GATEWAY TO THE SOLAR SYSTEM AND BEYOND

GATEWAY TO THE SOLAR SYSTEM AND BEYOND

James Essig

ISBN:	Softcover	979-8-3694-3857-2
	eBook	979-8-3694-3049-1

Print information available on the last page.

Rev. date: 02/19/2025

To order additional copies of this book, contact:
Xlibris
844-714-8691
www.Xlibris.com
Orders@Xlibris.com
862276

CONTENTS

Book Description

Would you be interested in understanding a wide variety of different starship propulsion concepts using only high school and even middle school math? If so, then this book is for you.

By purchasing and reading a copy of this book, you will be amazed at how much you can understand about the physics of starships using only grade-school math.

For example, there is something assuring about Special Relativity in that the subject includes a lot of formulas that you can understand using only high school algebra.

When you read this book, you will in a sense have graduated into having learned many concepts about starships. Having done so, you will have in a sense obtained an elite status and will be given the tools to work on your own starship concepts.

The mathematical and scientific presentation of the subject of starship propulsion in this book are thus easily accessible.

For those who are schooled in higher math, physics, and engineering, you will find this book refreshingly easy to wade thru and will likely be amazed at how the simple presentation provides you strong evidence for the possibilities of extreme forms of space travel.

There is something comforting about notions of reactionary or other forms of impulse space travel. At a time when wormholes and warp-drives are common topics of discussion among many advocates of advanced spacecraft propulsion, the notion that the speed of light may be an inviolable limit can be comforting. Well, in this book I have your comfort covered in many ways.

This book provides simple mathematical descriptions of how we will likely travel about the solar system and beyond. The book is mostly non-mathematical but the math included in this book is at the level of high-school algebra. It is my opinion that much of the developing science of manned space-flight can be explained via relatively easy math and formulations.

Impulse or reactionary propulsion at near the speed of light has profound applications including but not limited to potential hyper spatial travel. However, the focus of this book is more near-term and is highly relevant toward 21st Century applications.

For those adverse to any mathematical content, you likely will also find this book a fun read. As such, I include much prosaic explanations along with very simple arithmetic to back up the propositions made in this book.

Back in the days while I was a teenager, I attended a private school. The school psychologist was a consecrated Catholic religious brother with dark hair and a dark beard who used to let me ride in his fancy Ford Thunderbird. The car had a black exterior and interior.

Well, at about the same time, the sitcom, "The Jefferson's" was popular and the show theme song had a refrain that went like "Well we're movin' on up. (Movin' on up). To the east side.".

Even back then I was interested in interstellar travel concepts.

To make a long story short, I associated the school psychologist and rides in his Thunderbird with my internalized mantra of Movin' on up, to the future, at near light-speed. Thus, I became more hooked on special relativistic space travel and time dilation. The fire of my imagination for near light-speed travel was lit just as assuredly as the black Ford Thunderbird resembled the eternal black cosmic void. I knew then the ramifications of infinite time dilation, infinite forward time travel, and infinite travel distances through space made mathematically plausible for light speed impulse travel.

So, if you have the courage to delve into this book, or even only study select portions thereof and wade through the math, you will likely if not already also become intrigued with Movin' on up into the future with Special Relativity. As we now have a space travel industry, we have set before us the seas of infinity. Sailing these seas is what this book is all about.

The Torch of Science!

I have a golden torch
The torch of Science!
To drive out the darkness of ignorance

To burn out the burning hunger
To hell with death knell for malnutrition
To live disease free
To know the world, to rule the world
To be on Earth and to fly into space!

This is an ever-burning Science Olympics Torch!
Just bring your unlit torches
I shall torch them with Science!
Let the light spread everywhere
Long live Science. Long live the people!

Science Express

Get in, come aboard. Science express is ready
For a thrilling journey into science
Unraveling the mysteries of the world
Nature, flora and fauna, biology
Physics, chemistry, math's side by side
One after the other chugging along merrily

Throwing light on the origins of the universe
Formation of the first atoms, stars, planets, origin of life,
evolution, tracing discoveries, inventions of
vehicles, communicating devices, computers, net, ships,
planes, rockets, bacteria, virus, DNA and what not
in ever growing, endless list, sphere of knowledge

Travelling back into the primordial past,
Running through the present world,
Running on to the future in advance
Ahoy! It is great journey, grand, journey
Be involved, contribute to the science engines,
Bogies to be everlasting part of the express

It's untiring, ceaseless journey

As new frontiers keep expanding

More new ones keep beckoning

Stay safe, travel ahead, enriching yourself

Enriching the Science, upgrading the express

Baam, baam! Go, go, go, chug, chug, chug!

Billions and Billions Bulbed the Universe

Billions and billions bulbed the universe,
Yet darkness pervades in between.
Black holes growl, white and neutron stars dance,
Photons emitted on an endless journey.
Gamma ray, X-rays all abound.
Universe, the great universe!

Billions and billions eye see the skies
Thousands and thousands telescopes scan the space
Millions views spot the spectacles
Stars forming, stars dying
Galaxies spreading, lively blinking
Planets circling the stars
Moons circling the planets
Universe, the great universe!

Big Bang sprawled the universe away
Bang's remnants still on a journey
Journeying photons carrying the initial images
Dark matter pushing the universe away and away
Anti-matter lurking to destroy the matter

Yet the universe surviving to tell the tale
It's live story, ever expanding growth story
Universe, the great Universe!

Spectacles galore, yet the mysteries remain
Neutrinos, bosons, mesons, leptons, quarks
The endless sub particles thrive
The mortals make out a few, miss out a lot
Universe, the great Universe!

Grandeur universe, mammoth universe
Our life time goes out before fathoming it, grasping it
Yet the Universe lives on as we fade away
Long live the Universe, the great universe!

Poems by Mohan Sanjeevan from his book, *The Torch of Science! Science*.

Chapter 1

Going Interplanetary in the Next Thirty Years

This chapter includes a description of rocket propulsion techniques that are realizable within the next few decades for manned missions farther out into our planetary solar system. Included are descriptions of current technologies as well as those that may, in principle, be fairly quickly developed provided funding is made available.

A) Solar Thermal Rockets

Solar thermal rockets provide a specific impulse of as high as 900 seconds using hydrogen as the propellant where the hydrogen is indirectly heated by concentrated solar energy. Tantalum carbide and hafnium carbide are materials typically considered for the mechanism of the direct heating method where hydrogen is pumped into a heat-absorbing chamber containing conduits to carry hydrogen. For the direct heating method, a specific impulse as high as 1,200 seconds is possible (Wikipedia, 2011). Direct heating involves direct exposure of hydrogen or other fuel to concentrated solar radiation.

Consider a liquid hydrogen direct solar thermal rocket having a mass ratio of 1,000 such as might be accomplished using a large tank where the tank mass to fuel mass ratio is 0.0005, and where the remainder of the vehicle is crew quarters, radiation shielding, rocket engines, and life support supplies.

We first consider the case where the indirect heating method is used.

The nonrelativistic rocket equation is as follows:

$\Delta v = v_{ex} \ln (M_0/M_1)$ where $v_{ex} = (I_{SP})(g_0)$

Thus, the above spacecraft would obtain the following terminal velocity:

$$\Delta v = (I_{SP})(g_0) \ln (M_0/M_1)$$

$$= (900 \text{ s})[9.81 \text{ m/s}^2] \ln (1{,}000)$$

$$= [8{,}829 \text{ m/s}] (6.90775) = 60.9885 \text{ km/s}$$

The transit time of the spacecraft to Mars at Mars's closest approach to Earth would be about 963,000 seconds, or about 1.6168 weeks, provided the spacecraft could accelerate to this velocity in under one-half of a day.

If reverse rocket thrust is the mechanism for craft deceleration, then the terminal velocity is equal to:

$$\Delta v = (I_{SP})(g_0) \ln (M_0/M_1)$$

$$= (900 \text{ s})[9.81 \text{ m/s}^2] \ln (10^{1.5})$$

$$= [8{,}829 \text{ m/s}] (3.4538) = 30.49 \text{ km/s}$$

The transit time of the spacecraft to Mars at Mars's closest approach to Earth would be about 1,928,000 seconds, or about 3.234 weeks.

Here, Δv is the change in spacecraft velocity, I_{sp} is the specific impulse in seconds, g_0 is 9.81 m/s^2.

Such a large craft could use its fuel tanks as aerobraking chutes in the Martian atmosphere, although decelerations would amount to several tens of Gs for the crew members. The aerobraking mechanisms could be insulated from the hot plasma of reentry by magnetic and/or electrical field configurations.

Alternatively, the aerobrake could include a capillary-like network containing a high latent heat of vaporization, or high latent heat of ionization, carbonaceous liquid residue. The vaporization or ionization of carbon upon aerobraking would help cool the outer surface of the chute.

Another method of cooling the aerobrake surface would involve a network-like arrangement of cryogenic hydrogen-, oxygen-, nitrogen-, or helium-containing capillaries. The liquefied gases would be pumped to the outer surface layer of the chute through the capillary network. Perhaps the resulting superheated gases could be exhausted in a forward direction, thus providing reverse thrust to help slow the craft.

The crew members could be enclosed in a smart fabric type of pressure suits such as might be constructed from rheo-elastic or other electro-elastic materials.

Solar thermal hydrogen rockets are natural for use in bringing the craft back to Earth because such hydrogen can be made easily from water ice on or near the Martian surface.

The cost of getting the hydrogen to low Earth orbit would be high. However, perhaps 1,000 such launches could provide enough fuel to enable the above vehicle performance, provided that a heavy lift booster could be designed and assembled at low cost, including the possibility of reusable stages that would parachute back to Earth.

The uppermost rocket stage might optionally have wings so that it could glide back to Earth in a similar way that the space shuttle did.

Assume that such heavy lift boosters could be assembled at a cost of $100 million per vehicle and that each vehicle could be launched 100 times with quick mission turnaround. The cost of the assembled fuel launch hardware per Mars mission could be as low as $1 billion. One hundred missions could be launched using only $100 billion for fuel booster construction.

Obviously, the cost of booster flight fuel, flight operations, discarded stage collection, and any reconditioning of stages would add to the cost.

Such large mass ratios for interplanetary vessels could also enable practical travel to the Jovian and Saturnian systems. For cases where powered gravitational assists are utilized, such as with the sun or with Jupiter or Saturn, timely travel of human crews to the edge of the planetary solar system could be accomplished.

With solar electric rocket technology alone, we could explore and colonize any habitable or terraformable planet or moon within our solar system. We could mine the asteroids in situ and use the metals, alloys, and, perhaps, concretes thus produced to build elegant rotating space colonies that could be distributed throughout the solar system. The potential to support a simultaneous population of trillions of human beings becomes possible.

Include robotic or intelligent machine-based construction efforts and the assembly of the required habitats to make this dream a reality would become much easier.

Single stage to orbit fuel carrying craft such as space planes would help further reduce mission cost. However, the development of space planes has been a tough nut to crack. Perhaps the new commercial space hardware development companies such as SpaceX, Virgin Galactic, and Bigelow Aerospace can crack it since these companies are new and very nimble and thus may be very capable of fresh ideas.

For the direct heating method, we will assume a specific impulse of 1,200 seconds.

Consider a liquid hydrogen direct solar thermal rocket having a mass ratio of 1,000 such as might be accomplished using a large tank where the tank mass to fuel mass ratio is 0.0005, and where the remainder of the vehicle is crew quarters, radiation shielding, rocket engines, and life-support supplies.

Thus, the above spacecraft would obtain the following a terminal velocity:

$$\Delta v = (I_{sp})(g0) \ln (M_0/M_1)$$

$$= (1{,}200 \text{ s})[9.81 \text{ m/s}^2] \ln (1{,}000)$$

$$= [11{,}772 \text{m/s}] (6.90775) = 81.32 \text{ km/s}$$

The transit time of the spacecraft to Mars at Mars' closest approach to Earth would be about 723,000 seconds, or about 1.21 weeks provided the spacecraft could accelerate to this velocity in under 1/2 of a day.

If reverse rocket thrust is the mechanism for craft deceleration, then the terminal velocity is equal to:

$$\Delta v = (I_{sp})(g0) \ln (M_0/M_1)$$

$$= (1{,}200 \text{ s})[9.81 \text{ m/s}^2] \ln (10^{3/2})$$

$$= [11{,}772 \text{ m/s}] (3.4538) = 40.66 \text{ km/s}$$

Provided such a spacecraft could accelerate to this velocity in under a 1/2 of a day, the transit time of the spacecraft to Mars at Mars' closest approach to Earth would be a about 1,446,000 seconds, or 2.43 weeks.

Such large mass ratios for the proposed interplanetary vessels could also enable practical travel to the Jovian and Saturnian systems. For cases where powered gravitational assists are utilized such as with the sun, Jupiter, or Saturn, timely travel of human crews to the edge of the planetary solar system could be accomplished.

Note that the computed flight times to Mars assume that the entire distance is traveled at terminal velocity while neglecting the extra rocket thrust energy required to overcome the gravity well of the sun's and Earth's orbits, for cases where the missions are launched from low Earth orbit. However, the above-computed transit times are close to those obtainable for real systems because the compounded effects of these three mechanisms increase the travel time by only a relatively small percentage.

B) A Holdout for Chemical Rockets

Existing Technology

Ethanol + 25% water LOX produces a specific impulse equal to 269 seconds at sea level with combustion chamber pressure of Pc = 68 atm (1000 PSI). The performance of the rocket in the vacuum of space is close to that at sea level.

Consider an ethanol + 25% water LOX rocket having a mass ratio of 1,000 such as might be accomplished using a large tank where the tank mass to fuel mass ratio is 0.0005, and where the remainder of the vehicle is crew quarters, radiation shielding, rocket engines, and life-support supplies.

Thus, the above spacecraft would obtain a terminal velocity of 18.22 km/s. The transit time of the spacecraft to Mars at Mars's closest approach to Earth would be about 3.23 million seconds, or roughly five weeks, provided the spacecraft could accelerate to this velocity in under a few days.

Note that the term LOX is the abbreviated form of "liquid oxygen." Also note that a mass ratio of 1,000 and a fuel mass to tank mass ratio of 0.0005 is assumed for the remaining examples for ordinary chemical fuels.

Hydrazine and liquid oxygen produce a specific impulse equal to 303 seconds at sea level with combustion chamber pressure of Pc = 68 atm (1000 PSI).

The respective spacecraft would obtain a terminal velocity of 20.51 km/s. The transit time of the spacecraft to Mars at Mars's closest approach to Earth would be about 3 million seconds, or roughly 5 weeks, provided the spacecraft could accelerate to this velocity in under a few days.

Kerosene and liquid oxygen produce a specific impulse equal to 289 seconds at sea level with combustion chamber pressure of Pc = 68 atm (1000 PSI).

The respective spacecraft would obtain a terminal velocity of = 19.58 km/s.

Space Propulsion Group, Inc. has developed a paraffin LOX rocket that delivers a specific impulse of 349 seconds.

The respective spacecraft would obtain a terminal velocity of 23.65 km/s.

The transit time of the spacecraft to Mars at Mars's closest approach to Earth would be about 2.49 million seconds, or roughly 4.2 weeks provided the spacecraft could accelerate to this velocity in under a few days.

Liquid hydrogen and liquid oxygen fuels produced a specific impulse equal to 424 seconds in the vacuum of space for the Rocketdyne J-2 engine used in the upper stage of the Saturn V Apollo Program moon rocket.

The respective spacecraft would obtain a terminal velocity of 28.73 km/s.

The transit time of the spacecraft to Mars at Mars's closest approach to Earth would be about 2.05 million seconds, or roughly 3.43 weeks.

Liquid methane and liquid oxygen produces a specific impulse equal to 299 seconds at sea level with combustion chamber pressure of Pc = 68 atm (1000 PSI). The performance of the rocket in the vacuum of space is essentially identical to that at sea level.

The respective spacecraft would obtain a terminal velocity of 20.26 km/s.

The transit time of the spacecraft to Mars at Mars's closest approach to Earth would be about 2.9 million seconds, or roughly 5 weeks.

All of the above chemical fuels have at least demonstrated success in the laboratory, and most of these fuels have been used in modern rocket applications.

Such large chemical rocket–powered craft could use its fuel tanks as aerobraking chutes in the Martian atmosphere, although decelerations would amount to several tens of Gs for the crew members. The aerobraking mechanisms could be insulated from the hot plasma of reentry by magnetic and/or electrical field configurations.

Alternatively, the aerobrake could include a capillary-like network containing a high latent heat of vaporization or high latent heat of ionization carbonaceous liquid residue. The vaporization or ionization of the carbon upon aerobraking would cool the outer surface of the balloon.

Another method of cooling the aerobrake surface entails an arrangement of cryogenic hydrogen, oxygen, nitrogen, or helium that would be pumped to the outer surface layer of the shield through a capillary type of network. Perhaps the resulting superheated gases could be exhausted in a forward direction, thus providing reverse thrust to help slow the craft.

The crew members may be enclosed in smart fabric types of whole-body pressure suits such as might optionally be constructed from rheo-elastic or other electro-elastic materials.

Some chemical rocket fuels are natural for use in bringing the craft back to Earth because these fuels can be made from the Martian atmosphere and from water ice on or near the Martian surface.

The cost of getting the fuel(s) to low Earth orbit would be high. However, provided that a heavy lift booster could be designed and assembled at low cost with the possibility of reusable stages that would parachute back to Earth, perhaps 1,000 such launches could provide enough fuel to enable the above vehicles performance criteria. The upper rocket stage might optionally have wings so that it could glide back to Earth in a similar way that the Space Shuttle did.

Assume that such heavy lift boosters could be assembled at a cost of $100 million per vehicle and that each vehicle could be launched a hundred times with quick mission turnaround. The cost of the assembled fuel launch hardware per Mars mission could be as low as $1 billion. One hundred missions could be launched using only $100 billion for fuel booster construction. Obviously, the cost of booster flight fuel, flight operations, discarded stage collection, and any reconditioning of stages would add to the cost.

Such large-mass ratios for the proposed interplanetary vessels could also enable practical travel to the Jovian and Saturnian systems. Timely travel of human crews to the edge of the planetary solar system could be accomplished for cases where powered gravitational assists are utilized such as with the sun and with Jupiter or Saturn.

With chemical rocket technology alone, we could explore and colonize any habitable or terraformable planet or moon within our solar system. We could mine the asteroids in situ and use the metals, alloys, and, perhaps, concretes thus produced to build elegant, rotating space colonies that could be distributed all over the solar system. The potential to support a simultaneous population of trillions of human beings becomes possible. Include robotic or intelligent machine-based construction efforts and the assembly of the required habitats to make this dream a reality would become much easier.

A single stage to orbit fuel–carrying craft such as space planes would help further reduce mission costs; however, the development of space planes has been a tough nut to crack. Perhaps the new commercial space hardware development companies such as SpaceX, Virgin Galactic, and Bigelow Aerospace can crack it as these companies are new and seem very open to new ideas. Chemical-fueled rockets are good candidates because we have used such rockets ubiquitously to power our spacecraft.

A good scheme would include the incorporation of the fuel-carrying ferries into the interplanetary ships' construction. The rocket engines used to loft supply ferries into low Earth orbit may be repurposed to provide rocket thrust for the interplanetary ship. The fuel tanks carrying the ships' fuel could be repurposed and used as crew modules—or, optionally, as interplanetary rocket fuel modules.

Note that the computed flight times to Mars assume that the entire distance is traveled at terminal velocity while neglecting the extra rocket thrust energy required to overcome the gravity well of the sun, and Earth's orbit for cases where the missions are launched from low Earth orbit. However, the computed transit times are close to those obtainable for real systems because the compounded effects of these three mechanisms increase travel time by only a relatively small percentage.

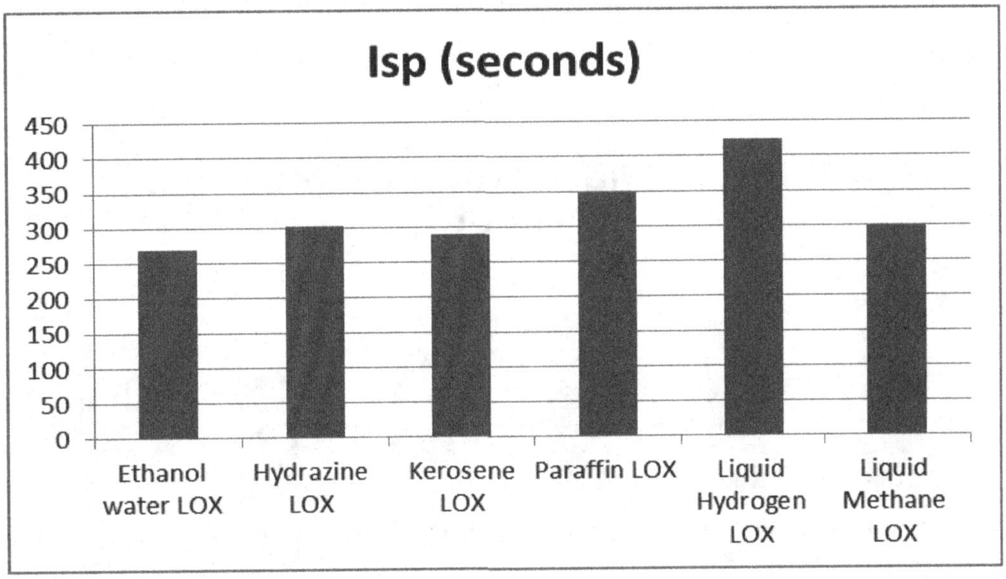

GRAPH 1 is a graph of specific impulse in seconds (y-axis) for various chemical fuels.

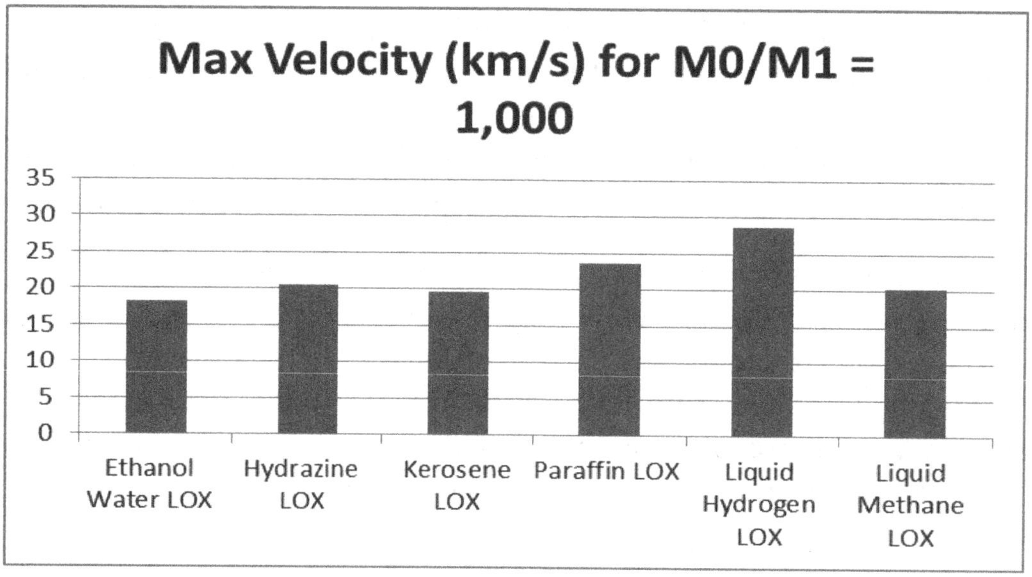

GRAPH 2 is a graph of maximum spacecraft velocity in km/s (y-axis) for various chemical fuels and for a mass ratio of 1,000.

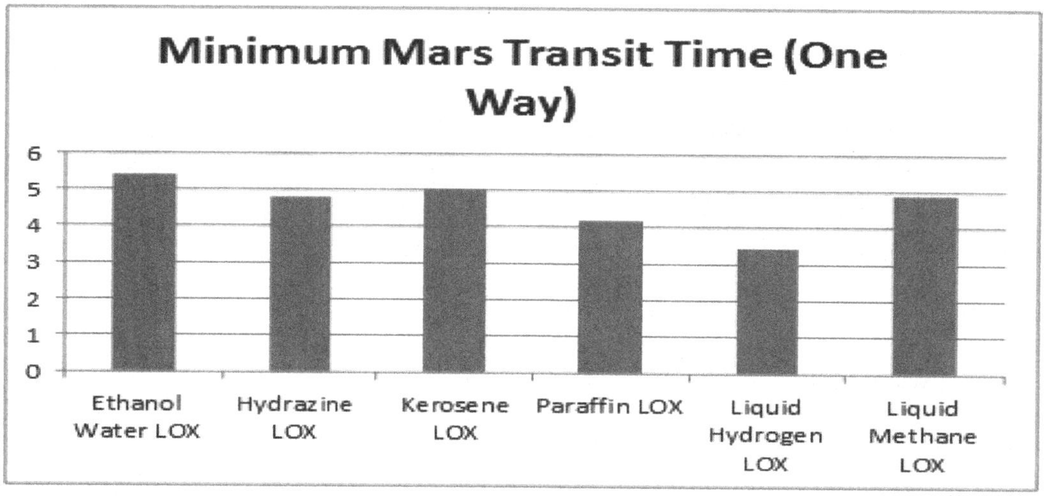

GRAPH 3 is a graph of one way Mars trip times for various chemical fuels in weeks (y-axis) for a mass ratio of 1,000.

In each of the preceding three graphs, rocketless deceleration is assumed. The transit time would be considerably longer where reverse rocket thrust was the primary deceleration mechanism.

Theoretical Chemistry

Free H radicals–based chemical rocket fuels have a maximum theoretical specific impulse of $I_{sp} = 2{,}130$ seconds.

Consider a free H-radicals–fueled rocket having a mass ratio of 1,000 such as might be accomplished using a large tank, where the tank mass to fuel mass ratio is 0.0005, and where the remainder of the vehicle is crew quarters, radiation shielding, rocket engines, and life-support systems.

The above spacecraft would obtain a terminal velocity of 144.34 km/s. The transit time of the spacecraft to Pluto would be about 33.9 million seconds, roughly 1.095 years, or about 13 months provided the spacecraft could accelerate to this velocity in under a few days.

Such a large craft might use magnetic sail–based braking to slow to Plutonian orbital velocity.

Now suppose we wanted to utilize reverse rocket thrust to slow a craft down to reach Mars. We will once again assume a mass ratio of 1,000 for the craft at the very beginning of its journey. However, for the accelerative phase of the trip, we will assume that the effective mass ratio of the craft is $10^{3/2}$.

Consequently, the terminal velocity of the craft will be 72.17 km/s. As a result, the craft will arrive on Mars in about 9.56 days, or about 1½ weeks. The leftover fuel could be used to slow the craft down, thus eliminating the need for aerobraking.

Hydrogen is a natural element choice for fabricating chemical fuels because it is ubiquitous throughout our solar system. What can be synthesized on Earth can be synthesized elsewhere.

We will assume a fully fueled initial mass ratio of 1,000 and a tank-mass-to-fuel-mass ratio of 0.0005 for the rest of the examples in this chapter.

Metastable helium chemical rocket fuels have a maximum theoretical specific impulse of $I_{sp} = 3{,}150$ seconds.

The respective spacecraft would obtain a terminal velocity of 213.46 km/s. The transit time of the spacecraft to Pluto would be about 22.9 million seconds, or about 0.74 years or about 8.9 months, provided the spacecraft could accelerate to this velocity in under a few days.

Such a large craft might use magnetic sail–based braking to slow to Plutonian orbital velocity.

The terminal velocity of the craft will be 106.73 km/s for cases where reverse thrust is used to slow the craft down, thus permitting it to arrive on Mars in about 6.47 days, or about 0.924 weeks.

Metastable helium is natural for deep solar system manned travel because craft powered by the fuel can refuel at locations such as the gas giant planets of our solar system. Converting the helium in situ into metastable helium is a different task. However, what can be done on Earth can be duplicated elsewhere.

Liquid ozone–liquid hydrogen fuel (LO_3 LH_2) has a specific impulse of 580 to 607 seconds.

The respective spacecraft would obtain a terminal velocity of 41.13 km/s. The transit time of the spacecraft to Mars at Mars's closest approach to Earth would be about 2.49 million seconds, or roughly 2.4 weeks, provided the spacecraft could accelerate to this velocity in under a few days

The terminal velocity of the craft will be 20.57 km/s for cases where reverse rocket thrust is used to slow the craft down. As a result, the craft will arrive on Mars in about 4.8 weeks.

Liquid F_2 and Liquid H_2 has a specific impulse of 546 to 703 seconds.

The respective spacecraft would obtain a terminal velocity of 47.64 km/s. The transit time of the spacecraft to Mars at Mars's closest approach to Earth would be about 1.234 million seconds, or about 2.07 weeks.

The terminal velocity of the craft will be 23.82 km/s in cases where reverse rocket thrust is used to slow the craft down. As a result, the craft will arrive on Mars in about 28.98 days, or about 4.14 weeks.

Be + O_2 (beryllium and oxygen fuel) has a specific impulse of 705 seconds.

The respective spacecraft would obtain a terminal velocity of 47.77 km/s. The transit time of the spacecraft to Mars at Mars's closest approach to Earth would be about 1.23 million seconds, or about 2.064 weeks, provided the spacecraft could accelerate to this velocity in under a few days.

The terminal velocity of the craft will be 23.89 km/s in cases where reverse rocket thrust is used. As a result, the craft will arrive on Mars in about 28.9 days, or about 4.128 weeks.

Tetrahedral N_4 (TdN_4) has a theoretical energy density of about 54 MJ/kg. I have not been able to locate the theoretical specific impulse of this fuel in open literature as of the time of this writing, so let us plug the energy density into the relativistic rocket equation to find the approximate velocity of rockets powered by this compound. First, we need to compute the specific impulse of the nitrogen compound.

In units of C,

$$I_{sp} = C\{[2n - n^2]^{11/2}\}$$

$$= C\{\{\{2[54 \text{ MJ}/ 90 \text{ billion MJ}]\} - \{[54 \text{ MJ}/ 90 \text{ billion MJ}]^2 \}\}^{1/2}\} = 0.00003461 \text{ C}$$

Here, *n* is the mass fraction of fuel mass converted into energy.

For a mass ratio of 1,000, the relativistic rocket equation yields the following:

$$\Delta v = C \text{ Tanh } \{[I_{sp} /C] \ln (M_0/M_1)\}$$

$$= C \text{ Tanh } \{[0.00003461 \text{ C}/ \text{C}] \ln (1,000)\}$$

$$= C \text{ Tanh } \{[0.00003461 \text{ C}/ \text{C}] (6.907755)\}$$

$$= C \text{ Tanh } [0.000239077] = 0.00023907699544496 \text{ C} \sim 0.000239 \text{ C} = 71.7 \text{ km/second}$$

For a mass ratio of 10,000:

$$\Delta v = C \text{ Tanh } \{[I_{sp} /C] \ln (M_0/M_1)\}$$

$$= C \text{ Tanh } \{[0.00003461 \text{ C}/ \text{C}] \ln (10,000)\}$$

$$= C \text{ Tanh } \{[0.00003461 \text{ C}/ \text{C}] (9.21034)\}$$

$$= C \text{ Tanh } [0.000239077] = 0.00031876986920281 \text{ C} \sim 0.0003188 \text{ C} = 95.64 \text{ km/second}$$

Thus, the respective spacecraft would obtain a terminal velocity of 71.7 km/s for a mass ratio of 1,000 and 95.64 km/s for a mass ratio of 10,000. The transit time of the spacecraft to Pluto would be about 68.32 million seconds, or roughly 2.2 years.

Provided such a spacecraft could accelerate to a terminal velocity of 95.64 km/s in under a few days, its transit time to Pluto would be about 51.22 million seconds, or roughly 1.65 years. Such a large craft might use magnetic sail–based braking to slow down to Plutonian orbital velocities.

Nitrogen is a natural element from which to synthesize rocket fuel because it is ubiquitous throughout our solar system.

Note that the above computed flight times assume that the entire distance is traveled at terminal velocity while neglecting the extra rocket thrust energy required to overcome the gravity well of the sun, and Earth's orbit for cases where the missions are launched from low Earth orbit. However, because the compounded effects of these three mechanisms increase the travel time by only a relatively small percentage, the computed transit times are close to those obtainable for real systems.

A good but brief Wikipedia reference on solar thermal rockets can be found under "Solar thermal rocket" (http://en.wikipedia.org/wiki/Solar_thermal_rocket).

A good short reference on chemical rocket fuels can be found here:

Rocket & Space Technology, "Rocket Propellants" by Robert A. Braeunig (2011). This reference is available at http://www.braeunig.us/space/propel.htm

Another good reference on chemical rocket fuels can be found in Orion's Arm, *Encyclopedia Galactica*, Chemical Rocket by M. Alan Kazlev, Richard Baker, David Dye, Mauk Mcamuk and Chris Shaeffer and is available at http://www.orionsarm.com/eg-article/493687ff373fd

I obtained most of the data on rocket fuel specific impulse values used in this chapter from the above three references.

A really interesting extremely speculative prospect would entail exothermic chemical reaction–based reification of the zero-point electromagnetic fields in a given volume of space, but in a controlled manner. Such a mechanism, if possible, might hold the potential for the production of rocket fuels having a super-relativistic energy density or a invariant mass specific energy yield greater than $E = mc^2$. How much greater would remain to be seen; however, the effective energy yield could be a huge multiple of the energy derived from an equal invariant mass of matter-antimatter fuel.

Perhaps such huge chemical fuel yields would ironically be potentially safer and less destructive than certain classical nuclear energy reactions. After all, nuclear reactions that could produce stable strangelets might be capable of propagating a chain reaction that would spread nonrecallable throughout our universe and even throughout our entire multiverse. Even more disastrous exotic nuclear energy scenarios can be considered, and have been discussed, by safety-minded high energy physicists.

However, let us not get carried away with the unknowns of nuclear energy. After all, nature within our very universe has been doing nuclear fusion for 13.75 billion years. We would not be here to contemplate interstellar space travel if the atoms that make up much of our bodily mass were not synthesized in stellar interiors and in supernova explosions by the continuous and sudden tremendous releases of nuclear energy associated with such natural systems. Thus, nature has used nuclear energy in a very safe manner for at least 13.75 billion years. We can do so too, even for the propulsion systems of our first starships.

I can safely say that nuclear energy is a good buddy of mine and is highly life oriented. I say this because nuclear energy can get us to the stars, and perhaps even beyond. This is a personal hope and dream of mine and of millions of other folks who have looked up at the night sky and wished upon a star.

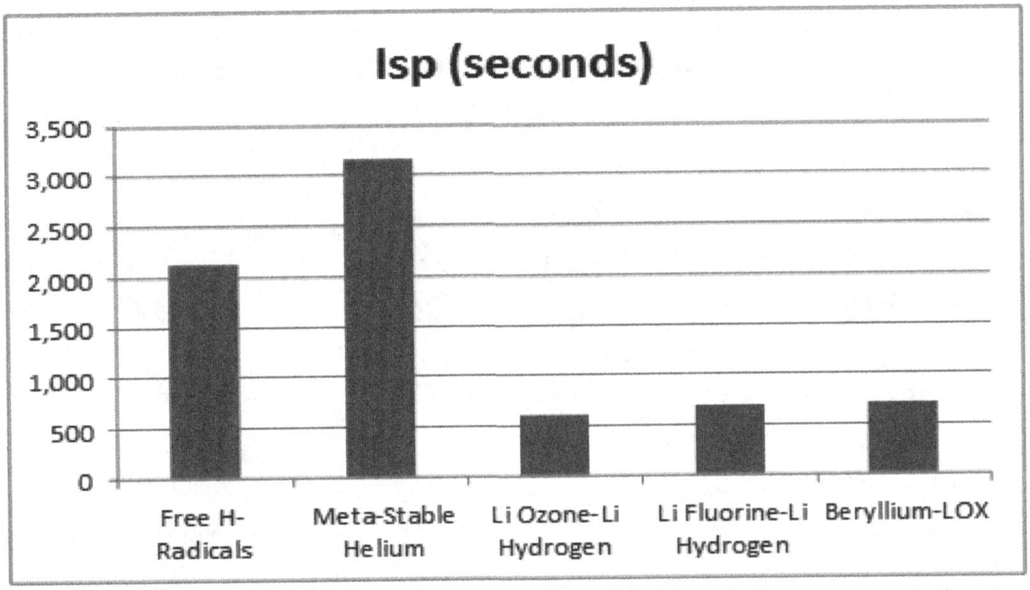

GRAPH 4 is a graph of specific impulse in seconds (y-axis) for various chemical fuels.

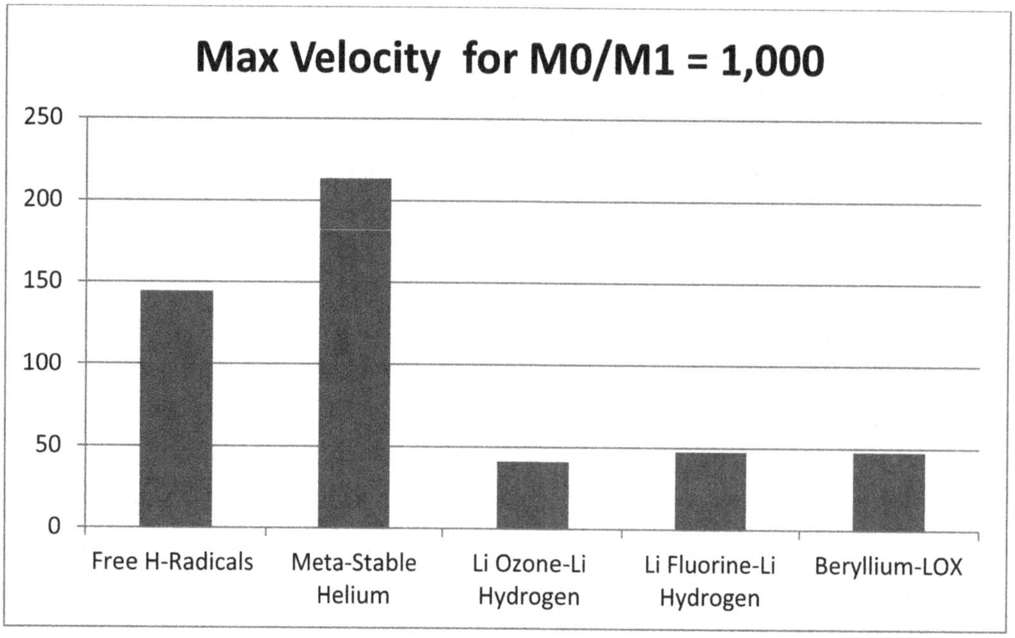

GRAPH 5 is a graph of maximum spacecraft velocity inkm/sec (y-axis) for various chemical fuels and for a mass ratio of 1,000.

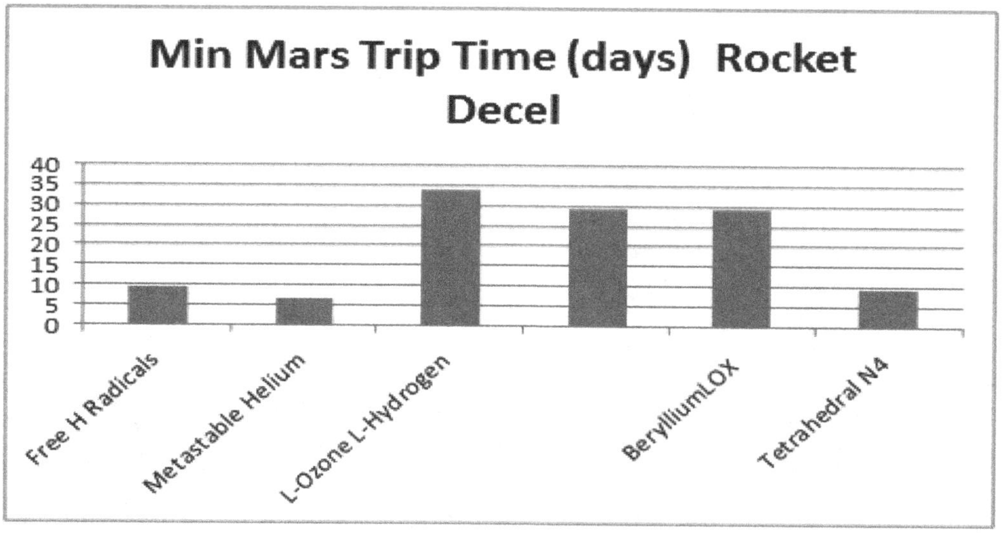

GRAPH 6 is a graph of one-way Mars trip times (y-axis in days) for various theoretical chemical fuels and reverse rocket thrust deceleration.

Chapter 2

Making Use of the Moon

This chapter includes speculative content on the practicality of the acquisition of lunar resources for facilitating manned interstellar space travel. The moon is known to contain resources that may have a vital role to play in the initial phases of manned interstellar exploration.

A) Mining the Moon for Basic Mineral Resources

Water has been discovered within the upper layers of the lunar soil. This means that there is water that can be collected and used for drinking, bathing, food preparation, green-house-based agriculture, and other industrial processes.

Since water molecules are composed of two hydrogen atoms and one oxygen atom, lunar water can be disassociated into its atomic components and used for rocket fuel. Liquid oxygen and diatomic hydrogen-based rocket fuel has among the highest specific impulse of any known chemical rocket fuels. Fuel for spacecraft can be manufactured on the moon, whereupon these spacecraft, perhaps in orbit around the moon, can be refueled and sent on missions into the far reaches of our planetary solar system. Thus, natural resources on the Moon and any initial infrastructure fielded thereon can facilitate manned excursions to Mars and, ultimately, lead to the development of lunar habitats, settlements, laboratories, factories, and mining operations.

So far, the quantity of water detected on the moon is very modest. In order to support extensive lunar colonies, it is likely that much additional water would need to be detected and made easily accessible. However, it is possible that improved water extraction methods can be used to collect enough water in cases where the amount of water per quantity of mined soil or lunar regolith would be very small.

Manned missions to the asteroid belt can also be staged from the moon. Such missions include the prospects of mining these bodies for the heavy—and, in some cases, rare and precious—metals they may contain.

Manned missions could be flown to some of the various moons of the gas giant planets, such as Jupiter, Saturn, Neptune, and Uranus, in cases where the radiation hazards would be minimal as our space-faring technology and infrastructure was further developed.

Manned missions could then be launched to explore the minor planets and, eventually, any Earth mass range Oort cloud objects.

Assume the most probably incorrect notion that manned spacecraft could travel at only Keplerian velocities through space, or at speeds of 1,000 km/sec or less. Manned expeditions ever farther from the sun could be launched, resulting in the colonization of the Oort cloud minor planets. The pioneers would eventually colonize the Oort cloud out to distances far enough removed from Earth that human civilization would become essentially interstellar in extent. At some point, we would colonize interstellar space into any likely existing Oort cloud analogues of adjacent stars. Over tens of millions of years, humanity could colonize the entire galaxy.

If mankind can just break the escape velocity of the Milky Way, which is less than 1,000 km/sec, we can travel to Andromeda, and perhaps to other galaxies of our local group, cluster, or supercluster. If mankind can break the escape velocity of our local supercluster—and again, 1,000 km/sec should suffice—we can travel to other superclusters and continually hop from cluster to cluster, supercluster to supercluster in highly insulated nuclear fusion–powered worldships or space Zonds.

Such world Zonds could be of a Dewar type, having a superconducting outer hull to keep in microwave, RF, and infrared electromagnetic emissions. The reason for it containing as much internal energy as possible involves issues of sustainability. Even with nuclear energy–based fuels or matter-antimatter–based fuels, world Zonds that are set on itineraries of cosmic durations will either need to conserve as much energy as possible or extract energy from the interstellar or intergalactic space. There are some merits to self-contained world Zonds that do not require in situ refueling or energy replacements.

One way or another, we can, and we will, go to the stars, even if it means outward travel at only Keplerian velocities and using chemical rocket–powered gravity assists via stellar gravitational fields.

The possibility of traveling unlimited distances through space and time presents itself even under the consideration of using chemical rocket–powered gravitational assists via other star systems.

I have hope and strongly believe that more advanced propulsion systems will be developed, but we know at least in theory, as ironic as it seems, good old-fashioned chemical rockets and gravitational assists can be our key to the great cosmos.

B) Mining the Moon for Uranium

Uranium has been detected on the moon. This fact should give individuals and research groups anticipating merely near-term nuclear fission–powered manned interplanetary and interstellar spacecraft a sense of hope for several reasons.

First, any bountiful source(s) of uranium on the moon would consist of the same uranium isotopes that exist on Earth. This implies that both U-238 and U-235 would be present on the moon. Both of these fuels can find application in nuclear fission reactors for powering spacecraft, direct nuclear fission fragment drive propulsion systems, and Project Orion–style nuclear bomb pulse drives.

Second, lunar uranium mining operations would not be a source of pollution on Earth. Such mining operations would do away with ground water, surface runoff, and atmosphere-polluting mining processes.

Third, in situ processing of uranium ores mined from the moon into ready-to-use fuels that could be optionally stored on the moon would alleviate the need for uranium-processing facilities on Earth.

Fourth, loading spacecraft on the moon's surface, in lunar orbit, Earth orbit, or solar orbit, with uranium fuels and uranium-derived fuels produced on the moon, would pose very little risk of fuel portions contaminating Earth's biosphere.

Fifth, launching nuclear-powered spacecraft constructed or assembled on the moon's surface, in lunar orbit, Earth orbit, or solar orbit, would be much easier than launching such craft from Earth's surface because of the lack of a lunar atmosphere and also the much smaller lunar gravity well compared to that of Earth.

Sixth, fueling or refueling of spacecraft with uranium fuels mined and processed into usable forms on the moon would be easier and safer than doing so with uranium fuels originating from Earth, which has a much deeper gravity well and requires more propulsive thrust and energy to loft supplies into Earth's orbit. Launching radioactive and chemically toxic uranium fuels from Earth's surface presents a greater risk to Earth's biosphere than similar launches from the lunar surface.

Seventh, nuclear fission-powered spacecraft could be launched from remote locations in solar orbit with exhaust plumes directed safely away from the direction of Earth. This can be an important consideration for very large spacecraft that produce very high levels of thrust.

Eighth, assembling atomic fission charges on the moon and then loading such devices onto spacecraft parked in lunar orbit, higher Earth orbit, or solar orbit, reduces the risk of such devices being introduced in proximity to low Earth orbit. Thus, the risk of weaponization of such devices within low Earth orbit and within Earth's atmosphere, oceans, or lithosphere is reduced.

Thus the mantra, "Atoms for Peace" can be more appropriately achieved by lunar mining of uranium in all isotopic forms present on the moon. This should give anti-nuclear activists, taxpayers, those in the voting booth, and civil and military leaders more confidence that nuclear fission–powered systems can be made safe and practical for use in manned interstellar travel. For those concerned about the dangers of employing high-yield atomic fission devices for propelling large Project Orion–style vehicles, the possibilities for antimatter catalyzed bomblets should alleviate their concerns to some degree. It is possible that such mini-fission bombs having a mass well below one gram could be used to provide large amounts of thrust required for accelerating a large Project Orion–style vehicle in a timely manner. However, the frequency of detonations would need to be much higher for cases where such mini-bombs would be used.

C) Mining the Moon for Helium-3

Going back to the moon by 2020 or sooner could be a great facilitator of manned interstellar space travel, at least to our nearest stellar neighbors. The Saturn V rocket developed approximately 200 million horsepower using the simple Power = d (∫ F•dx)/dt calculation. This is about (200 million) x (746) watts, or about 150 gigawatts. Thus, mankind is already capable of producing 150-gigawatt propulsion systems.

Lunar lasers powered by ^{3}He reactors would indeed offer a great means to propel manned craft out into the Oort cloud and farther abroad to our stellar neighbors.

An experimental Strategic Defense Initiative project included testing the accuracy of a laser beam directed to a roughly one-foot-wide mirror attached to a space shuttle in orbit. The beam reportedly held its position for a significant amount of time from an Earth-based location that was at least hundreds of miles away, if not a thousand miles or so. The Airborne Laser aircraft previously under development was able to test fire its beam to down a test missile at a distance of roughly 100 miles by the destructive effects of intense laser light emitted from the craft's several-megawatt infrared laser.

We should be able to work out the aiming requirements for a 10-gigawatt laser, or even a 1-terawatt–10-terawatt laser powered by a ^{3}He fusion reactor system. At 10 terawatts, assuming the beam could be utilized with at least 33 percent efficiency, and assuming ten years of shine time, a total of 10,000 kg [C ^{2}] of kinetic energy could be delivered to the craft. For a 1,000-metric–ton spacecraft, this works out to a Lorentz factor of about 1.01 and a velocity of about 0.2 C.

Electrodynamic braking could slow the craft down, or perhaps a reverse thrust ion rocket could do the same, even an ion rocket powered by the laser.

The ship itself could be powered by

- an ion rocket
- an electron rocket
- a photon rocket
- a magneto-hydrodynamic-plasma-drive
- an electro-hydrodynamic-plasma-drive
- an electro-magneto-hydrodynamic-plasma-drive
- and a magnetic field effect propulsion system.

Given a 1,000-terawatt laser beam generated by lunar-based nuclear fusion reactors and an overall system efficiency of 33 percent, in only ten years, a 1,000-metric-ton spacecraft could reach a terminal Lorentz factor of 2 or a velocity of 0.867 C. A 100-terawatt laser could accelerate a 1,000-metric-ton spacecraft to a Lorentz factor of 2 in one hundred years' lunar time.

We should take seriously the opportunity that lunar ^{3}He-powered lasing facilities can provide us in our travel among the stars. Other sources of ^{3}He include the gas giant planets of Jupiter and Saturn. I cannot think of a better use for ^{3}He that might otherwise find us as nuclear weapons fusion material!

Consider that the possibilities for human life expectancy increases, perhaps to 120 years or more if researchers in the field of anti-aging medicine have their way. We may then plausibly be able to reach any of the stars within a 100 light-year radius of Earth in about sixty years' ship time, assuming the 1,000-terawatt, ten-year acceleration scenario described above. The number of stars within 100 light-years of Earth is about 15,000.

Before all of this would take place, we must first go back to the moon or visit an asteroid, or go to Mars. For those of you who are U.S. citizens like I am, I strongly recommend that you write letters to government officials, as I do, or otherwise advocate for manned space travel and the practicality of going back to the moon and beyond. For those who are EU citizens, lobby your governments with the same goals in mind. I say the same to all of you citizens of the great countries of Russia and China as well. We can do great things as the civilization of humanity if we can cast aside the differences that at times put us somewhat at odds with each other.

Other resources can be mined from the moon, including any breeder reactor feedstock and metals for use in spacecraft construction and radiation shielding. In addition, lunar rock and minerals can also be used as radiation shielding for relativistic spacecraft and for use in concretes produced at least in part from such materials.

Mars has similar resources.

Martian soil is very high in perchlorate compositions and thus has strong potential for the production of powerful solid chemical rocket fuels.

Solid rocket fuels can become a large industry on Mars. To see how, continue on through the following digression on potential oxidizer components of solid rocket fuels.

Regarding principally solid oxidizers, consider the following:

The nitrite ion having the NO^-_2 is symmetric with equal N–O bond lengths.

Since tables of nitrites are hard to find, I have included the following list of more notable nitrites.

- Ammonium nitrite NH_4NO_2
- Calcium nitrite $Ca(NO_2)_2$
- Lithium nitrite $LiNO_2$
- Nickel(II) nitrite $Ni(NO_2)_2$
- Potassium cobaltinitrite $K_3[Co(NO_2)_6]$
- Potassium nitrite KNO_2

- Silver nitrite $AgNO_2$
- Silver(I) hyponitrite $Ag_2N_2O_2$
- Sodium cobaltinitrite $Na_3Co(NO_2)_6$
- Sodium hyponitrite $Na_2N_2O_2$
- Sodium nitrite $NaNO_2$

These nitrites, along with other nitrites, can have applications in chemical rocket fuels in manners similar to those of nitrates provided above.

A perchlorate is a chemical compound containing the perchlorate ion, ClO^-_4.

The following are some of the notable perchlorates:

- Ammonium perchlorate NH_4ClO_4
- Barium perchlorate $Ba(ClO_4)_2$
- Cesium perchlorate, $CsClO_4$
- Calcium perchlorate $Ca(ClO_4)_2$
- Copper(II) perchlorate $Cu(ClO_4)_2 \cdot 6H_2O$
- Fluorine perchlorate $FClO_4$ or $FOClO_3$
- Hexaperchloratoaluminate ion

  The hexaperchloratoaluminate ion is a triple negative complex of perchlorate with aluminum. It is related to hexanitratoaluminate and tetraperchloratoaluminate, all of which are energetic highly oxidizing materials.

- Potassium hexaperchloratoaluminate (may or may not exist)
- Hydrazinium hexaperchloratoaluminate (can only be made in an impure form)

Lithium perchlorate $LiClO_4$,

Magnesium perchlorate $Mg(ClO_4)_2$

Methyl perchlorate CH_3ClO_4

Nitronium perchlorate, NO_2ClO_4

Of all perchlorates, Nitronium perchlorate is the most powerful oxidizer.

Perchloratoborate $[B(ClO_4)_4]^-$

Potassium perchlorate $KClO_4$

Rubidium perchlorate $RbClO_4$

Silver perchlorate $AgClO_4$

Sodium perchlorate $NaClO_4$

Strontium perchlorate $Sr(ClO_4)_2$

Tetraperchloratoaluminates

These are salts of the tetraperchloratoaluminate anion, $[Al(ClO_4)_4]^-$.

Titanium perchlorate $Ti(ClO_4)_4$

Urea perchlorate $CO(NH_2)_2{\cdot}HClO_4$ is used as an oxidizer in liquid explosives, including underwater blasting.

Vanadyl perchlorate or **Vanadyl triperchlorate** $VO(ClO_4)_3$

Zirconium perchlorate $Zr(ClO_4)_4$

Nitrates include the following compounds (although the following list is not exhaustive):

Aluminum nitrate $Al(NO_3)_3$

Ammonium nitrate NH_4NO_3

Barium nitrate $Ba(NO_3)_2$

Beryllium nitrate $Be(NO_3)_2$

Bismuth(III) nitrate $Bi(NO_3)_3{\cdot}5H_2O$

Cadmium nitrate $Cd(NO_3)_2$

Cesium nitrate $CsNO_3$

Calcium nitrate $Ca(NO_3)_2$

Ceric ammonium nitrate $H_8N_8CeO_{18}$

Cerium nitrate $Ce(NO_3)_3$

Chlorine nitrate $ClNO_3$

Chromium(III) nitrate $Cr(NO_3)_3$ (anhydrous)

Cobalt Nitrate $Co(NO_3)_2$

Cobalt(III) nitrate $Co(NO_3)_3$

Copper(II) nitrate, $Cu(NO_3)_2$

Ethylammonium nitrate $C_2NH_8NO_3$

Europium(III) nitrate $Eu(NO_3)_3$

Fluorine nitrate FNO_3

Gadolinium(III) nitrate $Gd(NO_3)_3$

Gallium nitrate $Ga(NO_3)_3$

Guanidine nitrate $CH_6N_4O_3$

Hydrazine nitrate $N_2H_4 \cdot HNO_3$

Hydroxylammonium nitrate NH_3OHNO_3

Iron(III) nitrate $Fe(NO_3)_3$

Lead(II) nitrate $Pb(NO_3)_2$

Lithium nitrate $LiNO_3$

Magnesium nitrate $Mg(NO_3)_2$

Manganese(II) nitrate $Mn(NO_3)_2$

Mercury(I) nitrate $Hg_2(NO_3)_2$ (anhydrous)

Mercury(II) nitrate $Hg(NO_3)_2$

Methylammonium nitrate $CH_6N_2O_3$

Nickel hydrazine nitrate $[Ni(N_2H_4)_3](NO_3)_2$

Nickel nitrate $Ni(NO_3)_2$

Palladium(II) nitrate $Pd(NO_3)_2$

Potassium nitrate KNO_3

Scandium(III) nitrate $Sc(NO_3)_3$

Silver nitrate $AgNO_3$

Sodium nitrate $NaNO_3$

Strontium nitrate $Sr(NO_3)_2$

Thallium(III) nitrate $Tl(NO_3)_3$

Thorium(IV) nitrate $Th(NO_3)_4$

Titanium nitrate $Ti(NO_3)_4$

Urea nitrate $CH_5N_3O_4$

Vanadyl nitrate $VO(NO_3)_3$

Xenon nitrate (xenon dinitrate) $Xe(NO_3)_2$)

It has not been isolated and characterized, but mononitrate: **xenon fluoride nitrate** has been made and studied).

Zirconium nitrate $Zr(NO_3)_4$

Of all known perchlorates, Nitronium perchlorate is the most powerful oxidizer. There are at least some conjectured perchlorates that may exist, but they have not been observed.

Perchlorates seem to be the compounds of choice for solid rocket fuels.

I hold out hope that there exist perchlorates that are more powerful oxidizers than Nitronium perchlorate. Nitronium is not to be confused with the material neutronium, out of which neutron stars are made. Neutronium has never been fabricated on Earth except in the trivial cases of composite particles made of two or a few neutrons. These particles decay very, very rapidly and so do not constitute true neutronium as contemplated in relativistic astronautics and in science fiction.

As for more solid oxidizers, consider the following analogues of nitrite, perchlorates, and nitrates:

The proposed **fnitrite** ion having the NF^{-}_2, is symmetric with equal N–F bond lengths.

Since tables of nitrites are hard to find, I have included the following list of more notable nitrites:

Ammonium fnitrite NH_4NF_2

Calcium fnitrite $Ca(NF_2)_2$

Lithium fnitrite $LiNF_2$

Nickel(II) fnitrite $Ni(NF_2)_2$

Potassium cobaltfnitrite $K_3[Co(NF_2)_6]$

Potassium fnitrite KNF_2

Silver fnitrite $AgNF_2$

Silver(I) hypofnitrite $Ag_2N_2F_2$

Sodium cobaltifnitrite $Na_3Co(NF_2)_6$

Sodium hypofnitrite $Na_2N_2F_2$

Sodium fnitrite $NaNF_2$

These fnitrites and other fnitrites can have applications in chemical rocket fuels in manners similar to those of fnitrates provided above.

A proposed **fperchlorate** is a chemical compound containing the fperchlorate ion ClF^{-}_4.

The following are some notable fperchlorates:

Ammonium fperchlorate NH_4ClF_4

Barium fperchlorate $Ba(ClF_4)_2$

Cesium fperchlorate $CsClF_4$

Calcium fperchlorate $Ca(ClF_4)_2$

Copper (II) fperchlorate $Cu(ClF_4)_2 \cdot 6H_2F$

Fluorine fperchlorate $FClF_4$ or $FFClF_3$

The **fhexaperchloratoaluminate** ion is a triple negative complex of fperchlorate with aluminum. It is related to hexanitratoaluminate and tetraperchloratoaluminate, which are energetic, highly oxidizing materials.

Potassium fhexaperchloratoaluminate may or may not exist.

Hydrazinium fhexaperchloratoaluminate might only be made in an impure form. **Lithium fperchlorate** $LiClF_4$

Magnesium fperchlorate $Mg(ClF_4)_2$

Methyl fperchlorate, Nitronium fperchlorate NF_2ClF_4. Ff

All are fperchlorates. Nitronium fperchlorate is theoretically the most powerful oxidizer.

Perchloratoborate $[B(ClF_4)_4]^-$

Potassium fperchlorate $KClF_4$

Rubidium fperchlorate $RbClF_4$

Silver fperchlorate $AgClF_4$

Sodium fperchlorate $NaClF_4$

Strontium fperchlorate $Sr(ClF_4)_2$

Ftetraperchloratoaluminates $[Al(ClF_4)_4]^-$

These are salts of the ftetraperchloratoaluminate anion.

Titanium fperchlorate $Ti(ClF_4)_4$

Urea fperchlorate $CF(NH_2)_2 \cdot HClF_4$

This might be used as an oxidizer in explosives including underwater blasting.

Vanadyl fperchlorate or **Vanadyl trifperchlorate** $VF(ClF_4)_3$

Zirconium fperchlorate $Zr(ClF_4)_4$

Proposed fnitrates include the following compounds (although the following list is not exhaustive):

Aluminium fnitrate $Al(NF_3)_3$, **Ammonium fnitrate** NH_4NF_3, **Barium fnitrate** $Ba(NF_3)_2$, **Beryllium fnitrate** $Be(NF_3)_2$, **Bismuth(III) fnitrate** $Bi(NF_3)_3 \cdot 5H_2F$, **Cadmium fnitrate** $Cd(NF_3)_2$, **Caesium fnitrate** $CsNF_3$, **Calcium fnitrate** $Ca(NF_3)_2$, **Ceric ammonium fnitrate** $H_8N_8CeF_{18}$, **Cerium fnitrate** $Ce(NF_3)_3$, **Chlorine fnitrate** $ClNF_3$ **Chromium(III) fnitrate** $Cr(NF_3)_3$ (anhydrous), **Cobalt Nitrate** $Co(NF_3)_2$, **Cobalt(III) fnitrate** $Co(NF_3)_3$, **Copper(II) fnitrate**, $Cu(NF_3)_2$, **Ethylammonium fnitrate** $C_2NH_8NF_3$, **Europium(III) fnitrate** $Eu(NF_3)_3$, **Fluorine fnitrate** FNF_3, **Gadolinium(III) fnitrate** $Gd(NF_3)_3$, **Gallium fnitrate** $Ga(NF_3)_3$, **Guanidine fnitrate** $CH_6N_4F_3$, **Hydrazine fnitrate** $N_2H_4 \cdot HNF_3$, **Hydroxylammonium fnitrate** NH_3FHNF_3, **Iron(III) fnitrate** $Fe(NF_3)_3$, **Lead(II) fnitrate** $Pb(NF_3)_2$, **Lithium fnitrate** $LiNF_3$, **Magnesium fnitrate** $Mg(NF_3)_2$, **Manganese(II) fnitrate** $Mn(NF_3)_2$, **Mercury(I) fnitrate** $Hg_2(NF_3)_2$ (anhydrous), **Mercury(II) fnitrate** $Hg(NF_3)_2$, **Methylammonium fnitrate** $CH_6N_2F_3$, **Nickel hydrazine fnitrate** $[Ni(N_2H_4)_3](NF_3)_2$, **Nickel fnitrate** $Ni(NF_3)_2$, **Palladium(II) fnitrate** $Pd(NF_3)_2$, **Potassium fnitrate** KNF_3, **Scandium(III) fnitrate** $Sc(NF_3)_3$, **Silver fnitrate** $AgNF_3$, **Sodium fnitrate** $NaNF_3$, **Strontium fnitrate** $Sr(NF_3)_2$, **Thallium(III) fnitrate** $Tl(NF_3)_3$, **Thorium(IV) fnitrate** $Th(NF_3)_4$, **Titanium fnitrate** $Ti(NF_3)_4$, **Urea fnitrate** $CH_5N_3F_4$, **Vanadyl fnitrate** $VF(NF_3)_3$, Xenon fnitrate (Xenon difnitrate ($Xe(NF_3)_2$) monofnitrate: **xenon fluoride fnitrate**, **Zirconium fnitrate** $Zr(NF_3)_4$.

As for solid oxidizers, consider the following:

The **cnitrite** ion having the NCl^-_2, is symmetric with equal N–Cl bond lengths.

Since tables of nitrites are hard to find, I have included the following list of more notable cnitrites:

Ammonium cnitrite NH_4NCl_2, **Calcium cnitrite** $Ca(NCl_2)_2$, **Lithium cnitrite** $LiNCl_2$, **Nickel(II) cnitrite** $Ni(NCl_2)_2$, **Potassium cobalticnitrite** $K_3[Co(NCl_2)_6]$, **Potassium cnitrite** $KNCl_2$, **Silver cnitrite** $AgNCl_2$, **Silver(I) hypocnitrite** $Ag_2N_2Cl_2$, **Sodium cobalticnitrite** $Na_3Co(NCl_2)_6$, **Sodium hypocnitrite** $Na_2N_2Cl_2$, **Sodium cnitrite** $NaNCl_2$.

These cnitrites and other cnitrites can have applications in chemical rocket fuels in manners similar to those of nitrates provided above.

A **perfluorate** is a chemical compound containing the perfluorate ion, $FO^{-}{}_4$.

Some notable perfluorates are as follows:

Ammonium perfluorate NH_4FO_4, **Barium perfluorate** $Ba(FO_4)_2$, **Caesium perfluorate**, $CsFO_4$, **Calcium perfluorate** $Ca(FO_4)_2$, **Copper(II) perfluorate** $Cu(FO_4)_2{\cdot}6H_2O$, **Fluorine perfluorate** $FClO_4$ or $FOClO_3$,

The **fhexaperchloratoaluminate** ion is a triple negative complex of perfluorate with aluminium. It is related to hexanitratoaluminate and tetraperchloratoaluminate, which are energetic, highly oxidizing materials.

Potassium hexaperchlorofluoratoaluminate may or may not exist.

Hydrazinium hexaperchlorofluoroatoaluminate might only be made in an impure form.

Lithium perfluorate $LiFO_4$, **Magnesium perfluorate** $Mg(FO_4)_2$,

Methyl perfluorate, Nitronium perfluorate, NO_2FO_4.Of all perfluorates, nitronium perfluorate is theoretically the most powerful oxidizer, **Perchloratoborate** $[B(FO_4)_4]^-$,

Potassium perfluorate KFO_4, **Rubidium perfluorate** $RbFO_4$, **Silver perfluorate** $AgFO_4$, **Sodium perfluorate** $NaFO_4$, **Strontium perfluorate** $Sr(FO_4)_2$. **Ftetraperchloratoaluminates** are salts of the ftetraperchloratoaluminate anion, $[Al(FO_4)_4]^-$, **Titanium perfluorate** $Ti(FO_4)_4$, **Urea perfluorate** may find usage as an oxidizer in explosives including underwater blasting. $CO(NH_2)_2 \cdot HFO_4$, **Vanadyl perfluorate** or **Vanadyl triperfluorate** $VO(FO_4)_3$, **Zirconium perfluorate** $Zr(FO_4)_4$.

Cnitrates include the following compounds (although the following list is not exhaustive):

Aluminium cnitrate $Al(NCl_3)_3$, **Ammonium cnitrate** NH_4NCl_3, **Barium cnitrate** $Ba(NCl_3)_2$, **Beryllium cnitrate** $Be(NCl_3)_2$, **Bismuth(III) cnitrate** $Bi(NCl_3)_3 \cdot 5H_2Cl$, **Cadmium cnitrate** $Cd(NCl_3)_2$, **Caesium cnitrate** $CsNCl_3$, **Calcium cnitrate** $Ca(NCl_3)_2$, **Ceric ammonium cnitrate** $H_8N_8CeCl_{18}$, **Cerium cnitrate** $Ce(NCl_3)_3$, **Chlorine cnitrate** $ClNCl_3$ **Chromium(III) cnitrate** $Cr(NCl_3)_3$ (anhydrous), **Cobalt cnitrate** $Co(NCl_3)_2$, **Cobalt(III) cnitrate** $Co(NCl_3)_3$, **Copper(II) cnitrate**, $Cu(NCl_3)_2$, **Ethylammonium cnitrate** $C_2NH_8NCl_3$, **Europium(III) cnitrate** $Eu(NCl_3)_3$, **Fluorine cnitrate** $FNCl_3$, **Gadolinium(III) cnitrate** $Gd(NCl_3)_3$, **Gallium cnitrate** $Ga(NCl_3)_3$, **Guanidine cnitrate** $CH_6N_4Cl_3$, **Hydrazine cnitrate** $N_2H_4 \cdot HNCl_3$, **Hydroxylammonium cnitrate** $NH_3ClHNCl_3$, **Iron(III) cnitrate** $Fe(NCl_3)_3$, **Lead(II) cnitrate** $Pb(NCl_3)_2$, **Lithium cnitrate** $LiNCl_3$, **Magnesium cnitrate** $Mg(NCl_3)_2$, **Manganese(II) cnitrate** $Mn(NCl_3)_2$, **Mercury(I) cnitrate** $Hg_2(NCl_3)_2$ (anhydrous), **Mercury(II) cnitrate** $Hg(NCl_3)_2$,

Methylammonium cnitrate $CH_6N_2Cl_3$, **Nickel hydrazine cnitrate** $[Ni(N_2H_4)_3](NCl_3)_2$,, **Nickel cnitrate** $Ni(NCl_3)_2$, **Palladium(II) cnitrate** $Pd(NCl_3)_2$, **Potassium cnitrate** $KNCl_3$, **Scandium(III) cnitrate** $Sc(NCl_3)_3$, **Silver cnitrate** $AgNCl_3$, **Sodium cnitrate** $NaNCl_3$, **Strontium cnitrate** $Sr(NCl_3)_2$, **Thallium(III) cnitrate** $Tl(NCl_3)_3$, **Thorium(IV) cnitrate** $Th(NCl_3)_4$, **Titanium cnitrate** $Ti(NCl_3)_4$, **Urea cnitrate** $CH_5N_3Cl_4$, **Vanadyl cnitrate** $VCl(NCl_3)_3$, Xenon cnitrate (Xenon dicnitrate ($Xe(NCl_3)_2$) and monocnitrate: **xenon fluoride cnitrate**, **Zirconium cnitrate** $Zr(NCl_3)_4$.

As for solid oxidizers, consider the following:

The **brnitrite** ion having the $NBr^-{}_2$, is symmetric with equal N–Br bond lengths.

Since tables of Nitrites are hard to find, I have inBr uded the following list of more notable cnitrites.

Ammonium brnitrite NH_4NBr_2, **Calcium brnitrite** $Ca(NBr_2)_2$, **Lithium brnitrite** $LiNBr_2$, **Nickel(II) brnitrite** $Ni(NBr_2)_2$, **Potassium cobaltibrnitrite** $K_3[Co(NBr_2)_6]$, **Potassium brnitrite** $KNBr_2$, **Silver brnitrite** $AgNBr_2$, **Silver(I) hypobrnitrite** $Ag_2N_2Br_2$, **Sodium cobaltibrnitrite** $Na_3Co(NBr_2)_6$, **Sodium hypobrnitrite** $Na_2N_2Br_2$, **Sodium brnitrite** $NaNBr_2$.

Brnitrates inBrude the following compounds (although the following list is not exhaustive):

Aluminium brnitrate $Al(NBr_3)_3$, **Ammonium brnitrate** NH_4NBr_3, **Barium brnitrate** $Ba(NBr_3)_2$, **Beryllium brnitrate** $Be(NBr_3)_2$, **Bismuth(III) brnitrate** $Bi(NBr_3)_3 \cdot 5H_2Br$, **Cadmium brnitrate** $Cd(NBr_3)_2$, **Caesium brnitrate** $CsNBr_3$, **Calcium brnitrate** $Ca(NBr_3)_2$, **Ceric ammonium brnitrate** $H_8N_8CeBr_{18}$, **Cerium brnitrate** $Ce(NBr_3)_3$, **Chlorine brnitrate** $BrNBr_3$ **Chromium(III) brnitrate** $Cr(NBr_3)_3$ (anhydrous), **Cobalt brnitrate** $Co(NBr_3)_2$, **Cobalt(III) brnitrate**

$Co(NBr_3)_3$, **Copper(II) brnitrate**, $Cu(NBr_3)_2$, **Ethylammonium brnitrate** $C_2NH_8NBr_3$, **Europium(III) brnitrate** $Eu(NBr_3)_3$, **Fluorine brnitrate** $FNBr_3$, **Gadolinium(III) brnitrate** $Gd(NBr_3)_3$, **Gallium brnitrate** $Ga(NBr_3)_3$, **Guanidine brnitrate** $CH_6N_4Br_3$, **Hydrazine brnitrate** $N_2H_4 \cdot HNBr_3$, **Hydroxylammonium brnitrate** $NH_3BrHNBr_3$, **Iron(III) brnitrate** $Fe(NBr_3)_3$, **Lead(II) brnitrate** $Pb(NBr_3)_2$, **Lithium brnitrate** $LiNBr_3$, **Magnesium brnitrate** $Mg(NBr_3)_2$, **Manganese(II) brnitrate** $Mn(NBr_3)_2$, **Mercury(I) brnitrate** $Hg_2(NBr_3)_2$ (anhydrous), **Mercury(II) brnitrate** $Hg(NBr_3)_2$, **Methylammonium brnitrate** $CH_6N_2Br_3$, **Nickel hydrazine brnitrate** $[Ni(N_2H_4)_3](NBr_3)_2$,, **Nickel brnitrate** $Ni(NBr_3)_2$, **Palladium(II) brnitrate** $Pd(NBr_3)_2$, **Potassium brnitrate** $KNBr_3$, **Scandium(III) brnitrate** $Sc(NBr_3)_3$, **Silver brnitrate** $AgNBr_3$, **Sodium brnitrate** $NaNBr_3$, **Strontium brnitrate** $Sr(NBr_3)_2$, **Thallium(III) brnitrate** $Tl(NBr_3)_3$, **Thorium(IV) brnitrate** $Th(NBr_3)_4$, **Titanium brnitrate** $Ti(NBr_3)_4$, **Urea brnitrate** $CH_5N_3Br_4$, **Vanadyl brnitrate** $VBr(NBr_3)_3$, Xenon brnitrate (Xenon dibrnitrate ($Xe(NBr_3)_2$) monobrnitrate: **xenon fluoride brnitrate Zirconium brnitrate** $Zr(NBr_3)_4$.

A **perfluorbromate** is a chemical compound containing the perfluorbromate ion, FBr^-_4.

Some notable perfluorbromates are as follows:

Ammonium perfluorbromate NH_4FBr_4, **Barium perfluorbromate** $Ba(FBr_4)_2$, **Caesium perfluorbromate**, $CsFBr_4$, **Calcium perfluorbromate** $Ca(FBr_4)_2$, **Copper(II) perfluorbromate** $Cu(FBr_4)_2 \cdot 6H_2Br$, **Fluorine perfluorbromate** $FClBr_4$ or $FBrFBr_3$, The **fhexaperchloratoaluminate** ion is a triple negative complex of perfluorbromate with aluminium. It is related to hexanitratoaluminate and tetraperchloratoaluminate all of which are energetic highly oxidizing materials, **potassium hexaperchloratoaluminate** may or may not exist, **hydrazinium hexaperchloratoaluminate** might only be made in an impure form, **Lithium perfluorbromate** $LiFBr_4$, **Magnesium perfluorbromate** $Mg(FBr_4)_2$, **Methyl perfluorbromate, Nitronium perfluorbromate**, NBr_2FBr_4. Of all perfluorbromates, nitronium perfluorbromate is theoretically the most powerful oxidizer, **Fperchloratoborate** $[B(FBr_4)_4]^-$, **Potassium perfluorbromate** $KFBr_4$,

Rubidium perfluorbromate $RbFBr_4$, **Silver perfluorbromate** $AgFBr_4$, **Sodium perfluorbromate** $NaFBr_4$, **Strontium perfluorbromate** $Sr(FBr_4)_2$, **Tetraperchlorobromoatoaluminates** are salts of the tetraperchloratoaluminate anion, $[Al(FBr_4)_4]^-$, **Titanium perfluorbromate** $Ti(FBr_4)_4$, **Urea perfluorbromate** might find usage as an oxidizer in explosives including underwater blasting. $CBr(NH_2)_2{\cdot}HFBr_4$, **Vanadyl perfluorbromate** or **Vanadyl triperfluorbromate** $VBr(FBr_4)_3$, **Zirconium perfluorbromate** $Zr(FBr_4)_4$.

A **perbromochlorate** is a chemical compound containing the perbromochlorate ion, $ClBr^-_4$.

Some notable perbromochlorates are as follows:

Ammonium perbromochlorate NH_4ClBr_4, **Barium perbromochlorate** $Ba(ClBr_4)_2$, **Caesium perbromochlorate**, $CsClBr_4$, **Calcium perbromochlorate** $Ca(ClBr_4)_2$, **Copper(II) perbromochlorate** $Cu(ClBr_4)_2{\cdot}6H_2Br$, **Fluorine perbromochlorate** $FClBr_4$ or $FBrClBr_3$, The **hexaperchlorobromoatoaluminate** ion is a triple negative complex of perbromochlorate with aluminium. It is related to hexanitratoaluminate and tetraperchloratoaluminate all of which are energetic highly oxidizing materials, **potassium hexaperchloratoaluminate** may or may not exist, **hydrazinium hexaperchloratoaluminate** might only be made in an impure form, **Lithium perbromochlorate** $LiClBr_4$, **Magnesium perbromochlorate** $Mg(ClBr_4)_2$, **Methyl perbromochlorate, Nitronium perbromochlorate**, NBr_2ClBr_4. Of all perbromochlorates, nitronium perbromochlorate is theoretically the most powerful oxidizer, **Perchloratoborate** $[B(ClBr_4)_4]^-$, **Potassium perbromochlorate** $KClBr_4$, **Rubidium perbromochlorate** $RbClBr_4$, **Silver perbromochlorate** $AgClBr_4$, **Sodium perbromochlorate** $NaClBr_4$, **Strontium perbromochlorate** $Sr(ClBr_4)_2$, **Tetraperchlorobromoatoaluminates** are salts of the tetraperchlorobromoatoaluminate anion, $[Al(ClBr_4)_4]^-$, **Titanium perbromochlorate** $Ti(ClBr_4)_4$, **Urea perbromochlorate** might find usage as an oxidizer in explosives including underwater blasting. $CBr(NH_2)_2{\cdot}HClBr_4$, **Vanadyl perbromochlorate** or **vanadyl triperbromochlorate** $VBr(ClBr_4)_3$, **Zirconium perbromochlorate** $Zr(ClBr_4)_4$.

We can consider substantially solid analogues of perchlorates having any three or more of oxygen, chlorine, fluorine, bromine, and iodine.

A **perchlorofluorate:2,2** is a hypothetical chemical compound I have conceived containing the perchlorofluorate ion, Cl $F_2O_2^-$.

The following are some notable perchlorofluorate:2,2 s:

Ammonium perchlorofluorate:2,2 $NH_4ClF_2O_2$, **Barium perchlorofluorate:2,2** $Ba(ClO_2F_2)_2$, **Caesium perchlorofluorate:2,2**, $CsClO_2F_2$, **Calcium perchlorofluorate:2,2** $Ca(ClO_2F_2)_2$, **Copper(II) perchlorofluorate:2,2** $Cu(ClO_2F_2)_2{\cdot}6H_2O$, **Fluorine perchlorofluorate:2,2** $FClO_2F_2$ or $FOClO_3$

The **hexaperchlorofluoroatoaluminate:2,2** ion is a triple negative complex of perchlorofluorate:2,2 with aluminium. It is related to hexanitratoaluminate and tetraperchloratoaluminate all of which are energetic highly oxidizing materials. **Potassium hexaperchlorofluoroatoaluminate:2,2** may or may not exist, **hydrazinium hexaperchlorofluoroatoaluminate:2,2** might only be made in an impure form, **Lithium perchlorofluorate:2,2** $LiClO_2F_2$, **Magnesium perchlorofluorate:2,2** $Mg(ClO_2F_2)_2$, **Methyl perchlorofluorate:2,2, Nitronium perchlorofluorate:2,2**, $NO_2ClO_2F_2$. Of all perchlorofluorate:2,2 s, nitronium perchlorofluorate:2,2 is theoretically the most powerful oxidizer, **Perchloroflouroborate:2,2** $[B(ClO_2F_2)_4]^-$, **Potassium perchlorofluorate:2,2** $KClO_2F_2$, **Rubidium perchlorofluorate:2,2** $RbClO_2F_2$, **Silver perchlorofluorate:2,2** $AgClO_2F_2$, **Sodium perchlorofluorate:2,2** $NaClF_2O_2$, **Strontium perchlorofluorate:2,2** $Sr(ClO_2F_2)_2$, **Tetraperchloratoaluminates** are salts of the tetraperchlorofluoroatoaluminate:2,2 anion, $[Al(ClO_2F_2)_4]^-$, **Titanium perchlorofluorate:2,2** $Ti(ClO_2F_2)_4$, **Urea perchlorofluorate:2,2** might find usage as an oxidizer in explosives including underwater blasting. $CO(NH_2)_2{\cdot}HClO_2F_2$,

Vanadyl perchlorofluorate:2,2 or **vanadyl triperchlorofluorate:2,2** $VO(ClO_2F_2)_3$, **Zirconium perchlorofluorate:2,2** $Zr(ClO_2F_2)_4$.

A **perchlorofluorate:3,1** is a hypothetical chemical compound I have conceived containing the perchlorofluorate:3,1 ion, ClF_3O^-.

Some notable perchlorofluorate:3,1s are as follows:

Ammonium perchlorofluorate:3,1 NH_4ClF_3O, **Barium perchlorofluorate:3,1** $Ba(ClF_3O)_2$, **Caesium perchlorofluorate:3,1**, $CsClF_3O$, **Calcium perchlorofluorate:3,1** $Ca(ClF_3O)_2$, **Copper(II) perchlorofluorate:3,1** $Cu(ClF_3O)_2 \cdot 6H_2O$, **Fluorine perchlorofluorate:3,1** $FClF_3O$ or F_4OClO_3.

The **hexaperchlorofluoraluminate:3,1** ion is a triple negative complex of perchlorofluorate:3,1 with aluminium. It is related to hexanitratoaluminate and tetraperchloratoaluminate all of which are energetic highly oxidizing materials, **potassium hexaperchlorofluoroaluminate:3,1** may or may not exist.

Hydrazinium hexaperchlorofluoroaluminate:3,1 might only be made in an impure form, **Lithium perchlorofluorate:3,1** $LiClF_3O$, **Magnesium perchlorofluorate:3,1** $Mg(ClF_3O)_2$, **Methyl perchlorofluorate:3,1, Nitronium perchlorofluorate:3,1**, NO_2ClF_3O.

Of all perchlorates, nitronium perchlorate is theoretically the most powerful oxidizer, **Perchlorofluoroborate:3,1** $[B(ClF_3O)_4]^-$, **Potassium perchlorofluorate:3,1** $KClF_3O$, **Rubidium perchlorofluorate:3,1** $RbClF_3O$, **Silver perchlorofluorate:3,1** $AgClF_3O$, **Sodium perchlorofluorate:3,1** $NaClF_3O$, **Strontium perchlorofluorate:3,1** $Sr(ClF_3O)_2$, **Tetraperchlorofluoratoaluminates:3,1** are salts of the tetraperchlorofluoroatoaluminate:3,1 anion, $[Al(ClF_3O)_4]^-$, **Titanium perchlorofluorate:3,1** $Ti(ClF_3O)_4$, **Urea perchlorofluorate:3,1** may find use as an oxidizer in explosives including underwater blasting. $CO(NH_2)_2 \cdot HClF_3O$, **Vanadyl perchlorofluorate:3,1** or **vanadyl triperchlorofluorate:3,1** $VO(ClF_3O)_3$, **Zirconium perchlorofluorate:3,1** $Zr(ClF_3O)_4$.

A **perchlorofluorate:1,3** is a hypothetical chemical compound I have conceived containing the perchlorofluorate:1,3 ion, Cl $FO3^{-}$.

Some notable perchlorates are as follows:

Ammonium perchlorofluorate:1,3 NH_4ClFO_3, **Barium perchlorofluorate:1,3** $Ba(ClFO_3)_2$, **Caesium perchlorofluorate:1,3**, $CsClFO_3$, **Calcium perchlorofluorate:1,3** $Ca(ClFO_3)_2$, **Copper(II) perchlorofluorate:1,3** $Cu(ClFO_3)_2 \cdot 6H_2O$, **Fluorine perchlorofluorate:1,3** $FClFO_3$ or $FOClO_3$.

The **hexaperchlorofluoroaluminate:1.3** ion is a triple negative complex of perchlorofluorate:1,3 with aluminium. It is related to hexanitratoaluminate and tetraperchloratoaluminate all of which are energetic highly oxidizing materials, **potassium hexaperchlorofluoroaluminate:1,3** may or may not exist, **hydrazinium hexaperchlorofluoroaluminate:1,3** might only be made in an impure form, **Lithium perchlorofluorate:1,3** $LiClFO_3$, **Magnesium perchlorofluorate:1,3** $Mg(ClFO_3)_2$, **Methyl perchlorofluorate:1,3, Nitronium perchlorofluorate:1,3**, NO_2ClFO_3. Of all perchlorofluorate:1,3s, nitronium perchlorofluorate:1,3 is theoretically the most powerful oxidizer, **Perchloratoborate** $[B(ClFO_3)_4]^{-}$, **Potassium perchlorofluorate:1,3** $KClFO_3$, **Rubidium perchlorofluorate:1,3** $RbClFO_3$, **Silver perchlorofluorate:1,3** $AgClFO_3$, **Sodium perchlorofluorate:1,3** $NaClFO_3$, **Strontium perchlorofluorate:1,3** $Sr(ClFO_3)_2$, **Tetraperchlorofluoratoaluminates:1,3** are salts of the tetraperchloratoaluminate anion, $[Al(ClFO_3)_4]^{-}$, **Titanium perchlorofluorate:1,3** $Ti(ClFO_3)_4$, **Urea perchlorofluorate:1,3** might find usage as an oxidizer in explosives including underwater blasting. $CO(NH_2)_2 \cdot HClFO_3$, **Vanadyl perchlorofluorate:1,3** or **vanadyl triperchlorofluorate:1,3** $VO(ClFO_3)_3$, **Zirconium perchlorofluorate:1,3** $Zr(ClFO_3)_4$.

A **perchlorobromate:2,2** is a hypothetical chemical compound I have conceived containing the **perchlorofluorate:2,2** ion, Cl $Br_2O_2^{-}$.

Some notable perchlorates are as follows:

Ammonium perchlorobromate:2,2 $NH_4ClBr_2O_2$, **Barium perchlorobromate:2,2** $Ba(ClO_2Br_2)_2$, **Caesium perchlorobromate:2,2**,

$CsClO_2Br_2$, **Calcium perchlorobromate:2,2** $Ca(ClO_2Br_2)_2$, **Copper(II) perchlorobromate:2,2** $Cu(ClO_2Br_2)_2 \cdot 6H_2O$, **Bromine perchlorobromate:2,2** $BrClO_2Br_2$ or $BrOClO_3$.

The **hexaperchlorabromotoaluminate:2,2** ion is a triple negative complex of **perchlorobromate:2,2** with aluminium. It is related to hexanitratoaluminate and tetraperchloratoaluminate all of which are energetic highly oxidizing materials, **potassium hexaperchloratbromooaluminate:2,2** may or may not exist, **hydrazinium hexaperchlorabromotoaluminate:2,2** might only be made in an impure form, **Lithium perchlorobromate:2,2** $LiClO_2Br_2$, **Magnesium perchlorobromate:2,2** $Mg(ClO_2Br_2)_2$, **Methyl perchlorobromate:2,2, Nitronium perchlorobromate:2,2**, $NO_2ClO_2Br_2$. Of all perchlorobromate:2,2s, nitronium perchlorobromate:2,2 is theoretically the most powerful oxidizer, **Perchloratoborate** $[B(ClO_2Br_2)_4]^-$, **Potassium perchlorobromate:2,2** $KClO_2Br_2$, **Rubidium perchlorobromate:2,2** $RbClO_2Br_2$, **Silver perchlorobromate:2,2** $AgClO_2Br_2$, **Sodium perchlorobromate:2,2** $NaClBr_2O_2$, **Strontium perchlorobromate:2,2** $Sr(ClO_2Br_2)_2$. **Tetraperchlorabromotoaluminates:2,2** are salts of the tetraperchlorobromoatoaluminate:2,2 anion, $[Al(ClO_2Br_2)_4]^-$, **Titanium perchlorobromate:2,2** $Ti(ClO_2Br_2)_4$, **Urea perchlorobromate:2,2** might find usage as an oxidizer in explosives including underwater blasting. $CO(NH_2)_2 \cdot HClO_2Br_2$, **Vanadyl perchlorobromate:2,2** or **vanadyl triperchlorobromate:2,2** $VO(ClO_2Br_2)_3$, **Zirconium perchlorobromate:2,2** $Zr(ClO_2Br_2)_4$.

A **perchlorobromate:3,1** is a hypothetical chemical compound I have conceived containing the perchlorobromate:3,1 ion, $Cl\,Br_3O^-$.

Some notable perchlorobromate:3,1s are as follows:

Ammonium perchlorobromate:3,1 NH_4ClBr_3O, **Barium perchlorobromate:3,1** $Ba(ClBr_3O)_2$, **Caesium perchlorobromate:3,1**, $CsClBr_3O$, **Calcium perchlorobromate:3,1** $Ca(ClBr_3O)_2$, **Copper(II) perchlorobromate:3,1** $Cu(ClBr_3O)_2 \cdot 6H_2O$, **bromine perchlorobromate:3,1** $BrClBr_3O$ or Br_4OClO_3, The **hexaperchlorobromoatoaluminate:3,1** ion is a triple negative complex of **perchlorobromate:3,1** with aluminium. It is related to hexanitratoaluminate and tetraperchloratoaluminate all of which are energetic highly oxidizing materials, **potassium hexaperchlorobromoatoaluminate:3,1** may or may not exist, **hydrazinium hexaperchlorobromoatoaluminate:3,1** might only be made in an impure form, **Lithium perchlorobromate:3,1** $LiClBr_3O$, **Magnesium perchlorobromate:3,1** $Mg(ClBr_3O)_2$, **Methyl perchlorobromate:3,1, Nitronium perchlorobromate:3,1**, NO_2ClBr_3O. Of all perchlorobromate:3,1s, nitronium perchlorobromate:3,1 is theoretically the most powerful oxidizer, **Perchlorobromoatoborate:3,1** $[B(ClBr_3O)_4]^-$, **Potassium perchlorobromate:3,1** $KClBr_3O$, **Rubidium perchlorobromate:3,1** $RbClBr_3O$, **Silver perchlorobromate:3,1** $AgClBr_3O$, **Sodium perchlorobromate:3,1** $NaClBr_3O$, **Strontium perchlorobromate:3,1** $Sr(ClBr_3O)_2$, **Tetraperchlorobromoatoaluminates:3,1** are salts of the tetraperchlorobromoatoaluminate:3,1 anion, $[Al(ClBr_3O)_4]^-$, **Titanium perchlorobromate:3,1** $Ti(ClBr_3O)_4$, **Urea perchlorobromate:3,1** might find usage as an oxidizer in explosives including underwater blasting. $CO(NH_2)_2 \cdot HClBr_3O$, **Vanadyl perchlorobromate:3,1** or **vanadyl triperchlorobromate:3,1** $VO(ClBr_3O)_3$, **Zirconium perchlorobromate:3,1** $Zr(ClBr_3O)_4$.

A **perchlorobromate:1,3** is a hypothetical chemical compound I have conceived containing the perchlorobromate:1,3 ion, Cl $BrO3^-$.

Some notable perchlorobromate:1,3s are as follows:

Ammonium perchlorobromate:1,3 NH_4ClBrO_3, **Barium perchlorobromate:1,3** $Ba(ClBrO_3)_2$, **Caesium perchlorobromate:1,3**, $CsClBrO_3$, **Calcium perchlorobromate:1,3** $Ca(ClBrO_3)_2$, **Copper(II) perchlorobromate:1,3** $Cu(ClBrO_3)_2 \cdot 6H_2O$, **Bromine perchlorobromate:1,3** $BrClBrO_3$ or $BrOClO_3$, The **hexaperchlorobromoatoaluminate:1,3** ion is a triple negative complex of perchlorobromate:1,3 with aluminium. It is related to hexanitratoaluminate and tetraperchloratoaluminate all of which are energetic highly oxidizing materials, **potassium hexaperchlorobromoatoaluminate:1,3** may or may not exist, **hydrazinium hexaperchlorobromoatoaluminate:1,3** might only be made in an impure form, **Lithium perchlorobromate:1,3** $LiClBrO_3$, **Magnesium perchlorobromate:1,3** $Mg(ClBrO_3)_2$, **Methyl perchlorobromate:1,3, Nitronium perchlorobromate:1,3**, NO_2ClBrO_3. Of all perchlorobromate:1,3s, nitronium perchlorobromate:1,3 is theoretically the most powerful oxidizer, **Perchlorobromobromoatoborate:1,3** $[B(ClBrO_3)_4]^-$, **Potassium perchlorobromate:1,3** $KClBrO_3$, **Rubidium perchlorobromate:1,3** $RbClBrO_3$, **Silver perchlorobromate:1,3** $AgClBrO_3$, **Sodium perchlorobromate:1,3** $NaClBrO_3$, **Strontium perchlorobromate:1,3** $Sr(ClBrO_3)_2$, **Tetraperchlorobromoatoaluminates:1,3** are salts of the tetraperchlorobromoatoaluminate:1,3 anion, $[Al(ClBrO_3)_4]^-$, **Titanium perchlorobromate:1,3** $Ti(ClBrO_3)_4$, **Urea perchlorobromate:1,3** might find usage as an oxidizer in explosives including underwater blasting. $CO(NH_2)_2 \cdot HClBrO_3$, **Vanadyl perchlorobromate:1,3** or **vanadyl triperchlorobromate:1,3** $VO(ClBrO_3)_3$, **Zirconium perchlorobromate:1,3** $Zr(ClBrO_3)_4$.

Analogues are plausible with iodine replacing bromine, or iodine replacing fluorine. Additionally, analogues including any four of oxygen, chlorine, fluorine, bromine, and iodine are possible or all five of these elements.

Now infra-stoichiometric black powder solutions can be intentionally designed, for which liquid or gaseous oxidizers can be mixed in the combustion chambers. Some notable oxidizers include the following:

Liquid diatomic oxygen comes to mind as a mainstay oxidizer.

Other oxidizers may include liquid diatomic fluorine, liquid diatomic chlorine, monatomic oxygen, monatomic fluorine, monatomic chlorine, and ozone.

Compounds of two or more of oxygen, fluorine, chlorine, bromine, and iodine may also be useful as powerful oxidizers.

Chlorine dioxide has a standard enthalpy of formation 104.60 kJ/mol.

Dichlorine trioxide (Cl_2O_3) is a dark-brown solid discovered in 1967. The compound is explosive even below 0°C.

Dichlorine hexoxide (Cl_2O_6) is a very strong oxidizing agent and is stable at room temperature and explodes violently on contact with organic compounds.

Chlorine trioxide (ClO_3) is another powerful oxidizer.

Chlorine trioxide fluoride (ClO_3F) is another powerful oxidizer.

Chloryl fluoride has the formula ClO_2F.

Chlorine monofluoride (ClF) has a standard enthalpy of formation of −56.5 kJ/mol.

Dioxygen difluoride (O_2F_2) is an extremely powerful oxidizer with a standard enthalpy of formation of 19.2 kJ/mol.

Oxygen fluorides are compounds of oxygen and fluorine with the general formula O_nF_2, where $n = 1$ to 6. Many different oxygen fluorides are known:

oxygen difluoride (OF_2)

dioxygen difluoride (O_2F_2)

trioxygen difluoride or ozone difluoride (O_3F_2)

tetraoxygen difluoride (O_4F_2)

pentaoxygen difluoride (O_5F_2)

hexaoxygen difluoride (O_6F_2)

Chlorine pentafluoride (ClF_5) has a standard enthalpy of formation of −238.49 kJ/mol.

Chlorine trifluoride (ClF_3) has a standard enthalpy of formation of −158.87 kJ/mol.

Perchloryl Hypofluorite ($FClO_4$) has a standard enthalpy of formation of 9 kcal/mol.

Perchloryl fluoride (ClO_3F) is a powerful oxidizer with a standard enthalpy of formation of -21.42 kJ/mol.

Compounds of oxygen and iodine, oxygen and bromine, and oxygen and astatine exist.

Apparently, there are no definitively known interhalogen compounds containing atoms of three or more elements from the set of halogens. Some claims have been made for a couple of compounds as such. However, these claims are controversial and if the compounds claimed exist, they are so unstable and short lived so as to be all but trivial.

The halogens include: fluorine, chlorine, iodine, bromine, astatine, and tennessine. Astatine and tennessine are unstable and have short half-lives. So in their common isotopic forms, they would not make suitable rocket fuel components.

Bromine oxides include the following:

Dibromine monoxide (Br_2O)
Bromine dioxide (BrO_2)
Dibromine trioxide (Br_2O_3)
Dibromine pentoxide (Br_2O_5)
Tribromine octoxide (Br_3O_8)

An interhalogen compound contains two or more different halogen atoms (fluorine, chlorine, bromine, iodine, or astatine) and no elements from any other group. There are no definitely known interhalogen compounds containing three or more different halogens.

Diatomic interhalogens include:

Chlorine monofluoride (ClF)

Bromine monofluoride (BrF)

Iodine monofluoride (IF)

Bromine monochloride (BrCl)

Iodine monochloride (ICl)

Astatine monochloride (AtCl)

Iodine monobromide (IBr)

Astatine monobromide (AtBr)

Astatine monoiodide (AtI)

Tetratomic interhalogens include:

Chlorine trifluoride (ClF_3)

Bromine trifluoride (BrF_3)

Iodine trifluoride (IF_3)

Iodine trichloride (ICl_3)

Hexatomic interhalogens include:

Chlorine pentafluoride (ClF_5)

Bromine pentafluoride (BrF_5)

Iodine pentafluoride (IF_5)

Octatomic interhalogens include:

Iodine heptafluoride (IF_7)

Nanotech molecular manufacturing methods may be employed to fabricate the fuels considered herein.

The added order or arrangement of the molecules and atoms comprising these fuels can be chosen to add additional energy to the fuels as well as promote ideal stoichiometric reactions for a more complete combustion.

Note that all of the proposed oxidizers where not ordinarily possible to produce in stable forms might be produced using nano-technology methods of controlling molecular bonding. Also, some of my proposed oxidizers may already exis. However, it is plausible that novel bond angles can be designed to thus produce stable and novel oxidizers.

Chapter 3

The Asteroid Starship

This chapter includes concepts for hollowed out asteroids for use as starships.

As mentioned previously, Marc Millis, once the head of the former NASA Breakthrough Propulsion Physics Project has founded the 501(c)3 not-for-profit institute known as the Tau Zero Foundation, with the bold mission of advocating the cause for manned missions to the stars. Paul Gilster is the head of Tau Zero Centauri Dreams, the news forum for the Tau Zero Foundation, and posts new threads daily on the subject of manned interstellar space travel and related issues and draws an active readership as shown by the many comments he receives on the articles he posts.

One morning several months ago, Paul posted another one of his high-quality and inspirational articles on the subject of the very recent confirmation of water ice and organics on a large asteroid.

It is nice to know that such large asteroids have water and organics on or within them. I am also happy that such large asteroids exist. I did not previously know that asteroids came in such large sizes at two hundred kilometers across.

These asteroids may be an excellent source for minable materials including materials for use in the construction of large rotating interplanetary space stations.

The organics could be fashioned into carbon-graphite fiber composite materials for use in space station construction.

Since there are numerous large asteroids, the asteroids could be partially hollowed out and used as fusion rocket–powered spacecraft and sent out in droves all over the Milky Way galaxy, even if mass ratios for the craft are close to one, thus barely enabling the achievement of borderline relativistic velocities.

It would be very nice if we found an additional nuclear force. Even somehow obtaining the ability to decompose protons into unbound quark states would be nice. When released, the non-quark invariant mass energy of the protons would result in about 99 percent of the protons' mass being converted into energy. This would enable relativistic velocities to be obtained even with mass ratios that are close to one. Such a mechanism has been discussed in science fiction as a way to create so-called quark bombs. Provided the hadron fission process can include neutrons, upward of 99% of a portion of a bulk of materials based on the periodic table element(s) could be converted into pure energy.

Producing a self-scavenging asteroid with a diameter of two hundred kilometers using the above proton fission concept could permit extreme relativistic Lorentz factors and yet provide adequate shielding for a centrally located asteroidal metropolis, especially if strong shielding magnetic fields could be set up around the asteroid.

Assume a proton and neutron fission rocket's specific impulse is essentially equal to 1 C, and a mass ratio of 1,000, which would bring the diameter of the asteroid down to 20 kilometers after fuel burnup. The relativistic rocket equation yields a velocity of 0.999998 C yields $\gamma = 500$. The cosmic microwave background (CMBR) background radiation would be blue shifted by a factor of 500 and would thus appear to have a temperature of (500)(2.725 K), or 1,993°F. A hard, iron-nickel, natural-alloy asteroid should be able to withstand the blue-shifted CMBR especially if coated with a super-reflective, highly refractory, carbon-like material. The thickness of the asteroidal encasement should be more than adequate to stop the plasma jets produced by collisions with dust specks of up to 10 microns in diameter or with a mass of up to about 1 nanogram.

Note that the neutron fission process in this case is not the same as the decay process of isolated neutrons, which have a half-life roughly on the order of fifteen minutes. Commonly observed neutron decay does not liberate nearly as much energy as the ad hoc conjectural neutron fission process described above.

A more reasonable Lorentz factor of 200 would result in the CMBR being blue-shifted to a black body temperature of (200)(2.725) K, or 86°F. The asteroid could serve as a balmy late-spring-like temperature habitat where air-conditioning could be utilized as needed.

The above velocity is considered extremely relativistic, and might be accomplished without the use of matter-antimatter reactions and the risk of carrying large quantities of antimatter, along with the asteroid. If we need to rely on antimatter, in the end, I think we will rally to the occasion during the centuries, millennia, and eons to follow, but simple proton and neutron fission induced with low energy reactions would be a welcome breakthrough.

To achieve an I_{sp} of around 1 C, the energy produced by proton fission would need to be directed backward in a coherent stream such as a highly relativistic electron or ion beam, or perhaps as a photon beam. To accomplish this, the proton fissions would need to occur within a large thermal mass, and the fission reactions would need to release only a very small percentage of the fission energy in the form of neutrinos unless the neutrinos could somehow be emitted in a biased direction directly backward. The heat generated could then be used to drive turbo-electric steam cycle systems, preferably multi-cycle systems, which would then power lasing mechanisms or particle beam accelerators having an efficiency close to 100 percent. Any residual heat could be radiated nearly directly backward by a radiator system coupled to a highly reflective infrared mirror that is optimally shaped to direct the IR radiation in an as collimated configuration as possible.

The asteroid spacecraft could contain an electromagnetic Dewar type of shielding so as to prevent high-frequency IR radiation, as well as electrical equipment-based microwave and radiofrequency emissions from being lost to space in a useless manner. This electrodynamic energy could then be recycled to aid in ion, electron, or photon stream exhaust production.

The backwardly directed energy emissions in the form of collimated thrust would provide the thermodynamic gradient for the spacecraft to operate its internal machinery such as life-support systems and the like.

Now, how do we induce low-energy proton or neutron fissions, especially given the observed and theoretical realities that the constituent top and bottom quarks comprising the proton (2U, 1D) and the neutron (2D, 1U), cannot exist as stable and free particles? When these constituent quarks are pulled on by simple means, such as by high-energy particle collisions, the binding energy, or gluon field lines, holding the quarks together build up tension to a maximum level and then metaphorically snap with the resulting energy release taking the form of additional quarks, which are then bound to the original quarks at least in part. No doubt, quarks can decay or be transmuted into other particles such as by proton-antiproton interactions, ultra-high-energy proton-proton, proton-antiproton collisions, or other high energy reactions.

However, the energy of the decay products is essentially the same as the input energy, and thus very little net energy can be gained this way. Consider the possibility that individual proton or neutron fissions could be motivated using, say, only 1 million eV of perturbative energy per nucleon. Perhaps the yield factor would be about (936 MeV)/(1 MeV), where the numerator is about equal to the energy equivalence of the invariant mass of either the proton or the neutron.

We can always hope for invariant mass-specific reactions with energy releases greater than $E = MC^2$. We will have to see what the folks operating the Large Hadron Collider, the proposed TeV range electron-positron linac, the eventual

LHC upgrade to 40 TeV collision energies, the Facility for Rare Isotope Beams being constructed within the United States, and a proposed Muon accelerator can come up with. All I can say is expect the unexpected with these machines.

Regardless, we know we have nuclear fusion, and this may enable multigenerational ships formed from partially evacuated large asteroids. An optional mini-magnetosphere setup using permanent magnets or electromagnets could shield the crew from cosmic rays.

I am pleased that President Obama desires manned missions to near Earth asteroids. If we can mine asteroids for rare Earths or exotic elemental and isotopic materials, this would jump-start our society into a truly interplanetary civilization, and then onward to other star systems in perhaps relatively short order.

Paul also touched on the subject of the complete lack of detected monopoles, or exotic particles that, theoretically, would embody unipolar magnetic charges just as an electron embodies a unipolar electric charge. He mentions the prodigious energy that might be released when a negative magnetically charged monopole would be directed into a positive magnetically charged monopole.

Consider the hypothetical scenario where we would create isolated unipolar magnetically charged particles or monopoles and then combine them at will. Perhaps we will have the wherewithal to produce high-energy density materials like none that are theoretically known to exist. At the very least, materials that are many times higher in invariant mass specific yield than ordinary exothermically nuclear fusionable materials might be produced. See Gilster, The Enigma of Contact, April 26, 2010, at http://www.centauri-dreams.org/?p=12229

A perhaps more realistic scenario would involve hollowing out the asteroid and replacing 90 percent of its mass with nuclear-fusion fuel obtained from the gas giant planets of Jupiter and Saturn or from Neptune or Uranus. The latter two planets may be a more practical source of fusion fuel due to the planets' smaller gravity. Alternatively, cometary bodies, Pluto, other Kuiper Belt objects, or Oort cloud objects could be scavenged for primordial hydrogen fuel.

Note that care would be needed while extracting large quantities of hydrogen and helium from the gas giants or from any other planetary body in our solar system. The many body problem of the associated orbital mechanics would need to be solved fairly rigorously so that orbital destabilization does not occur. Restabilization of unstable orbits or stabilized mining of fuels from the gas giants could be achieved by properly timed extraction, as well as by dropping inert mass into the planets to compensate for fuel mass extraction. The inert mass could be captured from asteroidal bodies, comets, Kuiper Belt, and/or Oort cloud objects. In addition, spacecraft "tug boats" could gradually pull the mined planets into a more stable orbit.

Assume nearly 100 percent efficient fusion rocket propulsion systems and a mass ratio of 10. The asteroid could obtain a velocity of 0.13615 C, and also come to rest at the target star system using its fusion powered rocket mechanism. Assume nearly 100 percent efficient fusion rocket propulsion systems and a mass ratio of 10. The asteroid could obtain a velocity of 0.26734 C and come to rest at the target star system without rocket-assisted braking. Once again, some methods of rocketless braking include field effect electrodynamic systems and electrodynamic-hydrodynamic-plasma-drive systems. Electrodynamic braking systems can include linear induction magnetic braking, magnetic plasma bottle sails, and the like. Some forms of electrodynamic-hydrodynamic-plasma-drive systems include reverse electro-hydrodynamic-plasma-drives, reverse magneto-hydrodynamic-plasma-drives, reverse electromagneto-hydrodynamic-plasma-drives, reverse-thrust interstellar ramjets, and the like.

Just as occurred in a whimsical episode of the original made-for-TV Star Trek series, huge asteroidal spacecraft might be fashioned into highly relativistic semi worldships. The construction materials for these ships are already in place. The materials of construction need only be sequestered and an appropriately high nuclear energy density fuel be collected along with R&D and construction efforts to turn asteroids into huge interstellar spacecraft.

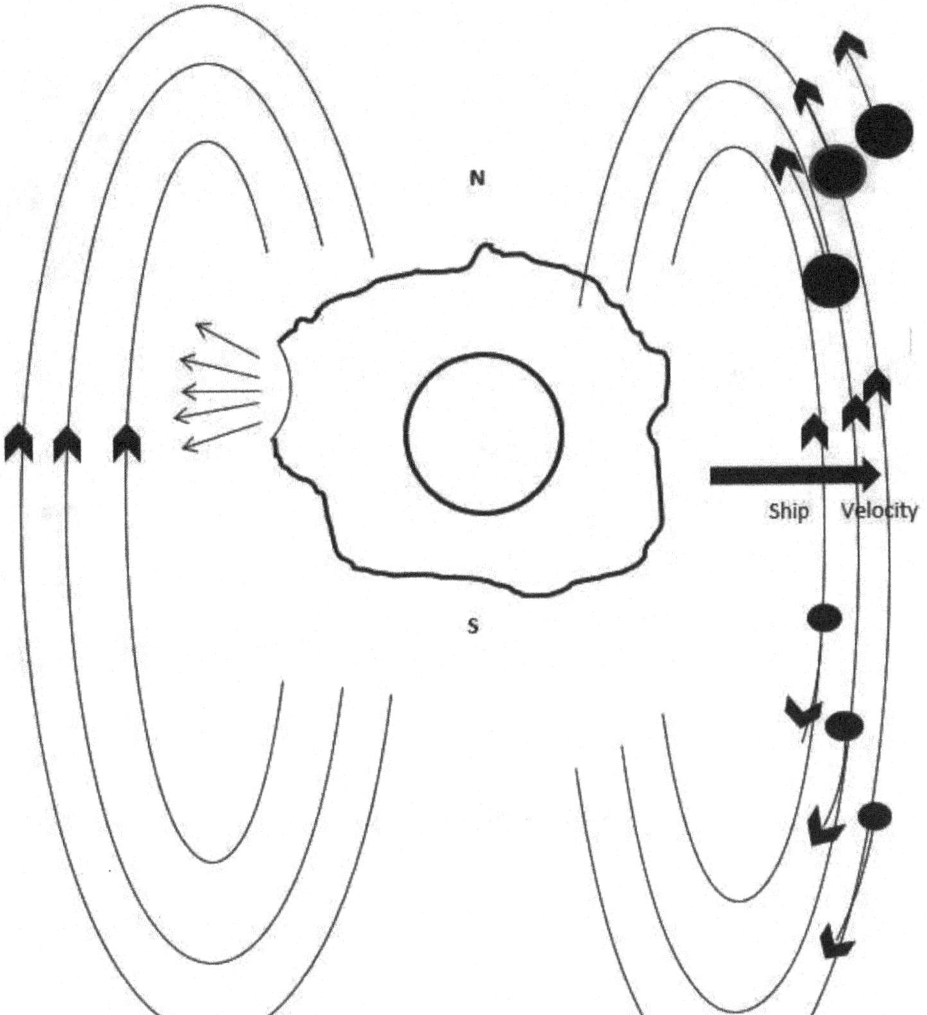

FIGURE 1 depicts a hollowed out asteroid propelled by a relativistic rocket mechanism. The fuel may plausibly include any one or all of the following: 1) matter and antimatter, 2) antimatter, 3) nuclear fusion fuel, and 4) hybrid matter-antimatter atomic fuels. The rays projected from the bottom of the image represent the relativistic exhaust.

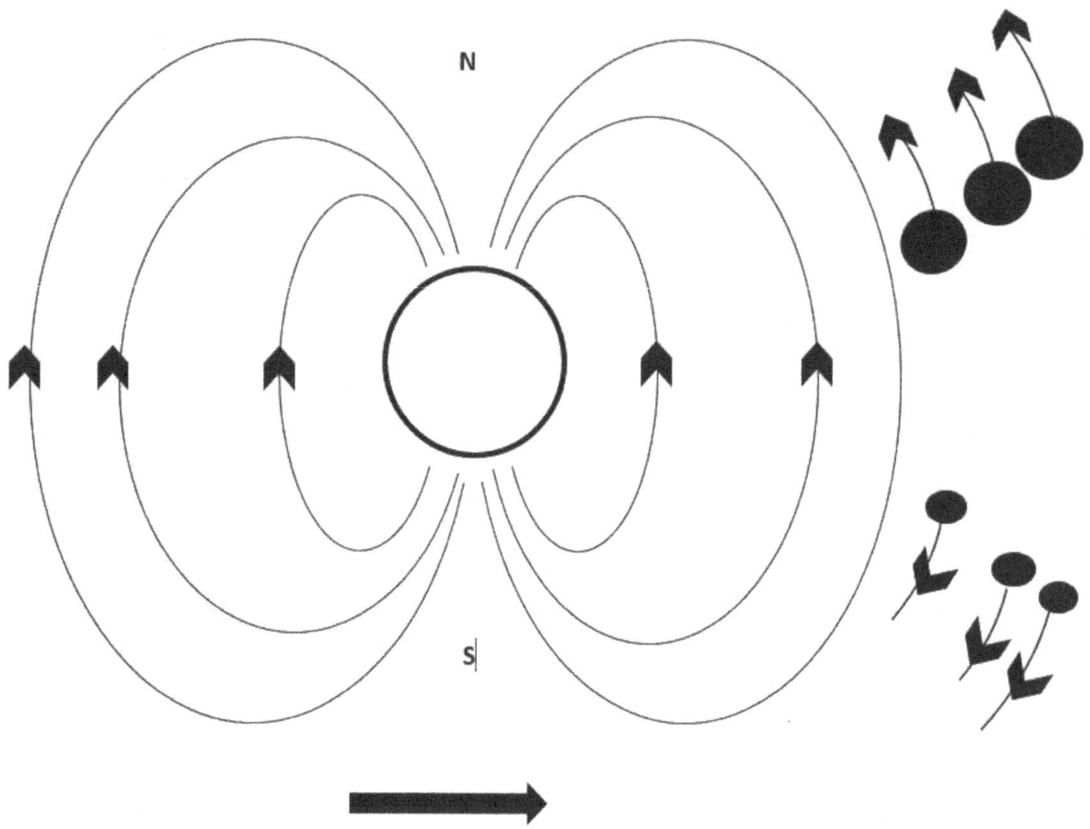

Spacecraft Velocity

FIGURE 2 depicts a mini-magnetosphere such as may be deployed around a generic relativistic spacecraft. The magnetic field indicated by the dashed arrowed lines is depicted as seen from the spacecraft reference frame. The larger dots with upwardly oriented arrows attached indicate positively charged particles, most likely protons and heavy metal ions, and the smaller dots represent negatively charged particles.

Chapter 4

Nuclear Comet Starships with Solar Power Augmentation

This chapter presents comets for use as relativistic rocket nuclear-fusion fuel.

Self-contained nuclear fusion starships may, in theory, attain velocities of 0.975 C, or a Lorentz factor of 4.5 when fabricated out of comet-sized hydrogen fuel balls or other fusion fuel ice balls, while allowing for a final payload of roughly the size and mass of a modern-day nuclear-powered aircraft carrier.

A comet's capture and its subsequent incorporation into a spacecraft might enable fusion rocket craft to obtain Lorentz factors close to 2, which is commensurate with velocities on the order of 0.867 C. A billion-metric-ton comet that was almost pure hydrogen ice might permit a mass ratio of 100,000. A velocity close to 0.867 C is obtainable according to the relativistic rocket equation, $\Delta v = C \text{ Tanh } [(I_{sp}/C)\ ln\ (M_0/M_1)]$, with a payload of 10,000 metric tons and where an I_{sp} expressed in units of C is assumed to be equal to the maximum value of 0.119 C for fusion fuel.

Consider a mass ratio of 1,000,000 utilizing a 10-billion-metric-ton invariant mass comet, and thus a final payload of 100,000 metric tons invariant mass, and an I_{sp} of 0.119 C. The resulting velocity is equal to 0.928035 C.

Consider a mass ratio of 10,000,000 utilizing a 1 trillion metric ton invariant mass comet and thus a final payload of 100,000 metric tons invariant mass, and an I_{sp} of 0.119 C. The resulting velocity is equal 0.957756 C.

Consider a mass ratio of 100,000,000 utilizing a 10-trillion-metric-ton-invariant mass comet and thus a final payload of 100,000 metric tons invariant mass, and an I_{sp} of 0.119 C. The resulting velocity is equal to 0.97536 C. In the latter case, the relativistic Lorentz factor is 4.5327.

For tabulations and graphs of mass ratio, B = v/C, and gamma for ideal fusion fueled systems having a specific impulse of 0.119, and for systems that are 80 percent and 60 percent efficient, refer to chapter 6.

We can use any of the previously listed electrodynamic braking mechanisms to slow down the craft.

I once read a paper regarding the fusion fuel ice ball concept including a mass ratio of 100,000 able to obtain a velocity in the range of 0.63 C. The comet-like ice ball consisted of a concentric arrangement of fusion fuel ice layers where each layer would be wrapped in some form of insulating and containment membranous material(s).

A friend of mine, Adam Crowl, provided me the following three URLs regarding the similar, if not nearly identical, concept known as the Enzmann Starship:

http://www.centauri-dreams.org/?p=1142

http://crowlspace.com/?p=589

http://enzmannstarship.com/

However, achieving 0.075 C to 0.1 C would be perfectly most welcome by me. After all, it was about forty years ago when the voyager spacecraft were launched and we are still receiving signals from these spacecraft. In about forty years, a 0.1 C class space probe or manned starship could reach Alpha Centauri. This realization gives me much hope.

It is disappointing that, so far, no gas giant planets have been found around Alpha Centauri. I still wonder about the prospects for finding two to five near Earth mass planets around Alpha Centauri around which might orbit 0.2 to 0.5 Earth mass moons. Perhaps even a 0.1 Earth mass moon could support life.

Regardless, when considering huge interstellar spacecraft, the fictional account in the first *Aliens* movie provides for some whimsical mental imagery and possible future directions of 0.9 C capable starships.

Assume that some form of volumetrically extremely dense storage of I_{sp} optimized fusion fuel could be developed. Also assume the development of extremely lightweight fusion fuel containments. Such containments might be fabricated from the following:

1. Monolithic multilayer graphene-based materials
2. Carbon nanotube materials
3. Boron nitride nanotube materials
4. Diamond fiber fabrics
5. Beta carbon nitride fiber fabrics
6. Graphene oxide paper
7. One atom-wide carbon chain-based fabrics
8. Combinations of the above

In addition, there are other low-density extremely strong materials. Thus, the above extreme Lorentz factors for self-contained fusion-powered starships become plausible.

The ship could start out with low acceleration at ~ 0.03 to ~0.1 G and ramp up acceleration to perhaps 1 G to 2 Gs after the mass of the fuel was mostly depleted in order to reduce mechanical stress. Any of the previously listed electrodynamic braking mechanisms can be used to slow the craft.

Even though we might not have a Pandora to aim for, science fiction still gives us plenty of other whimsical scenarios. The good news is that we have loads of fuel right here in our own solar system, and presumably, any stellar destinations would have the same. Such natural resources could provide fuel for highly relativistic fusion-powered starships that carry all their fuel on board from the start of the mission.

Assuming we can determine how to do suspended animation, or at least, nanotechnology controlled hibernation at body temperatures of perhaps 3°C or slightly lower, and maintain such states indefinitely, travel itineraries over truly cosmic spatial-temporal distances into the future and then become possible. In addition, if we assume that high relativistic Lorentz factors are possible for crews that remain awake for the journey, indefinite human life expectancy enhancement is possible, or huge relativistic world Zonds can be fielded, the universe is the limit.

Photovoltaics, thermoelectric, photothermal, turboelectric, and nuclear fission apparatus may be used to power mining operations.

Sunlight may optionally, as nonlimiting example, be concentrated for more mass-specific efficient power generation and process heat.

Many options exist for enhancing the effects of captured or intercepted solar energy for enabling relativistic rocket flight.

For example, as contemplated earlier in this text, photovoltaic, thermoelectric, thermomechanical, and combinations of any two or all three of the latter mechanisms can be used to power optimized chargon rockets.

In order to reduce the mass of spacecraft vehicles driven by electrical propulsion systems, large solar concentrators may be deployed as now presented.

First, solar concentrators may include simple gas-pressure-inflated reflectors. Accordingly, the reflectors would energize electrical power–producing mechanisms such as photovoltaic, thermoelectric, thermomechanical, and combinations of the any two or all three of the latter mechanisms.

These concentrators may include continuously curved surfaces or tiled or faceted surfaces. Faceted surfaces are useful for limiting light concentration. Faceting also permits superambient pressurization of a basic support mechanism on which membranous tiles can be attached at discreet or continuous portions around the tiles' effective periphery. The facets may be formed from one multiply connected reflective membrane or more than one membrane where the multiple membranes form an array.

Faceted reflector surfaces may also be formed by discretely corded attachment of two opposing gas or electrostatically superambient pressured membranes and/or by wall-like attachments between the two opposing membranes.

Another faceting mechanism would include corded or walled superambient pressured membranes on which rigid or rigidizable reflective plates would be attached. Such reflective plates may be initially permanently solid, or include resin-filled platelike pockets or envelopes that are hardened by ultraviolet light originating from the sun, or perhaps by manual and/or automated UV light chemical activation mechanisms.

For cases where the reflective membranes are deployed by means other than gas pressure (for example, such as by electrostatic, magnetostatic, cord ties, or wall bonds), the reflector can take the form of a net or grid of crossed reflective fibers for greatly reduced reflector mass.

Second, solar concentrators may include electrostatically deployed reflectors for which a gas-pressure-inflated ring supports two axially spaced and opposing membranes. One membrane would be reflective while the other membrane may be optionally reflective. The reflective membrane may be parabolic when deployed, or have other deployed configurations such as having cross-section of: spherical shape; oblate spherical shape, prolate spherical shape, ovular shape, elliptical shape, catenary in shape, and the like as nonlimiting examples.

Continuously curved membranes may be deployed by superambient or subambient gas inflation and/or electrostatic pressure.

Such reflective membranes when held at subambient pressure by gas pressure would need confinement within a pressurized vessel with higher internal gas pressure.

Simple superambient gas-pressured membranous reflectors can include versions for which the focal point, or focal points, lie outside of the pressurized cavity, on the surface of one of the pressure-deformed membranes, or within the pressurized encompassment.

Instead of using gas, an electrostatic or magnetostatic mechanism may be used to spread two or more membranes joined to form a pocket for which at least one of the membranes forms a surface that is concave with respect to the incident sunlight. However, in some cases where inflation gas used alone would be to massive, a combination of inflation gas, electrostatic repulsion, and/or magnetostatic repulsion may be employed.

Superambient pressured devices of any of the basic forms described above can optionally have internal and/or external membranous and/or rigid element-reflective facets. Moreover, the reflective facets may also include UV-hardened resin for rigidizement.

For cases where the reflective membranes are deployed by means other than gas pressure, for example, such as by electrostatic, magnetostatic, cord ties, or wall bonds, the reflector can take the form of a net or grid of crossed reflective fibers for greatly reduced reflector mass.

Third, non-pressure-deployed membranes may be attached to a concave assemblage of inflatable toroidal or other ring-type inflatable vessels. The concavity may be enabled by a concentric or staggered concentric series of rings.

For example, a series of nested inflatable rings of radially increasing thickness and diameter may be covered with a reflective membrane or thin flexible sheet of reflective material. The rings may have radially increasing thickness along one or both cross-sectional coordinate axes.

The reflector element attached to any such ring assemblage may be attached in a discretized manner of multiple continuous circular or ring-type attachments or, alternatively, be attached at discrete points.

Staggered concentric inflatable ring assemblages may include rings of identical or differing thickness. Rings of differing thickness may be nested in any arbitrary patterns and combinations of thickness such as may be most useful or cost effective for one or more given reflector applications.

The contour of such nested ring assemblage, where the reflector element is attached may include the following shapes as nonlimiting examples: a parabola, an oblate spheroid, a prolate spheroid, an oval, an ellipse, a sphere, a catenary, a cone, a pyramidal cone of three or any useful greater number of facets, and the like.

Other shape include concave surfaces having a wavy contour that can include continuity of curvature and/or step like or peal-like patterns having singularities.

Where facets are included in the reflective surface, the facets themselves may be concave and/or convex, or planar in surface contour.

For cases where the reflective membranes are deployed by means other than gas pressure, for example, such as by electrostatic, magnetostatic, cord ties, or wall bonds, the reflector can take the form of a net or grid of crossed reflective fibers for greatly reduced reflector mass.

Fourth, low mass concentrators may plausibly be deployed by inflatable truss works. Here, an assemblage of inflatable tubes would deploy by inflation extension. The reflective membranes attached to the tubes at select locations would be spread into a concave shape. The reflective membranes may be monolithic or, alternatively, a netlike grid, so long as the grid-lines were separated by distances no greater than slightly less than the length of the intended reflected electromagnetic radiation.

The concavity of the truss structure can be continuously curved or discretely tiled. Concave shapes can include a parabola, a spherical shape, an oblate spherical shape, a prolate spherical shape, an ovular shape, an elliptical shape, a catenary shape, a cone, a pyramid of arbitrary number of facets, and the like as nonlimiting examples.

The truss work may optionally include UV light–hardening resins so that the deployed truss upon inflation then becomes a rigid structure without further need of inflation pressure. However, a slight under-pressure as a result of the initial inflation gas may be employed to help support and stabilize the rigidized structures. Note that the UV-hardened resin can be used to rigidize any of the pressure deployable concentrators previously described.

For cases where the reflective membranes are deployed by means other than gas pressure, for example, such as by electrostatic, magnetostatic, cord ties, or wall bonds, the reflector can take the form of a net or grid of crossed reflective fibers for greatly reduced reflector mass.

Fifth, any of the above-mentioned space-based solar concentrators may be used as single units or employed as an array of units. Moreover, concentrators presented in more than one or all of the previous four general scenarios may be used as an array. Use as an array enables reduced quantities of required inflation gas because the effective thickness of the gas envelope of an array is reduced by roughly the power of two-thirds of the volume of the inflation gas required under the condition where only one large inflatable concentrator would be used but where the array of smaller concentrators would have the same light-capturing area as would the large single concentrator otherwise deployed.

Reflective membranes where monolithic include single sheets, multisheet laminates, fiber-reinforced single sheets, or fiber-reinforced multisheet laminates. The reflective portion of the reflective membranes may include internal or external metallic layers, which may be optionally instilled by chemical deposition, vapor deposition, ion beam deposition, electrolytic processes, or any other process by which reflective coatings are installed on reflective membranes.

For fiber-reinforced membranes, the fibers themselves may optionally but not necessarily compose of alloys or elemental metals with very high tensile strength or, optionally, include other high-strength materials such as carbon nanotubes, multiwalled carbon nanotubes, graphene fibers and/or strips, boron nitride nanotubes, graphene oxide paper strips, diamond fibers, carbon atom chains, beta-carbon-nitride fibers, and the like.

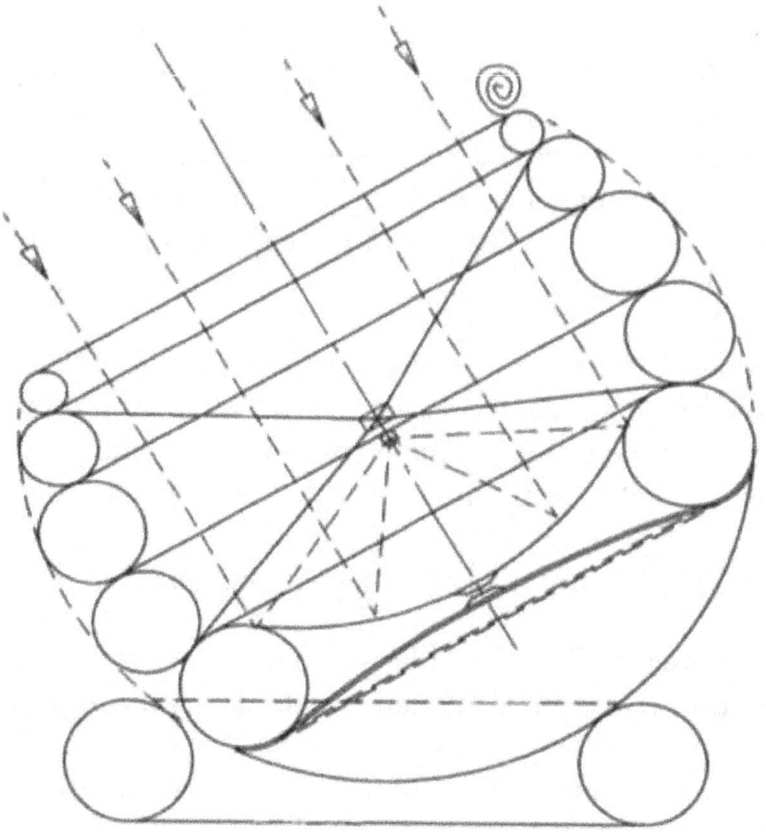

FIG. 1B

Fig. 1B. An excerpt from US Patent Application No. 10/729,145 of John R. Essig and James M. Essig illustrating a low-friction inflated ring-mounted solar concentrator that is herein interpreted to be pressure deformed by electrostatic charge.

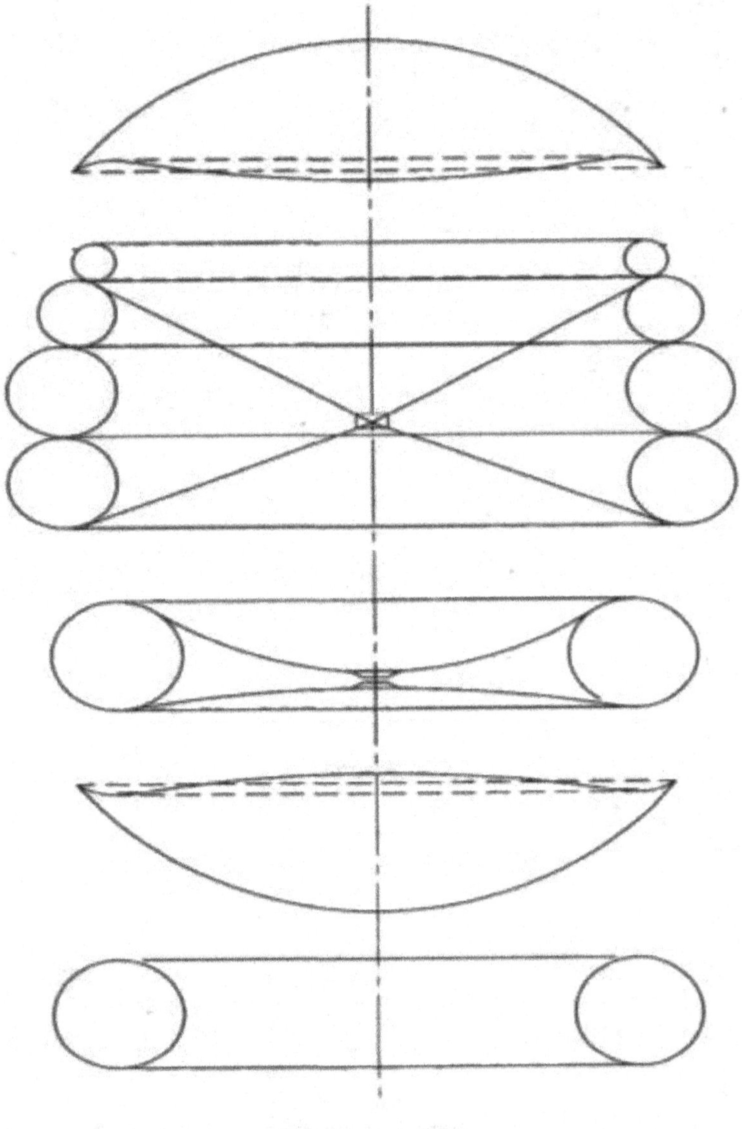

FIG. 1C

Fig. 1C. An excerpt from the US Patent Application No. 10/729,145 of John R. Essig and James M. Essig illustrating the portions of a low-friction inflated ring-mounted solar concentrator, which is herein interpreted to be pressure deformed by electrostatic charge.

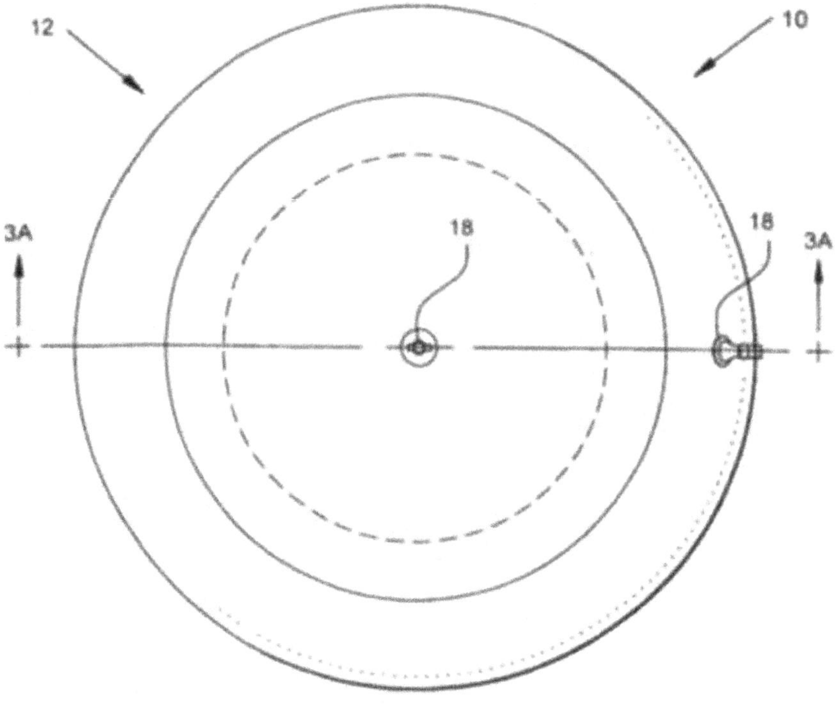

FIG. 2A

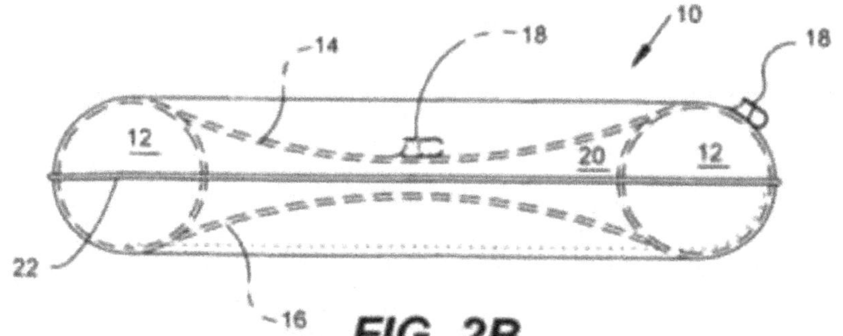

FIG. 2B

Fig. 2A–2B. An excerpt from the US Patent Application No. 10/729,145 of John R. Essig and James M. Essig illustrating a simple pressure-deployable solar concentrator, which is herein interpreted to be pressure deformed by electrostatic charge.

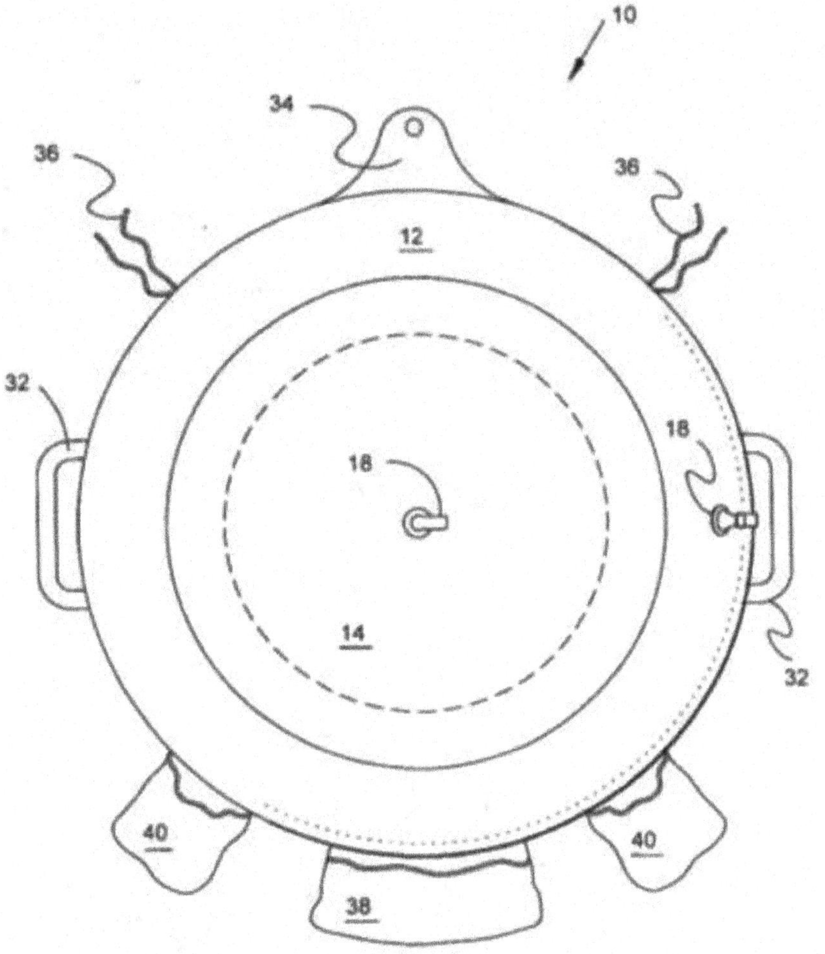

FIG. 2C

Fig. 2C. An excerpt from the US Patent Application No. 10/729,145 of John R. Essig and James M. Essig illustrating an inflated ring-mounted solar concentrator, which is herein interpreted to be pressure deformed by electrostatic charge and also showing various hardware attachment features.

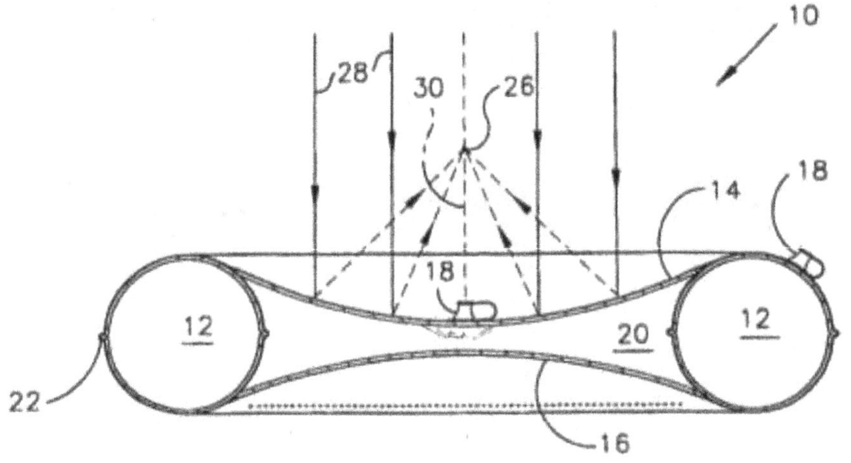

FIG. 3A

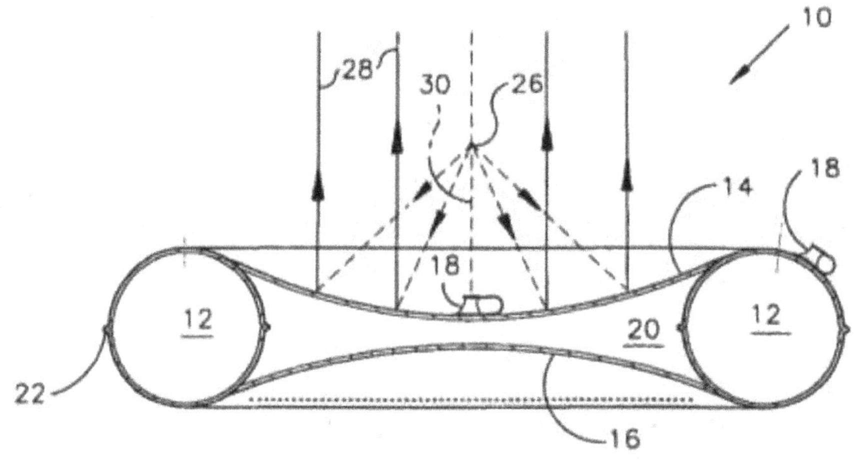

FIG. 3B

Fig. 3A–B. An excerpt from the US Patent Application No. 10/729,145 of John R. Essig and James M. Essig illustrating two pressure-deployed solar concentrators, which are herein interpreted to be pressure deformed by electrostatic charge.

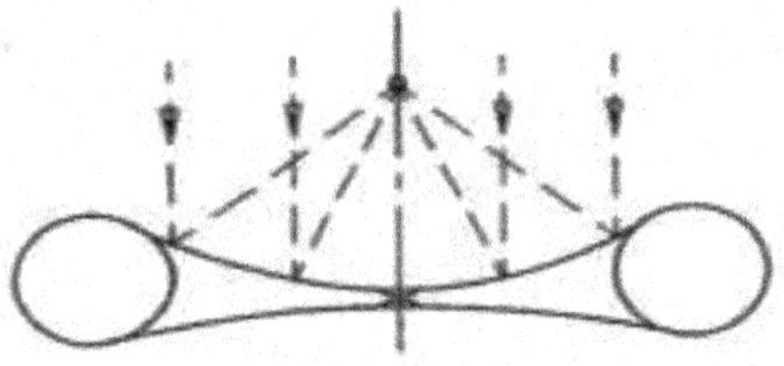

FIG. 3C

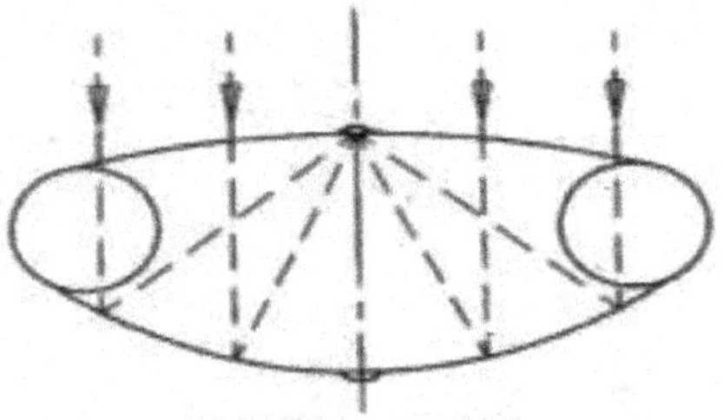

FIG. 3D

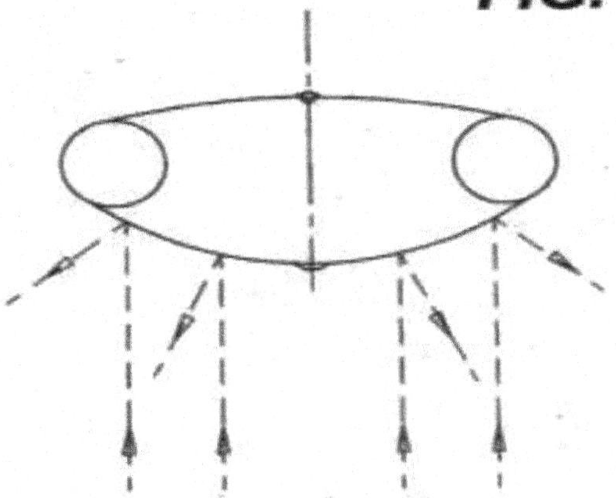

FIG. 3E

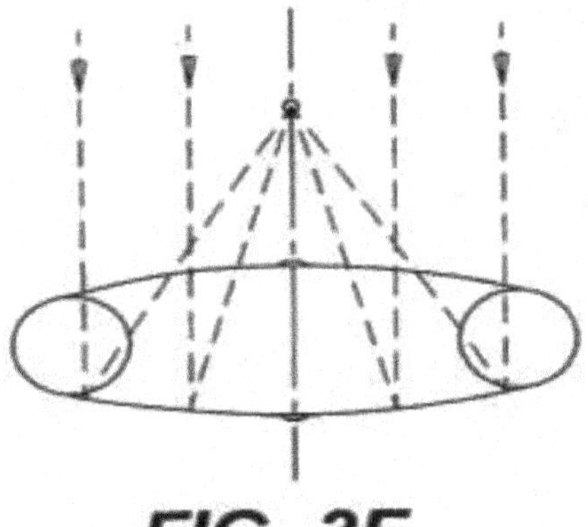

FIG. 3F

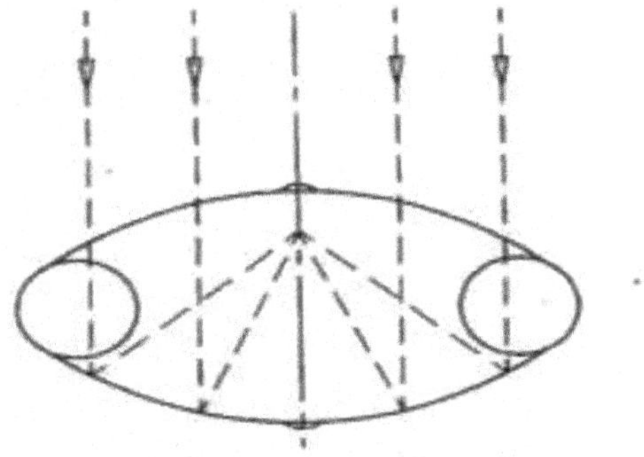

FIG. 3G

Fig. 3F–G. An excerpt from the US Patent Application No. 10/729,145 of John R. Essig and James M. Essig illustrating inflated ring-mounted solar concentrators, which are herein interpreted to be pressure deformed by electrostatic charge and/or inflation gas.

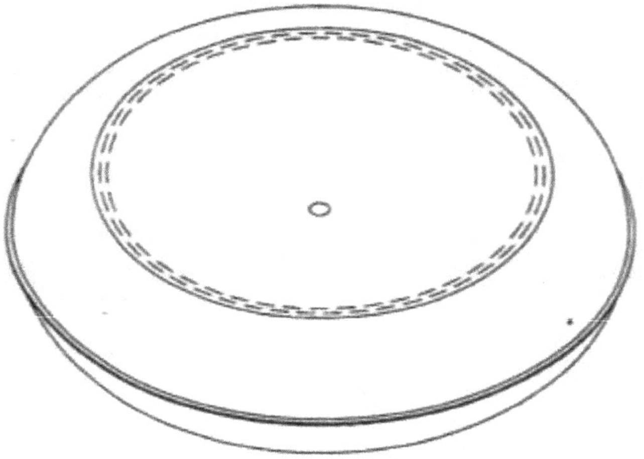

FIG. 5A

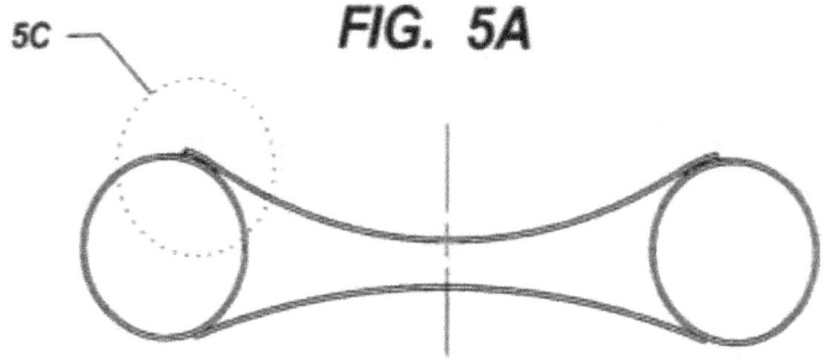

FIG. 5B

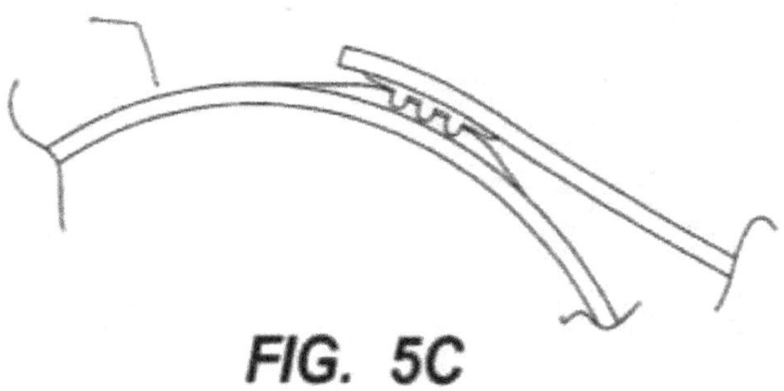

FIG. 5C

Fig. 5A–C. An excerpt from the US Patent Application No. 10/729,145 of John R. Essig and James M. Essig depicting a mechanism for attachment of a pressure-deployable solar concentrator membrane to an inflated ring.

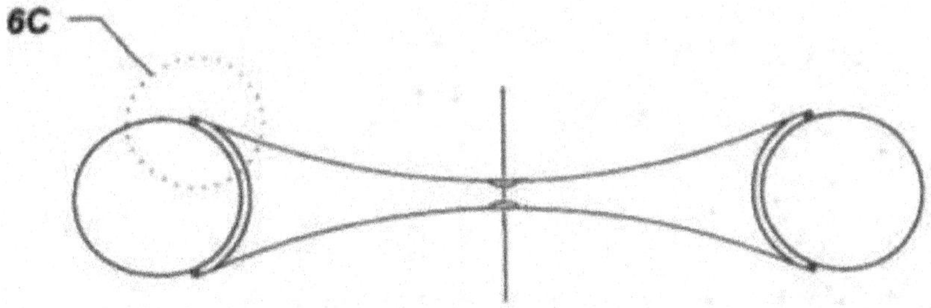

FIG. 6A

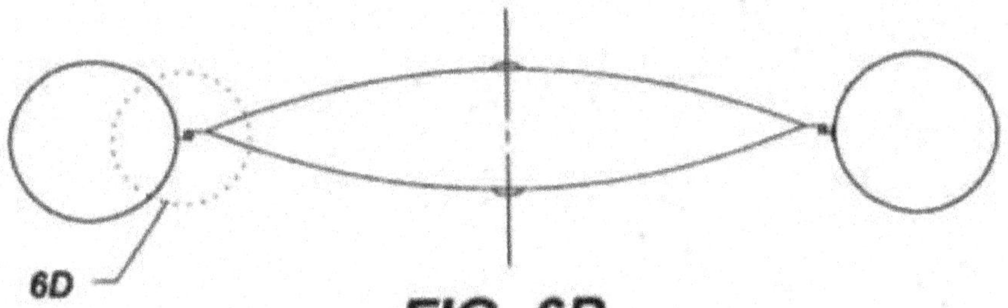

FIG. 6B

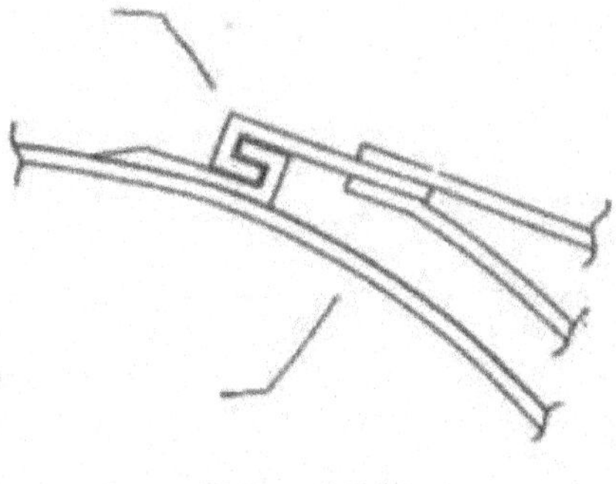

FIG. 6C

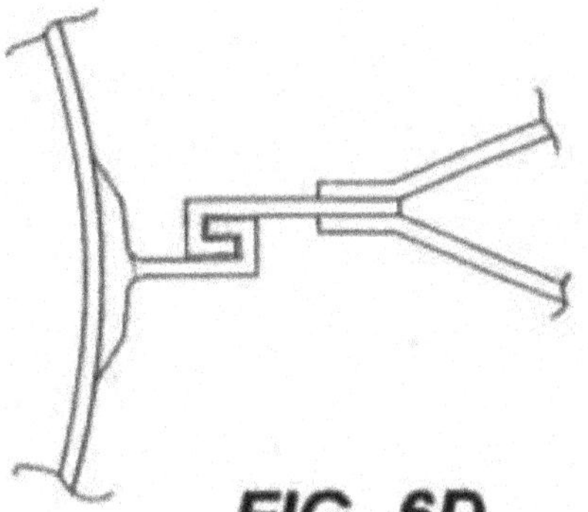

FIG. 6D

Fig. 5A–C. As excerpt from the US Patent Application No. 10/729,145 of John R. Essig and James M. Essig depicting mechanisms for the attachment of pressure-deployable solar concentrator membranes to inflated rings.

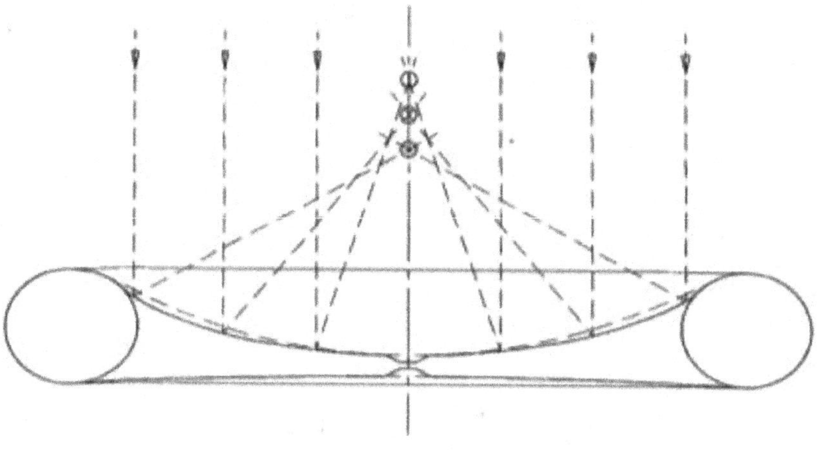

FIG. 7A

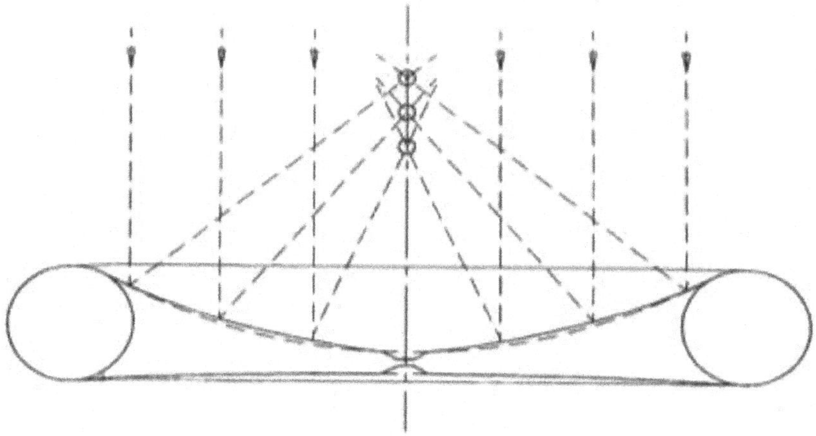

FIG. 7B

Fig. 7A–B. An excerpt from the US Patent Application No. 10/729,145 of John R. Essig and James M. Essig depicting conical collimated concentrated light patterns produced by a pressure-deployed solar concentrator membranes that are appropriately skewed from parabolic form.

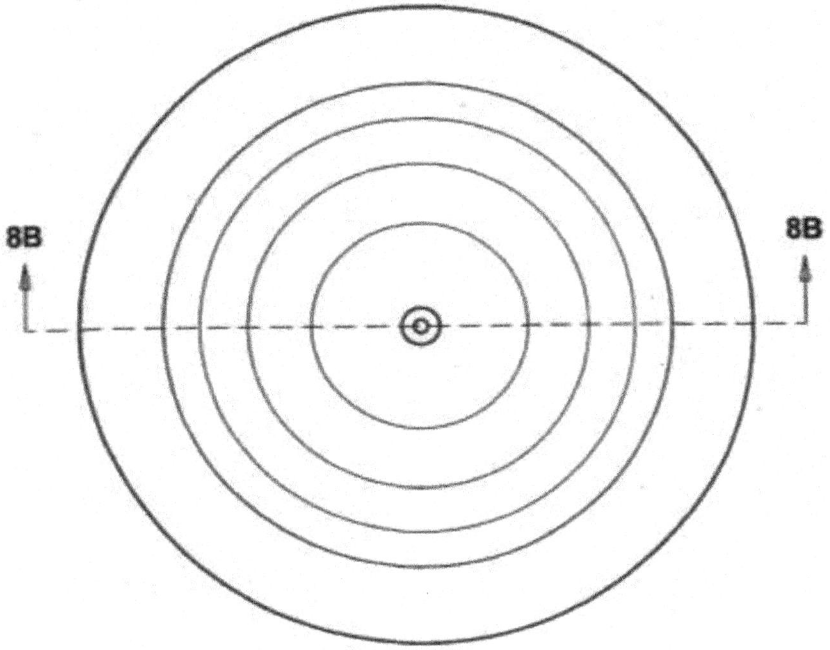

FIG. 8A

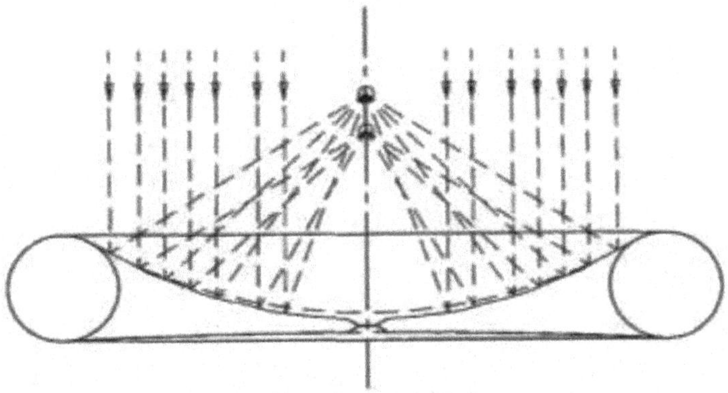

FIG. 8B

Fig. 8A–B. An excerpt from the US Patent Application No. 10/729,145 of John R. Essig and James M. Essig depicting a conical collimated concentrated light pattern produced by a pressure-deployed solar concentrator membrane that is appropriately skewed from parabolic form and having annular reflective facets of radially decreasing width.

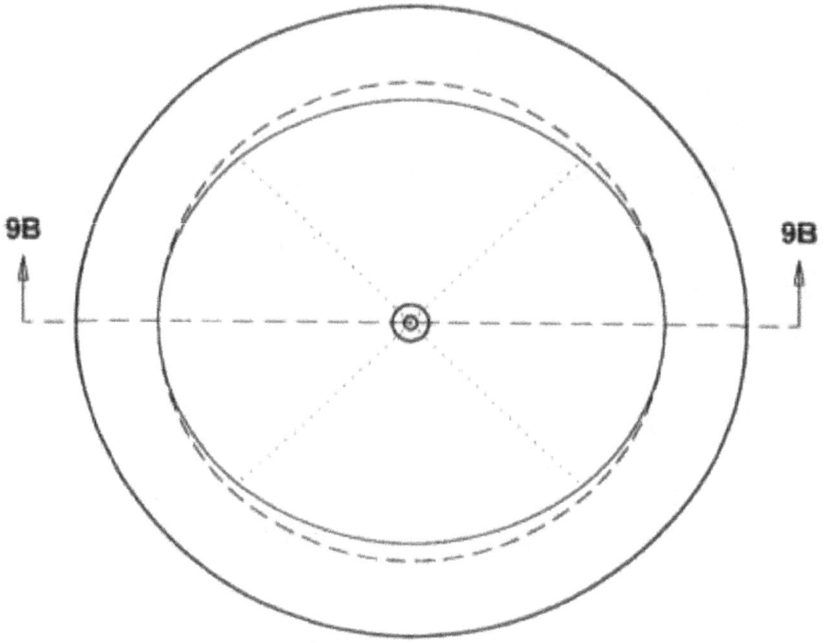

FIG. 9A

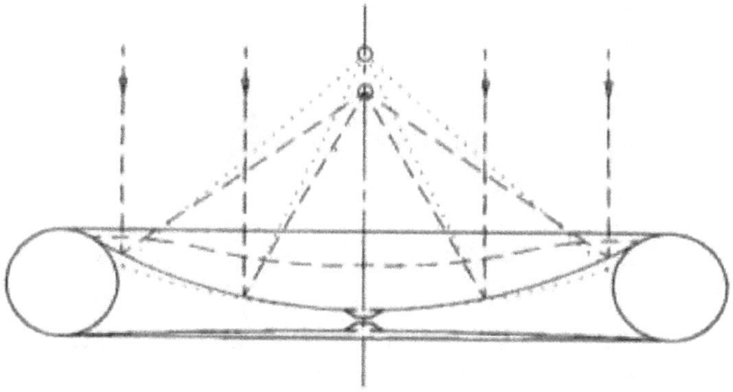

FIG. 9B

Fig. 9A–B. An excerpt from the US Patent Application No. 10/729,145 of John R. Essig and James M. Essig depicting a conical collimated concentrated light pattern produced by a pressure-deployed solar concentrator membrane that is appropriately skewed from parabolic form. Herein, nodes of undulation are shown.

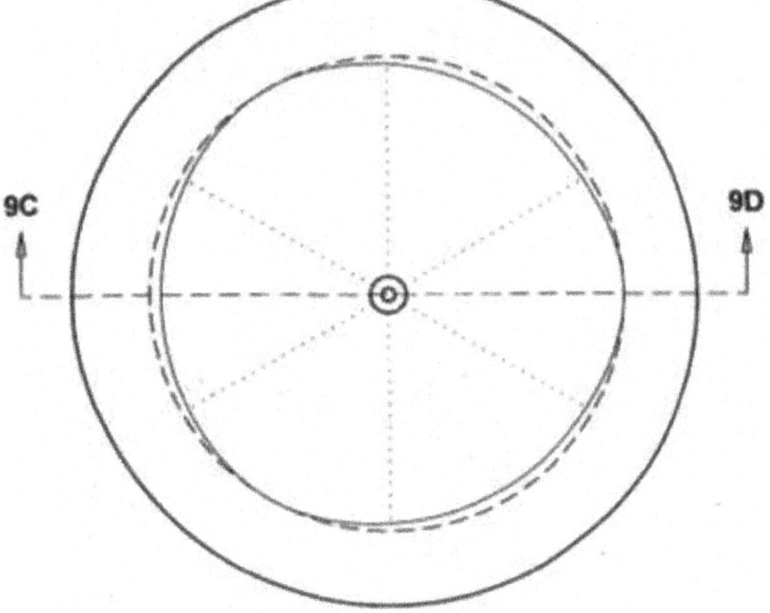

FIG. 9C

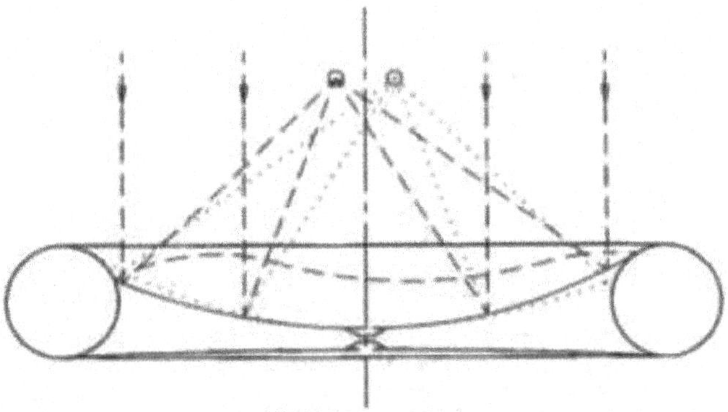

FIG. 9D

Fig. 9C–D. An excerpt from the US Patent Application No. 10/729,145 of John R. Essig and James M. Essig depicting a conical collimated concentrated light pattern produced by a pressure-deployed solar concentrator membrane that is appropriately skewed from parabolic form. Three nodes are shown, which produce a wavy, undulating ring like focal area.

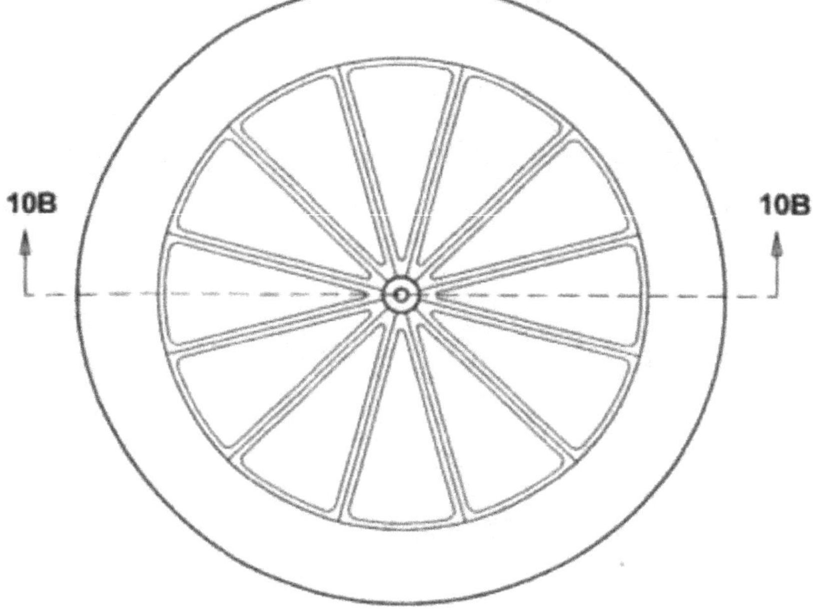

FIG. 10A

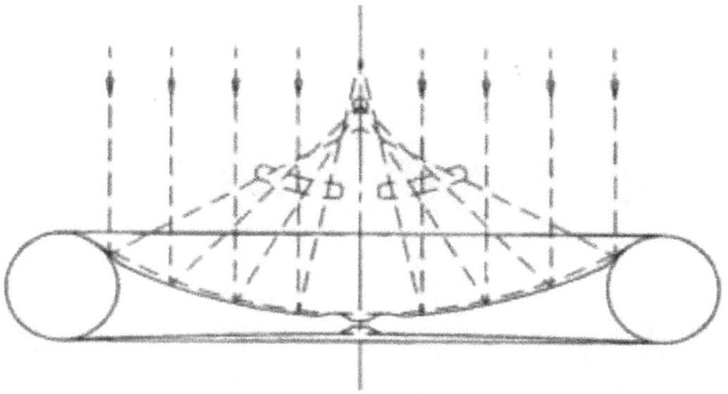

FIG. 10B

Fig. 10A–B. An excerpt from the US Patent Application No. 10/729,145 of John R. Essig and James M. Essig depicting a conical collimated concentrated light pattern produced by a pressure-deployed solar concentrator membrane that is appropriated skewed from parabolic form. The focal area has a segmented regular polygonal ring like shape.

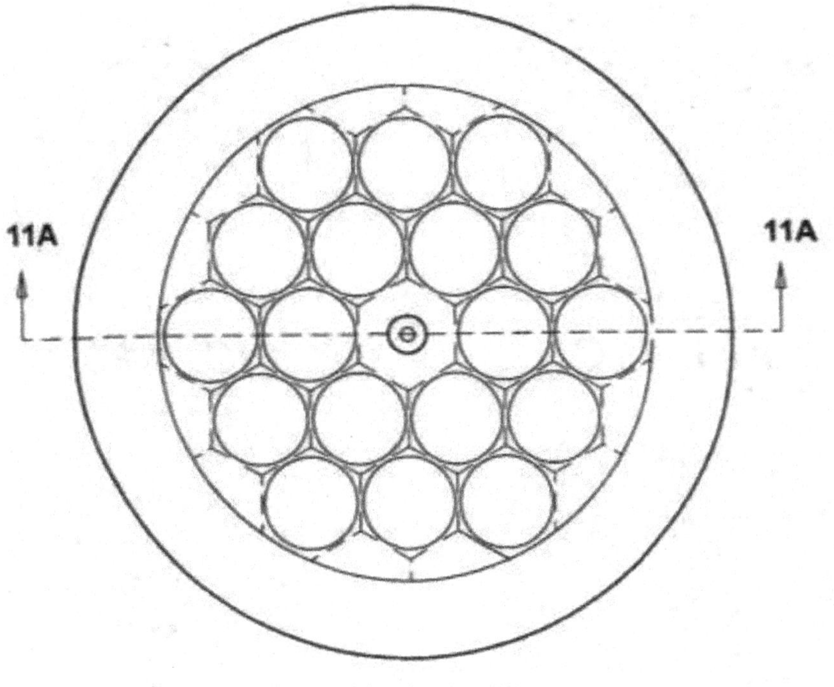

FIG. 11A

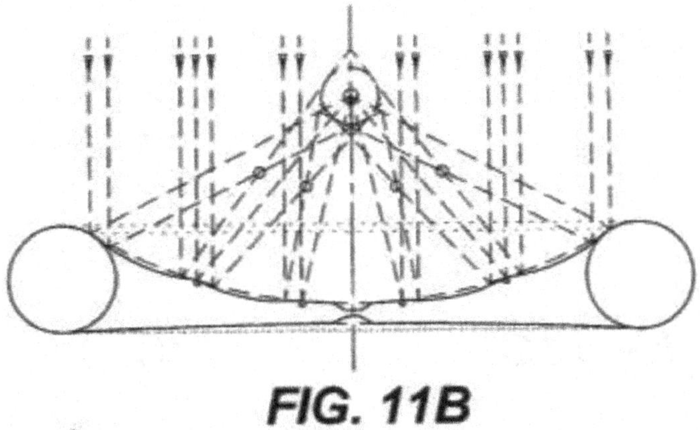

FIG. 11B

Fig. 11A–B. An excerpt from the US Patent Application No. 10/729,145 of John R. Essig and James M. Essig depicting a conical collimated concentrated light pattern produced by a pressure-deployed solar concentrator membrane that is appropriately skewed from parabolic form. The membrane contains substantially circular facets supported by a hexagonal wire or cable grid and produces a spherical focal area.

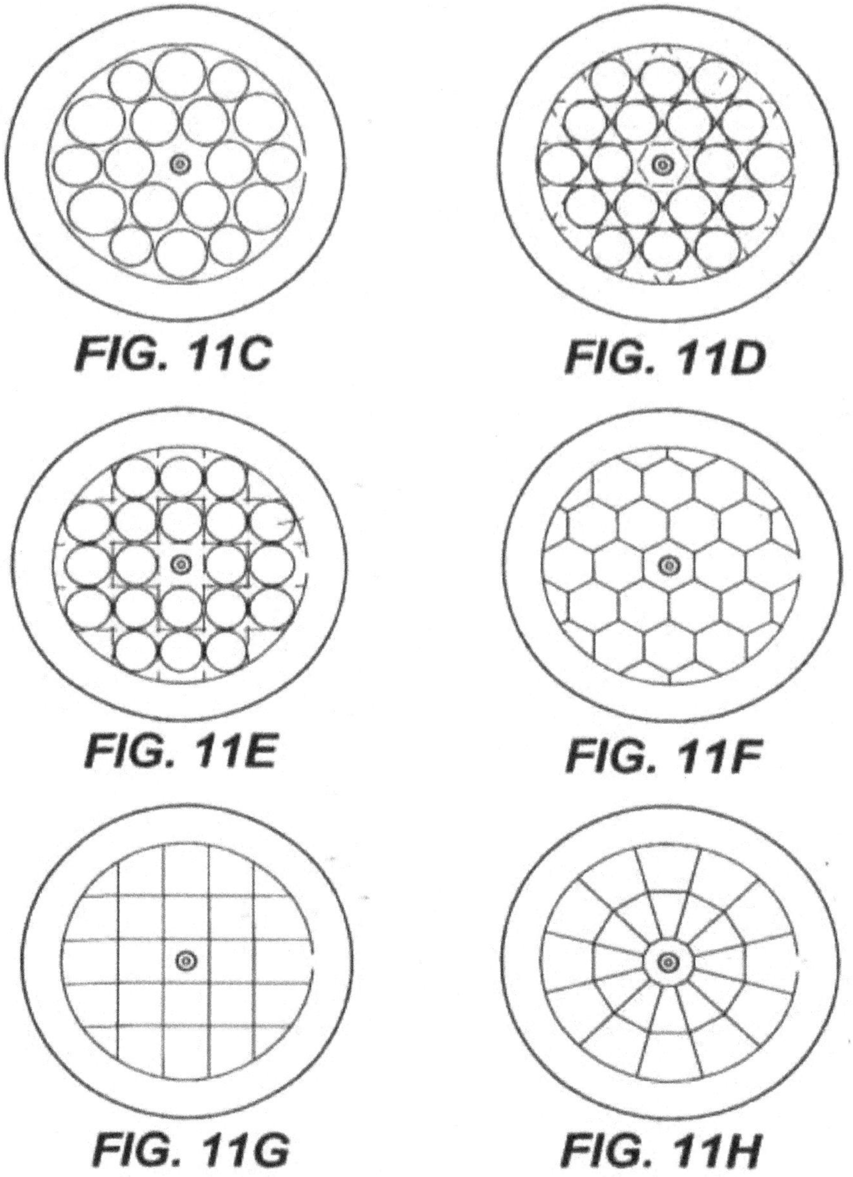

Fig. 11C–H. An excerpt from the US Patent Application No. 10/729,145 of John R. Essig and James M. Essig depicting faceted pressure-deployed solar concentrator membranes that are intended to limit solar or starlight concentration.

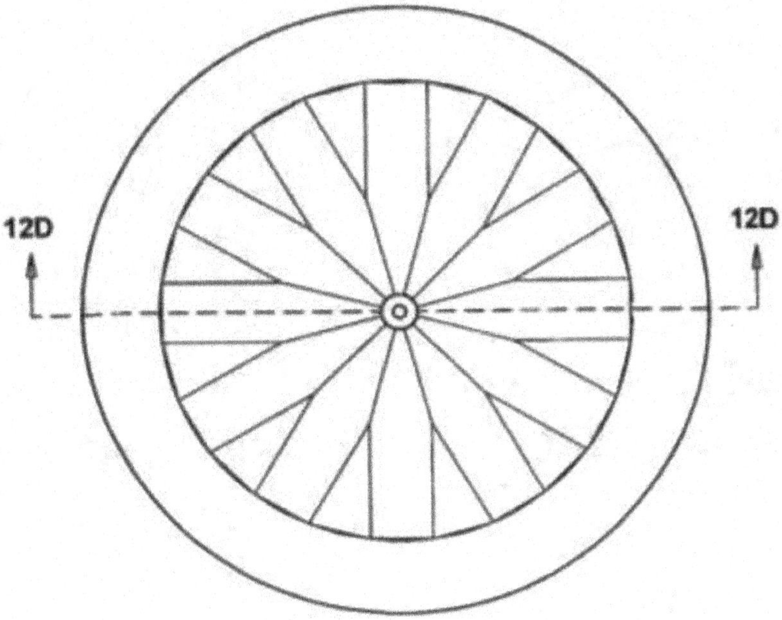

FIG. 12C

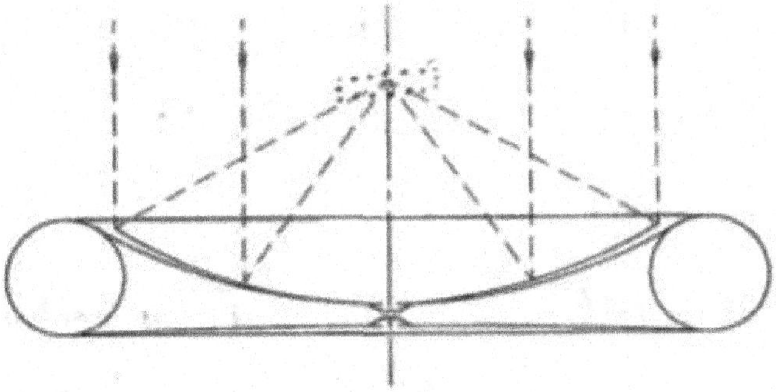

FIG. 12D

FIGURES 12 A & B as excerpted from U.S. Patent Application No. 10/729,145 of John R. Essig and James M. Essig depict a concentrated light pattern produced by a pressure deployed solar concentrator membrane which is appropriated skewed from parabolic form and having two shapes of distinct irregular polygonal facets.

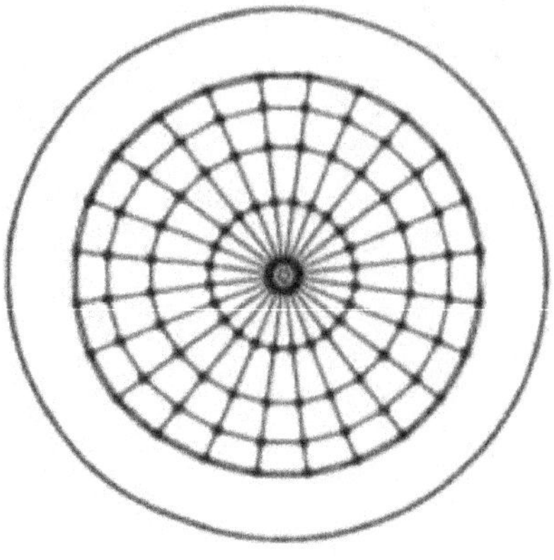

FIG. 14C

FIG. 14D

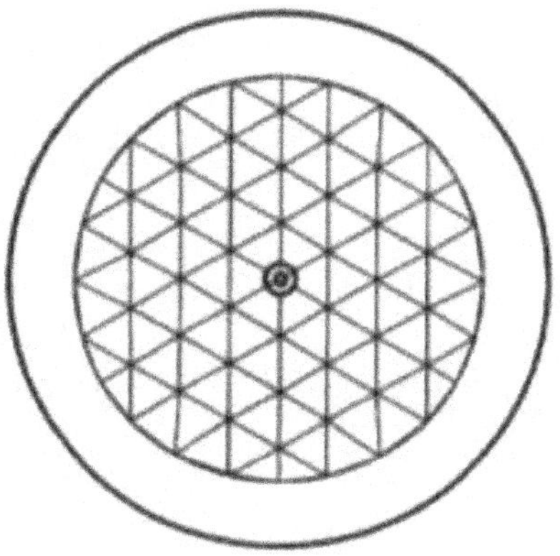

FIG. 14E

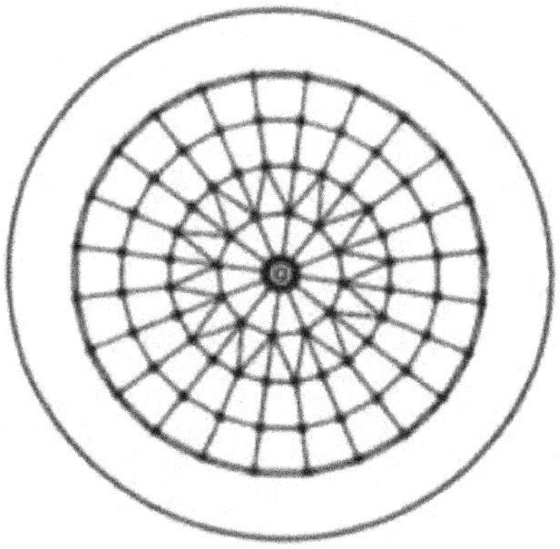

FIG. 14F

Fig. 14C–F. An excerpt from the US Patent Application No. 10/729,145 of John R. Essig and James M. Essig depicting faceted pressure-deployed solar concentrator membranes that are intended to limit solar or starlight concentration.

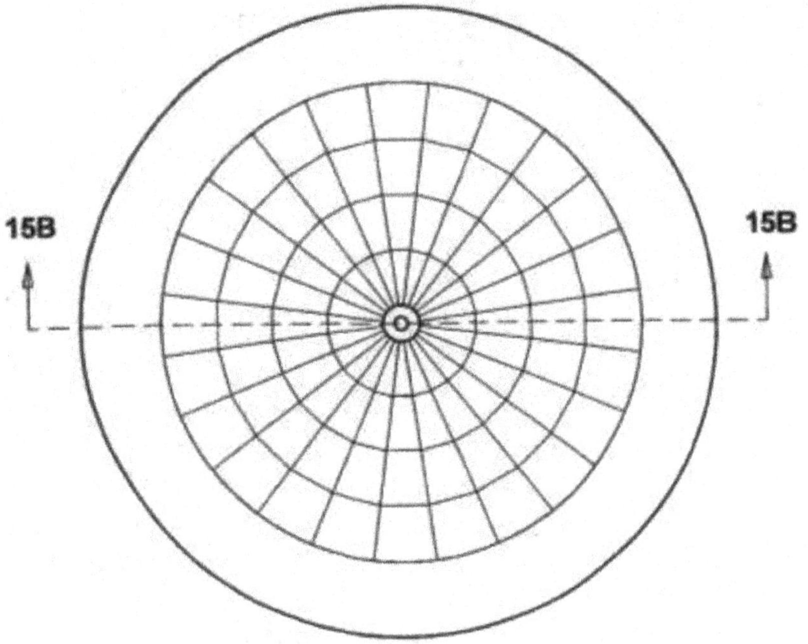

FIG. 15A

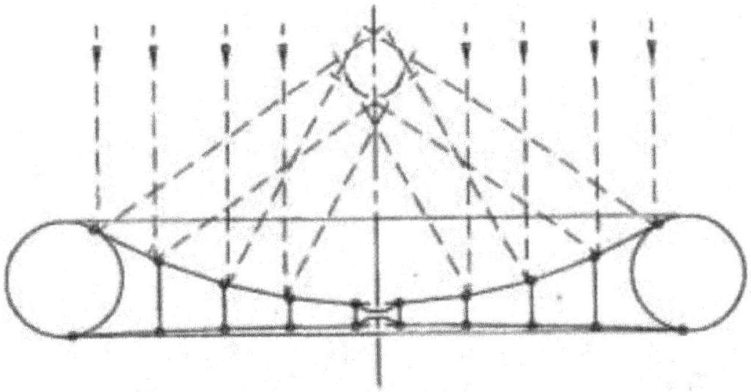

FIG. 15B

Fig. 15A–B. An excerpt from the US Patent Application No. 10/729,145 of John R. Essig and James M. Essig depicting a faceted pressure-deployed solar concentrator membranes that is intended to limit solar or starlight concentration. Here, the membrane is gas pressure deployed and/or electrostatically deployed, having a cord or wall-like holding mechanism for preventing reflector overextension.

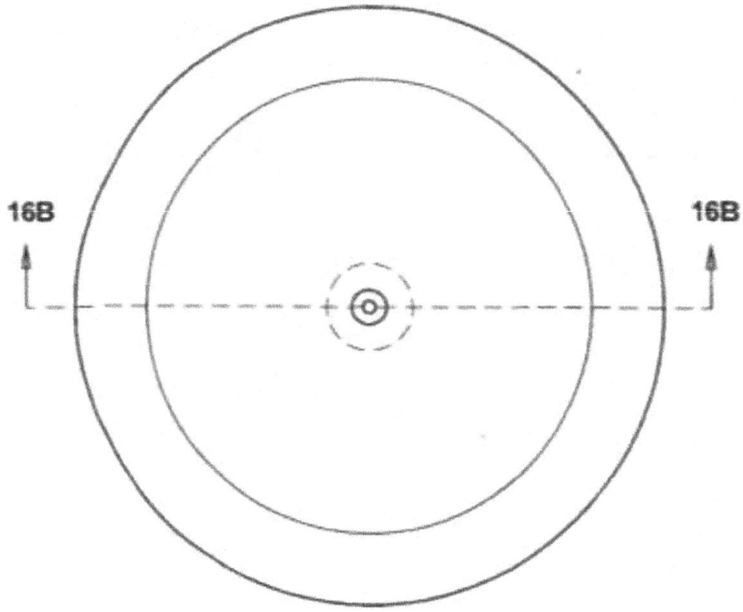

FIG. 16A

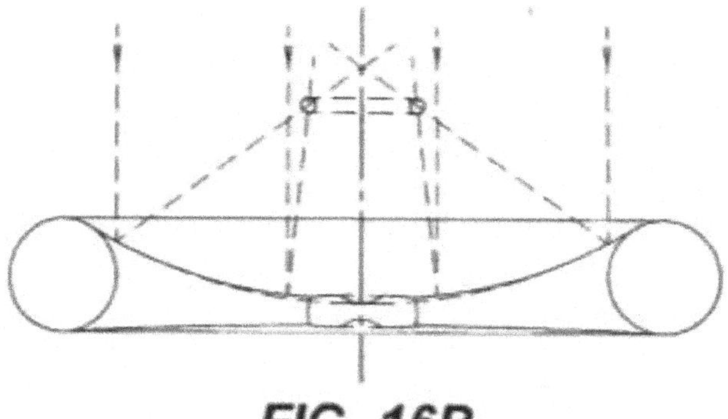

FIG. 16B

Fig. 16A–B. An excerpt from the US Patent Application No. 10/729,145 of John R. Essig and James M. Essig depicting a pressure-deployed solar concentrator membrane that is appropriately skewed from parabolic form as intended to limit solar or starlight concentration. Here, two distinct focal areas are provided as may be appropriate for certain applications.

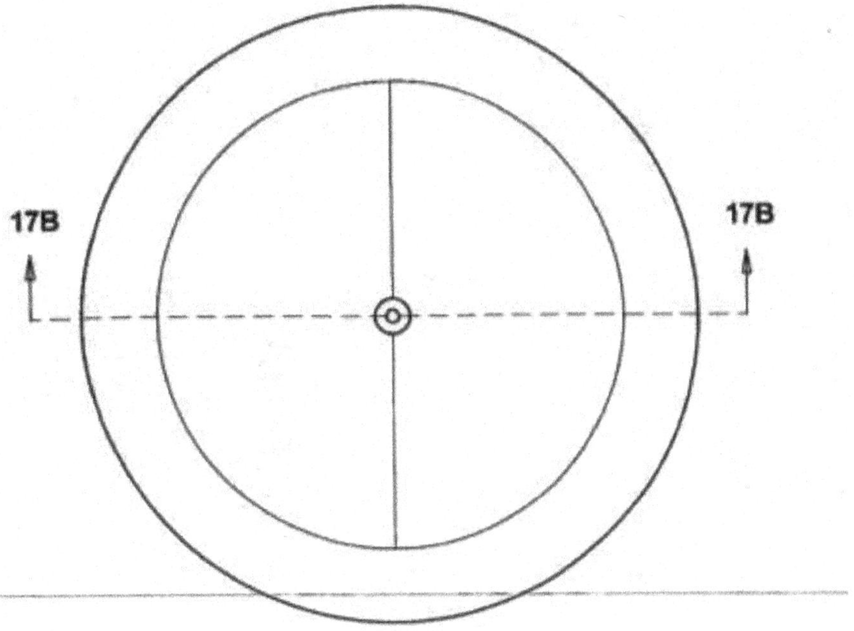

FIG. 17A

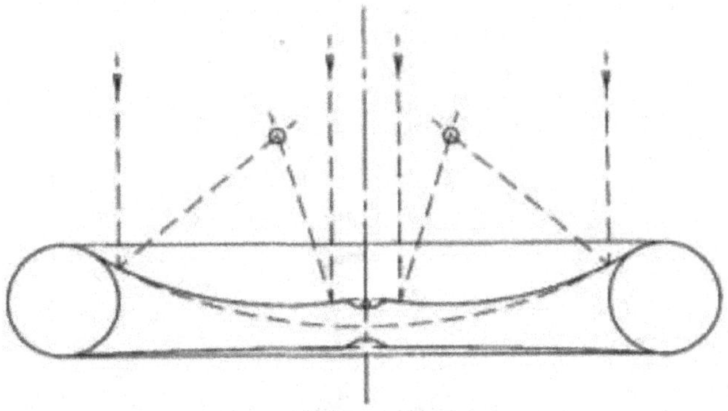

FIG. 17B

Fig. 17A–B. An excerpt from the US Patent Application No. 10/729,145 of John R. Essig and James M. Essig depicting a pressure-deployed solar concentrator membrane that is appropriately skewed from parabolic form as intended to limit solar or starlight concentration. Here, the bipartite membrane produces two distinct focal points.

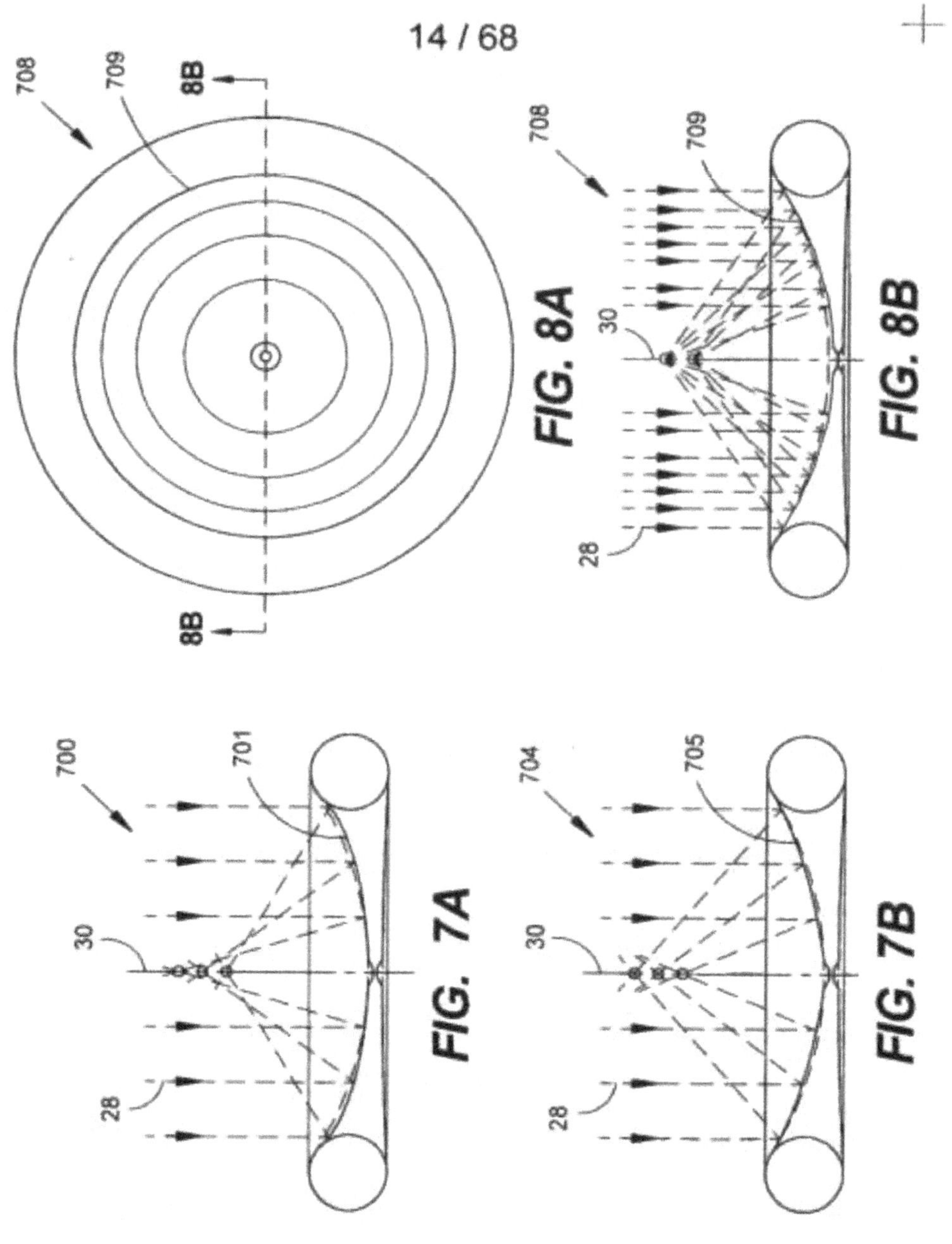

Fig. 7A–B and 8A–B. An excerpt from the US Patent Application No. 11/254,023 illustrate comparison of focal area of two pressure deployable reflectors where the conical focal areas are relatively inverted with respect to one another.

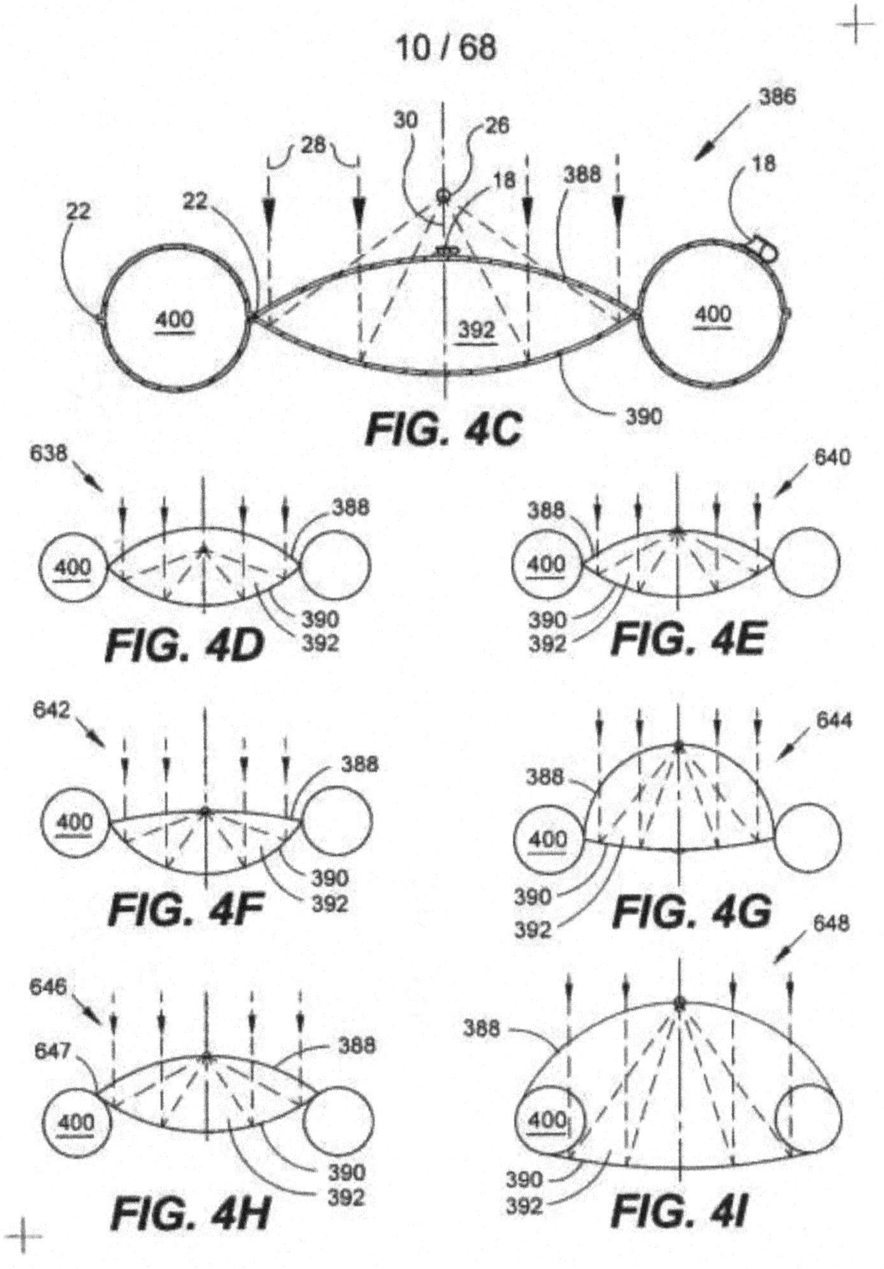

Fig. 4C–I. An excerpt from the US Patent Application No. 11/254,023 of John R. Essig and James M. Essig illustrate comparison of focal areas for gas-inflated superambient pressured reflectors of various shapes and focal lengths.

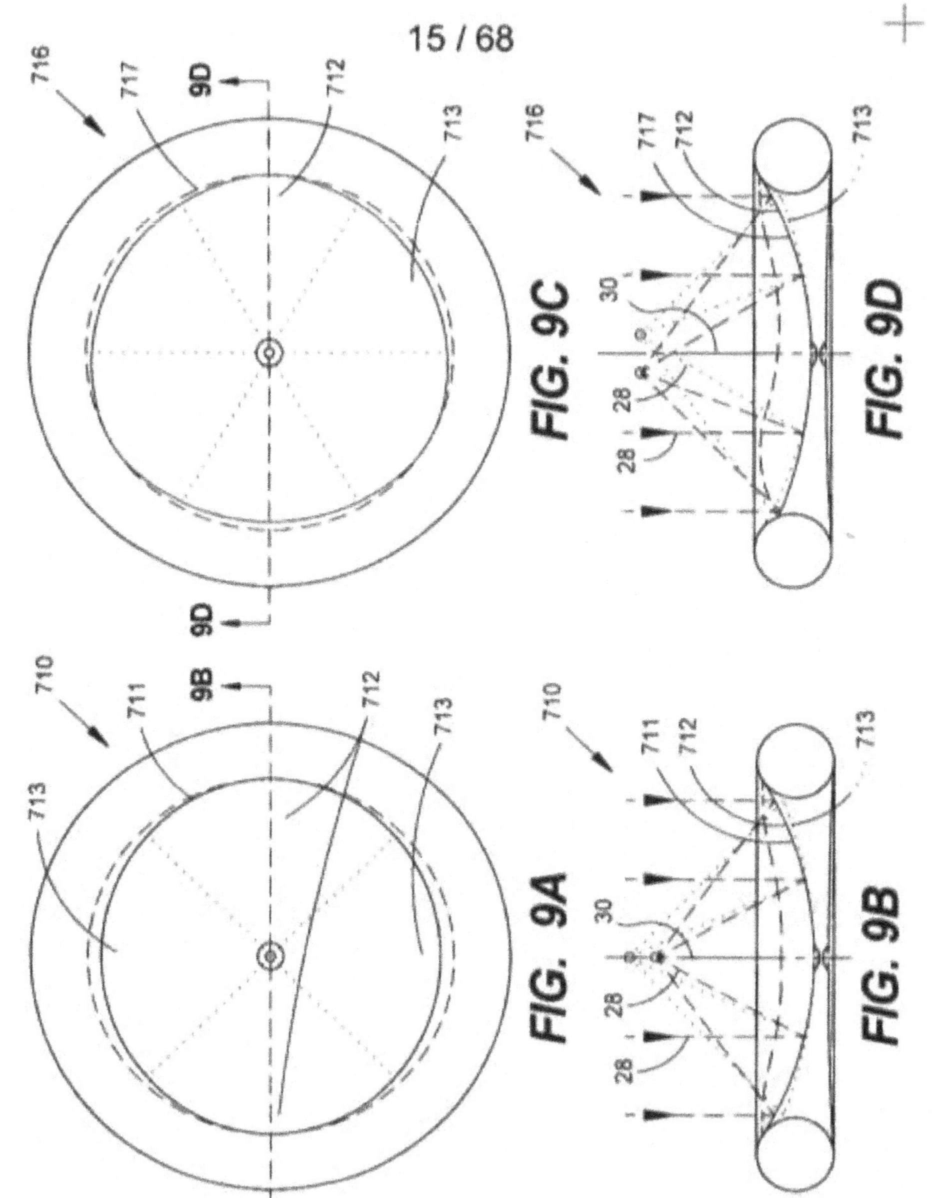

Fig. 9A–B. An excerpt from the US Patent Application No. 11/254,023 of John R. Essig and James M. Essig illustrating comparison of focal areas of two pressure-deployable reflectors having differing geometries, one with two nodes and one with three nodes of surface undulations.

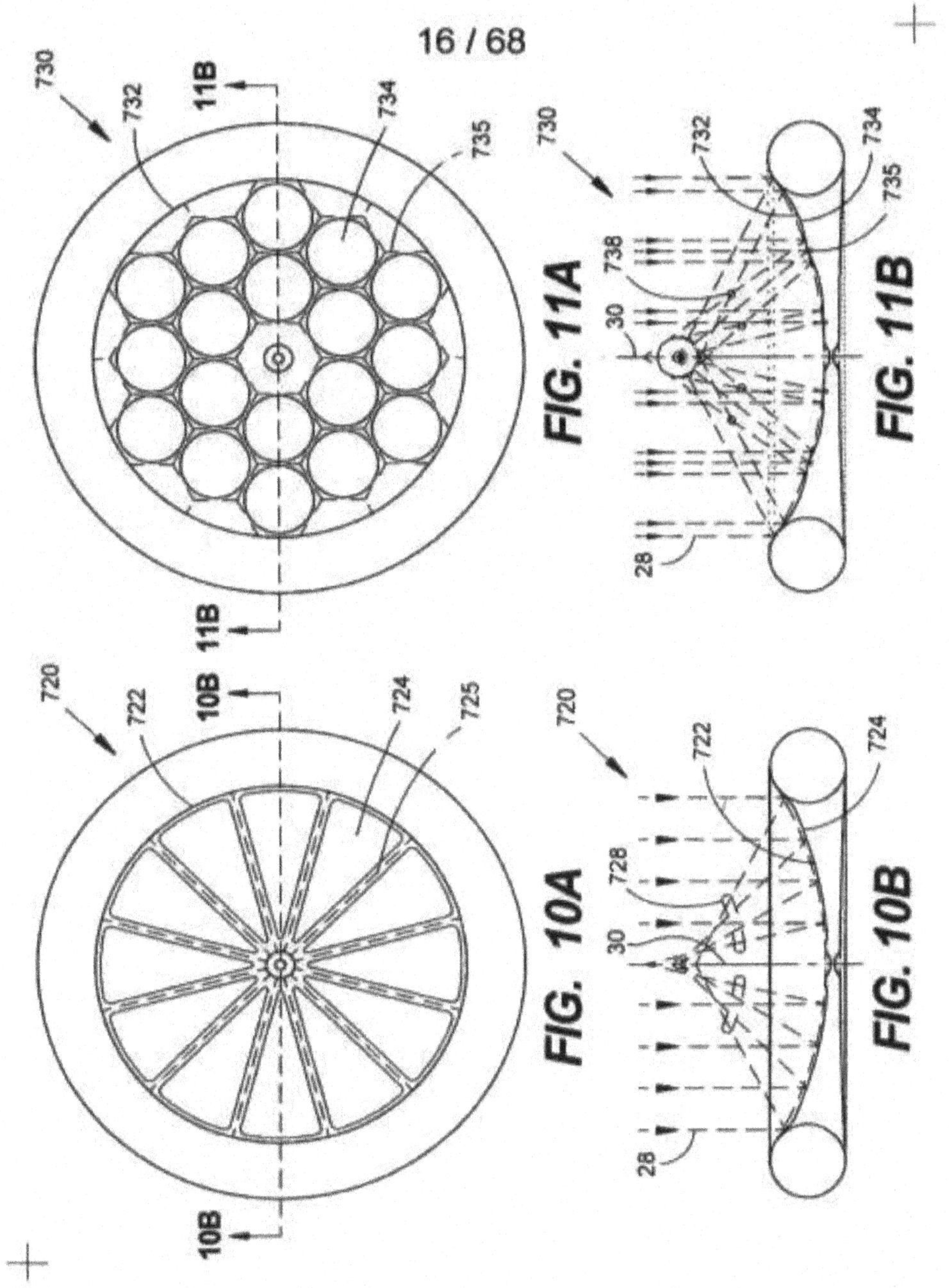

Fig. 10A–B and 11A–B. An excerpt from the US Patent Application No. 11/254,023 of John R. Essig and James M. Essig illustrating comparison of focal area of two pressure-deployable reflectors where the focal areas include discretized circular rings in segments and a spherical light concentration pattern, respectively.

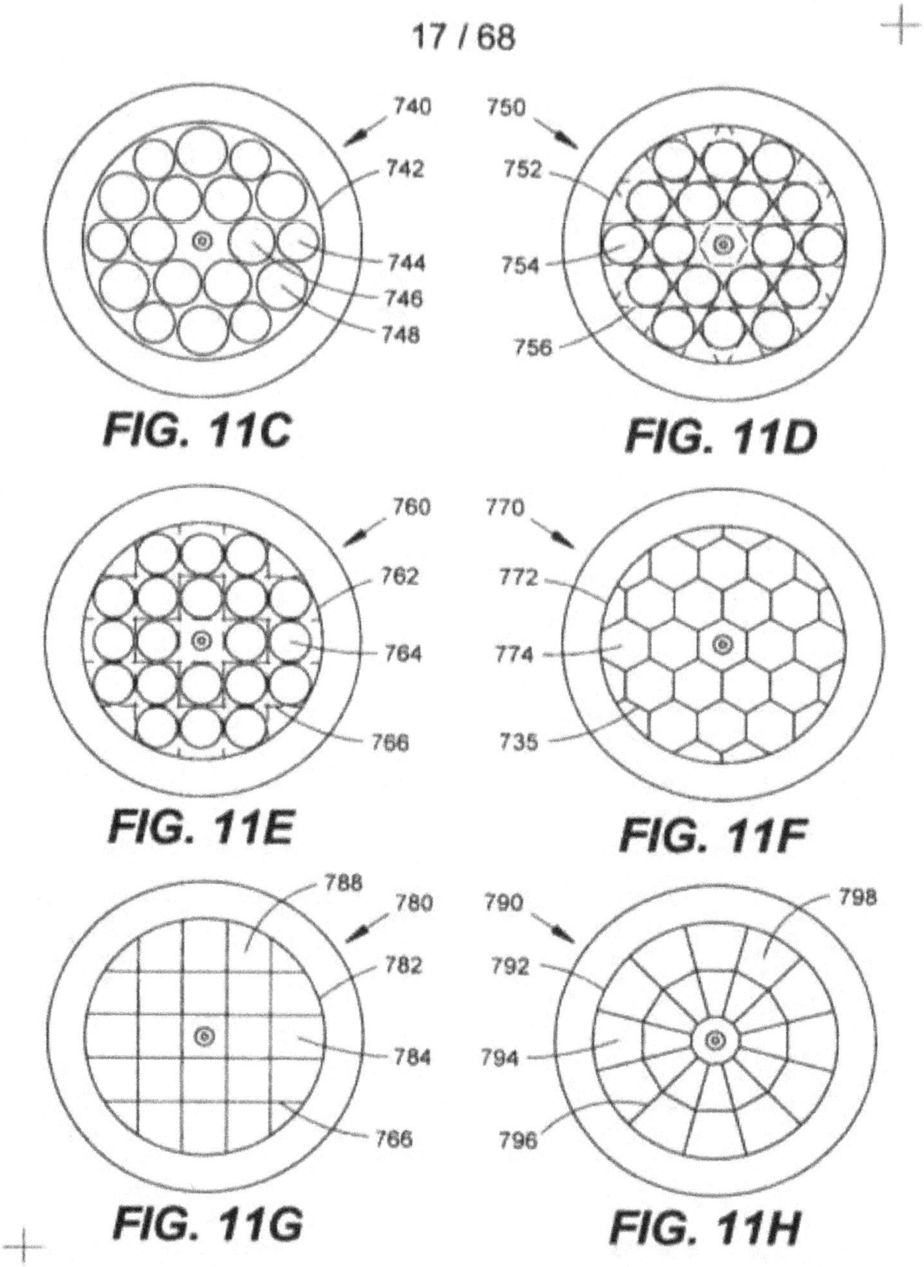

Fig. 11C–H. An excerpt from the US Patent Application No. 11/254,023 of John R. Essig and James M. Essig illustrating several distinct optional tiling patterns for reduced light concentration.

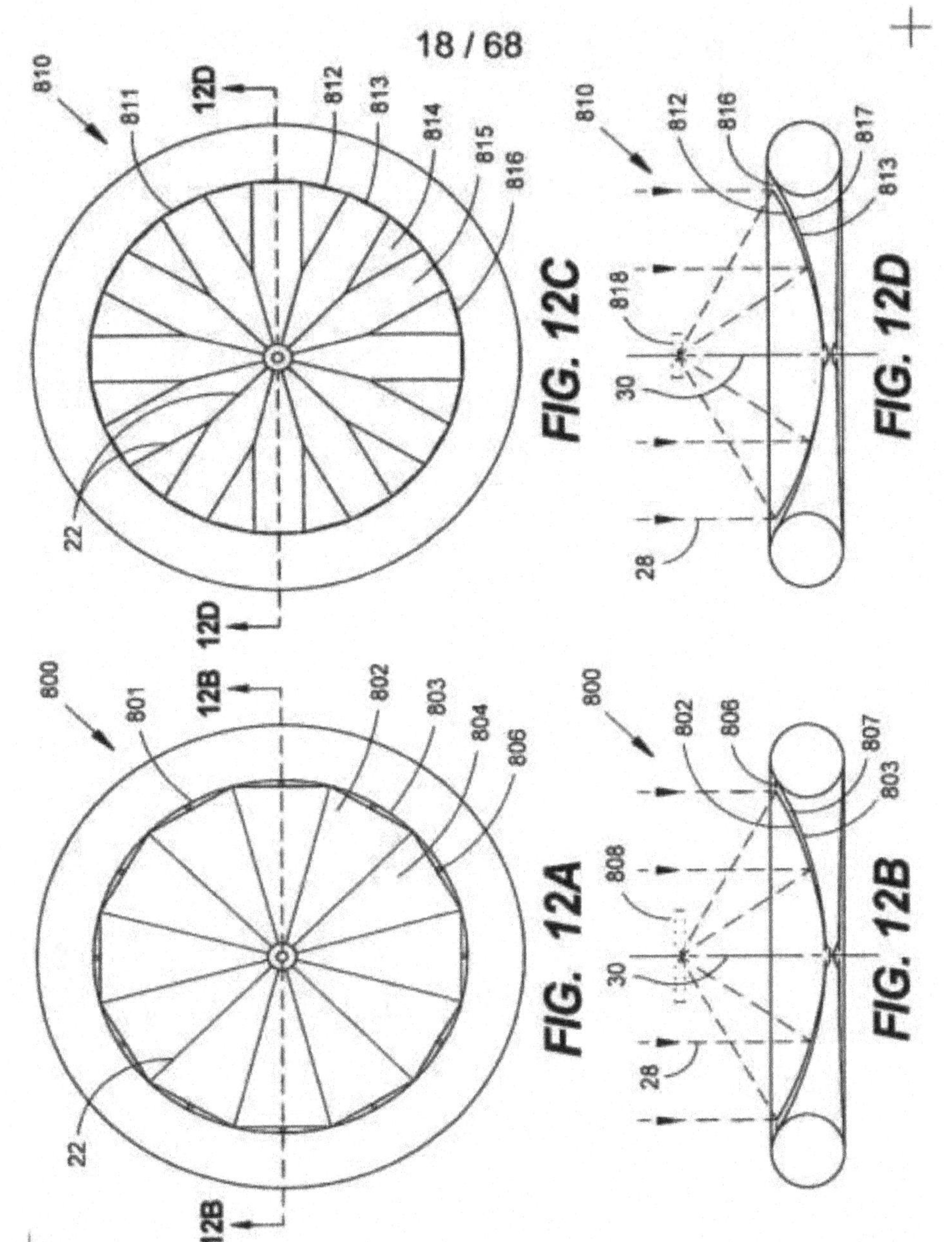

Fig. 12A–B and 12C–D. An excerpt from the US Patent Application No. 11/254,023 of John R. Essig and James M. Essig illustrating comparison of focal area of two pressure deployable reflectors where the conical focal areas have effectively dissimilar aspect ratios.

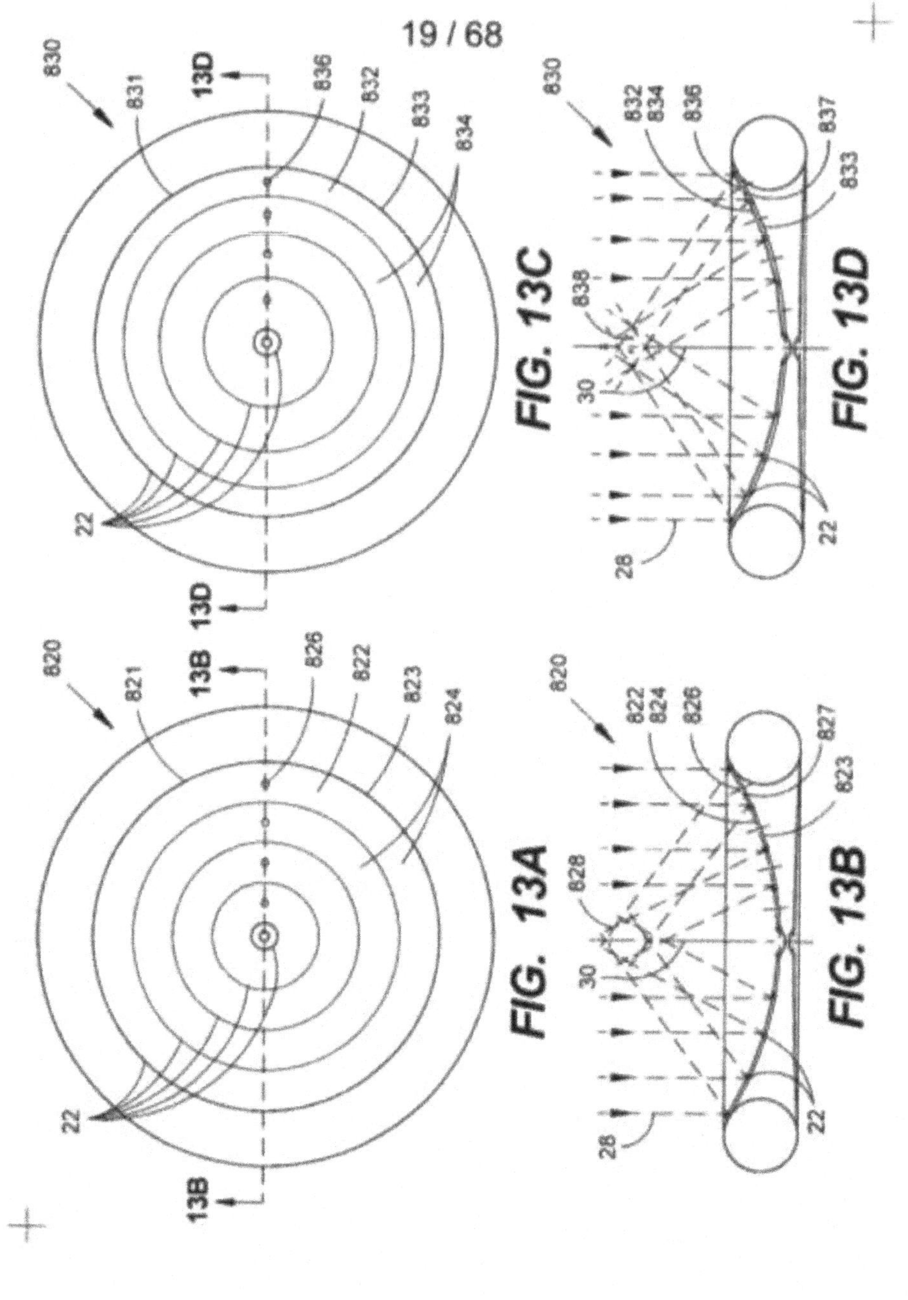

Fig. 13A–B and 13C–D. An excerpt from the US Patent Application No. 11/254,023 of John R. Essig and James M. Essig illustrating the comparison of focal area of two pressure deployable reflectors where focal areas are spherical in shape and of two differing sizes relative to the reflector sizes.

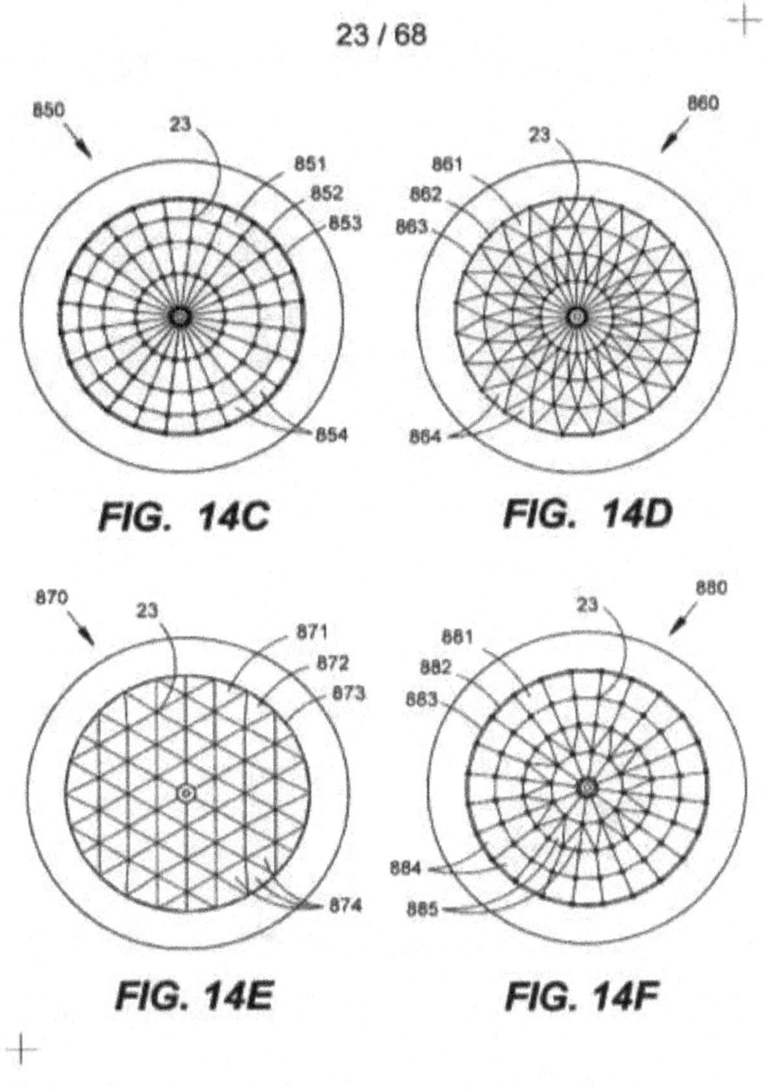

Fig. 14C–D. An excerpt from the US Patent Application No. 11/254,023 of John R. Essig and James M. Essig illustrating various optional reflector faceting patterns such as for safely reducing the concentration of sunlight.

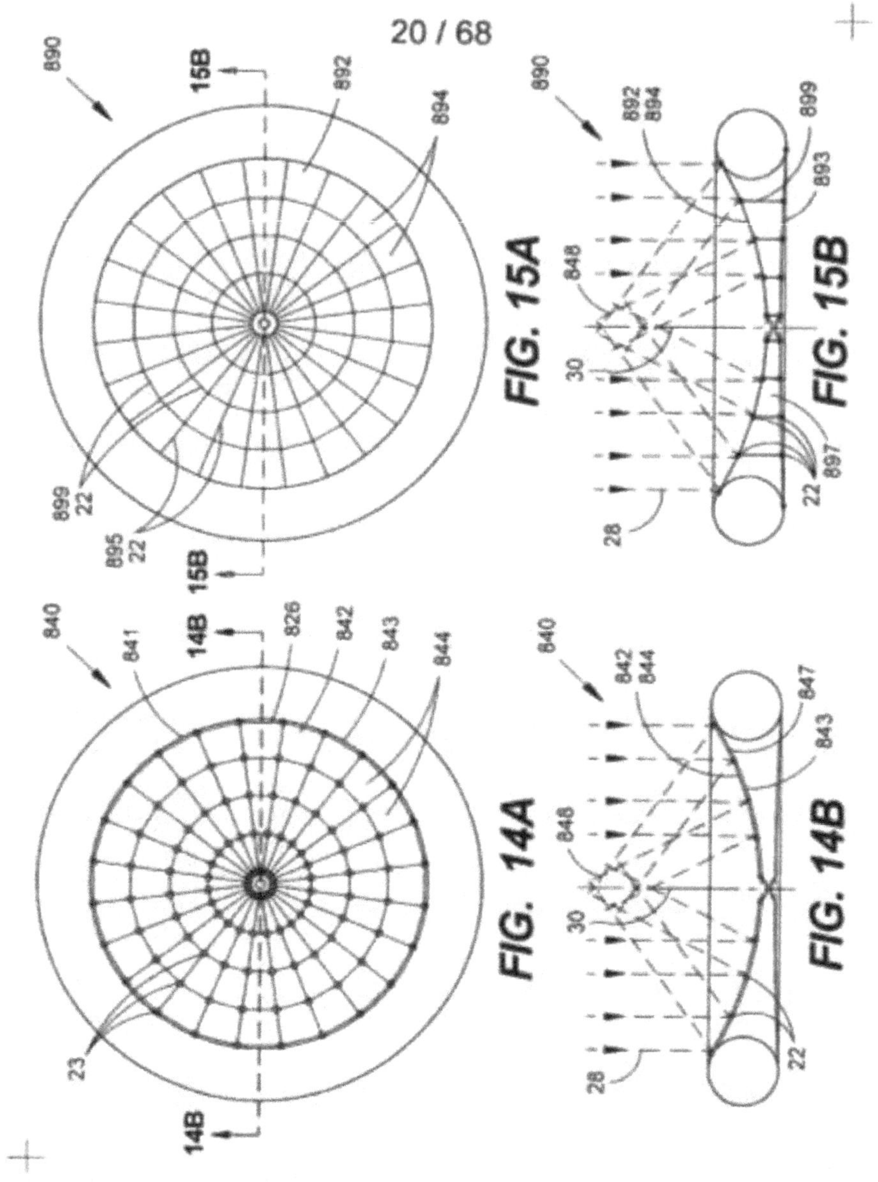

Fig. 14A–B and 15A–B. An excerpt from the US Patent Application No. 11/254,023 of John R. Essig and James M. Essig illustrating spherical focal areas of two pressure-deployable reflectors for which two distinct membrane-shaping mechanisms are deployed where one reflector membrane is spot-bonding to a substrate membrane and the other reflective membrane is cord-tied or wall-bonded to a substrate membrane, respectively.

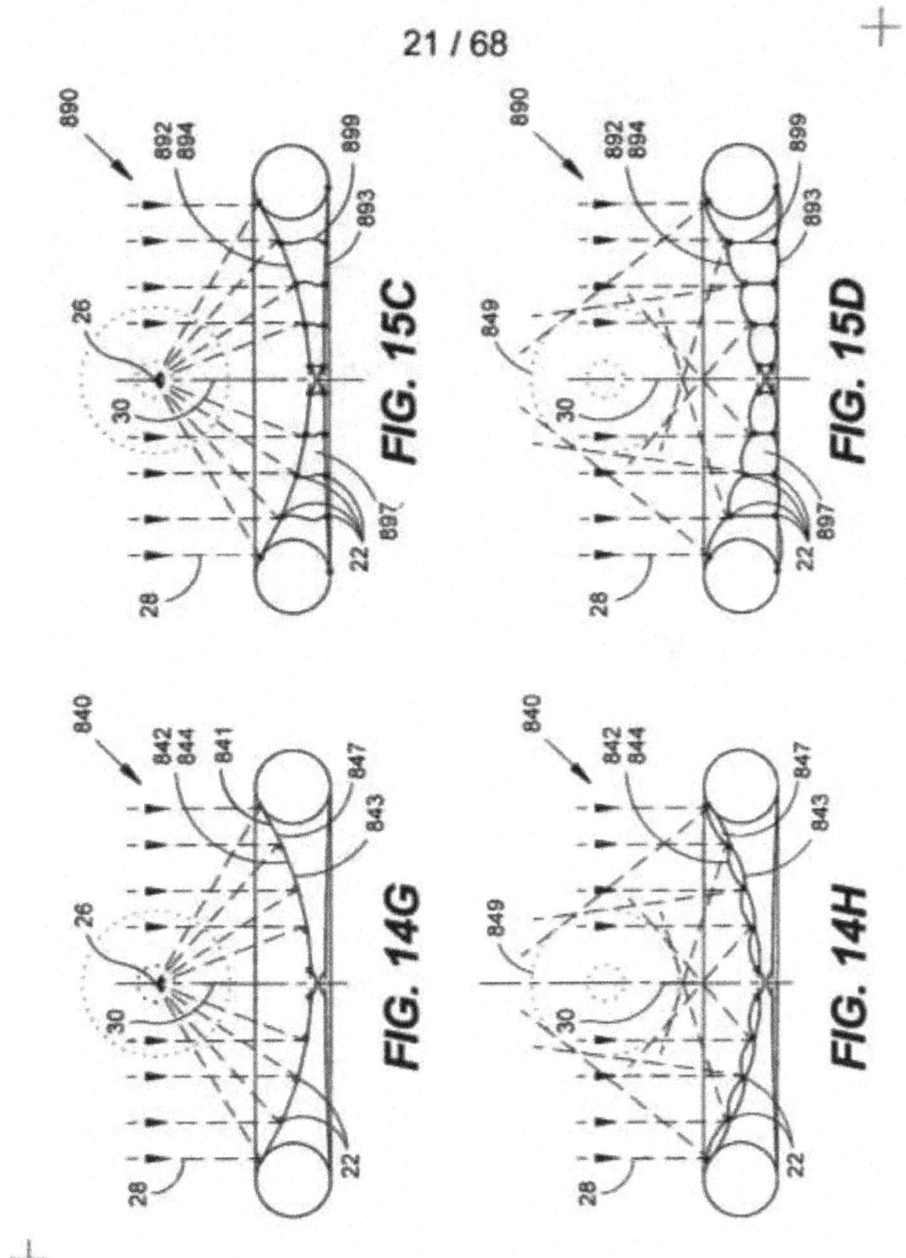

Fig. 14G–H and 15C–D. An excerpt from the US Patent Application No. 11/254,023 of John R. Essig and James M. Essig illustrating comparison of focal areas of four pressure-deployable reflectors .

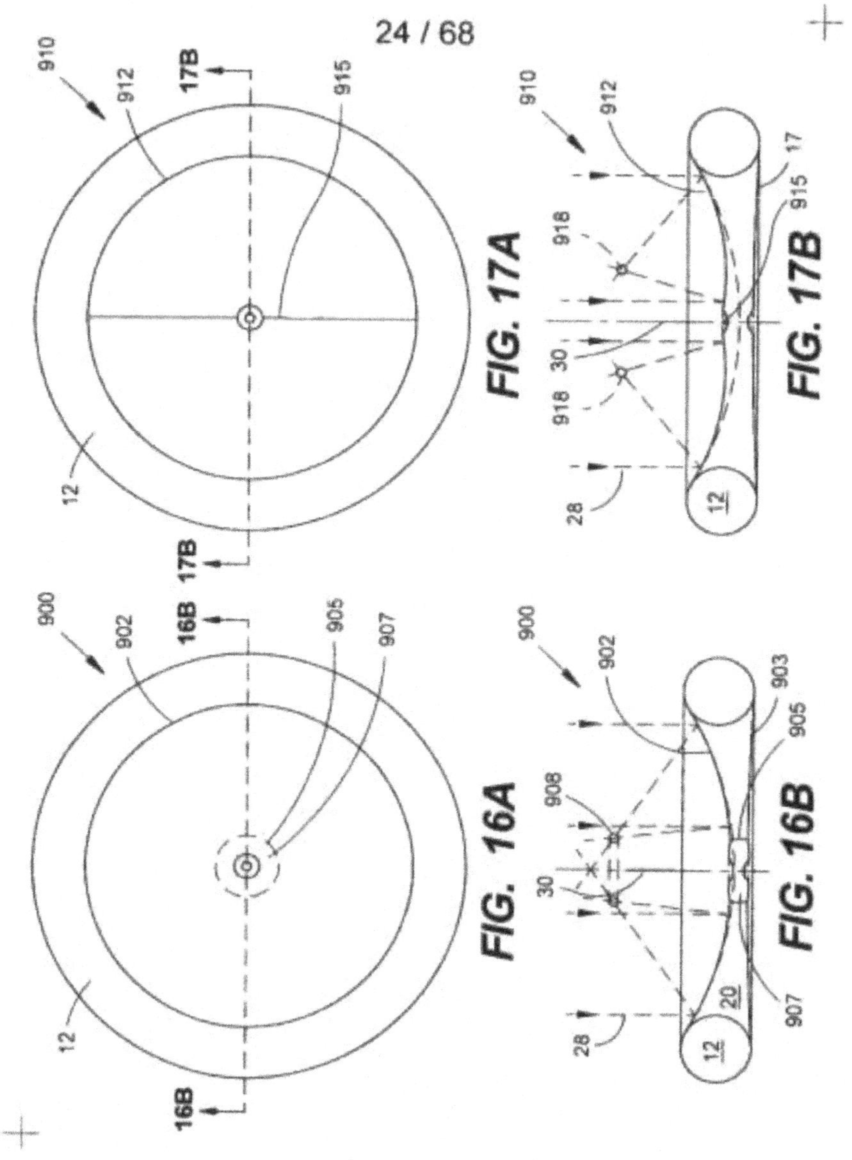

Fig. 16A–B and 17A–B. An excerpted from the US Patent Application No. 11/254,023 of John R. Essig and James M. Essig illustrating the comparison of focal area of two pressure-deployable reflectors where one of the reflectors has an air bladder modification and the other reflector has two focal points.

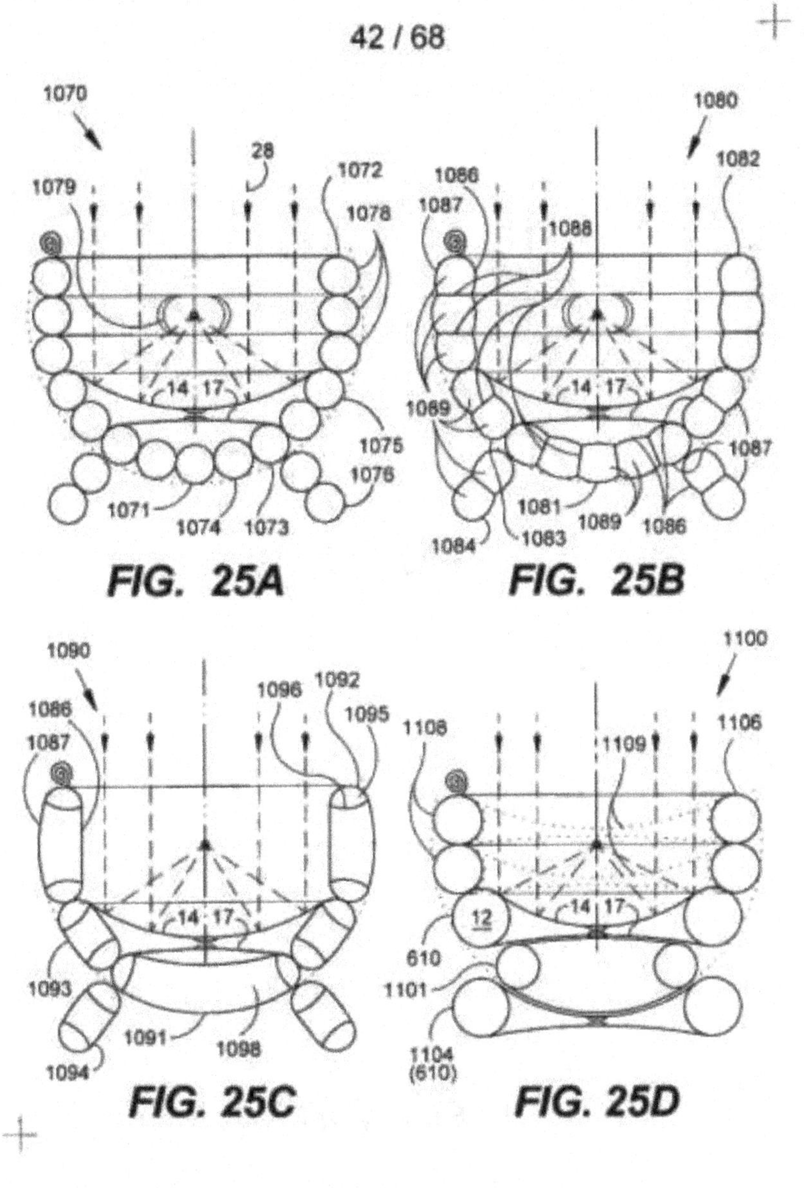

Fig. 25A–D. An excerpt from the US Patent Application No. 11/254,023 of John R. Essig and James M. Essig illustrating four distinct types of pressure-deployable reflector apparatus and methods of holding objects at the reflector focal points.

43 / 68

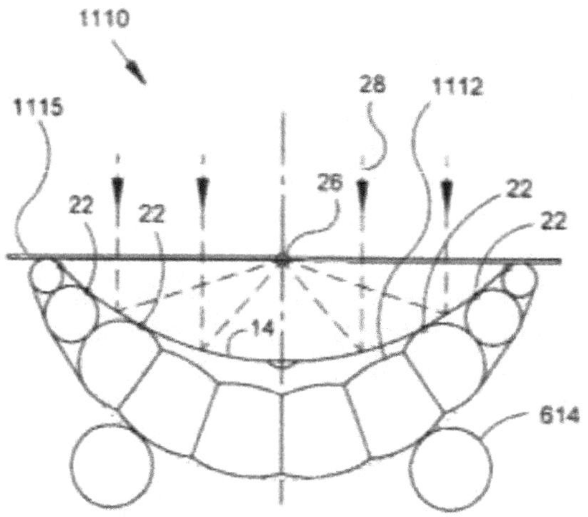

FIG. 26A

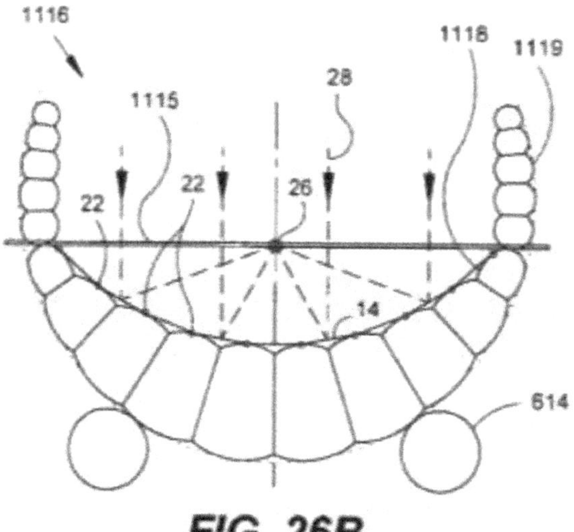

FIG. 26B

Fig. 26A–B. An excerpt from the US Patent Application No. 11/254,023 of John R. Essig and James M. Essig illustrating two distinct forms of purely superambient pressure gas–inflated light concentrators and mechanism for holding objects at the focal points.

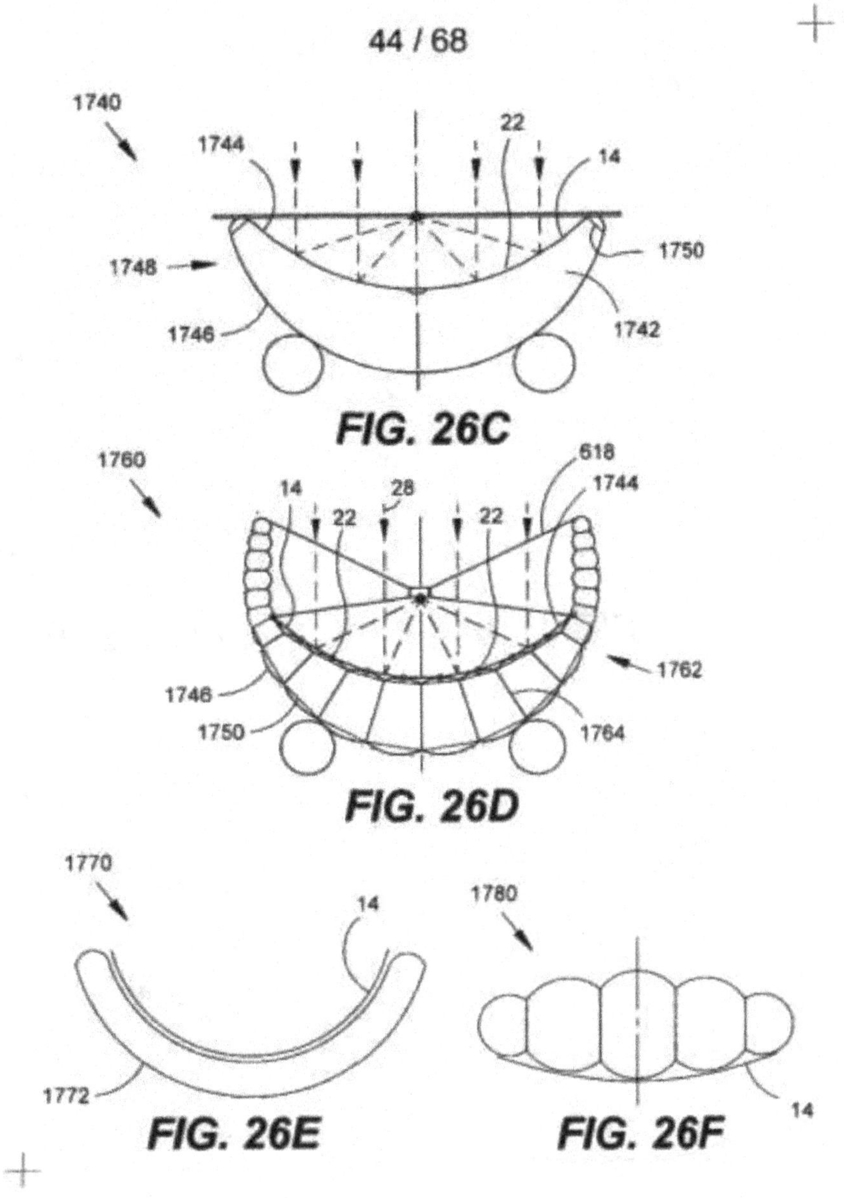

Fig. 26C–F. An excerpted from US Patent Application No. 11/254,023 of John R. Essig and James M. Essig illustrating four differing styles of superambient gas-pressured solar concentrators.

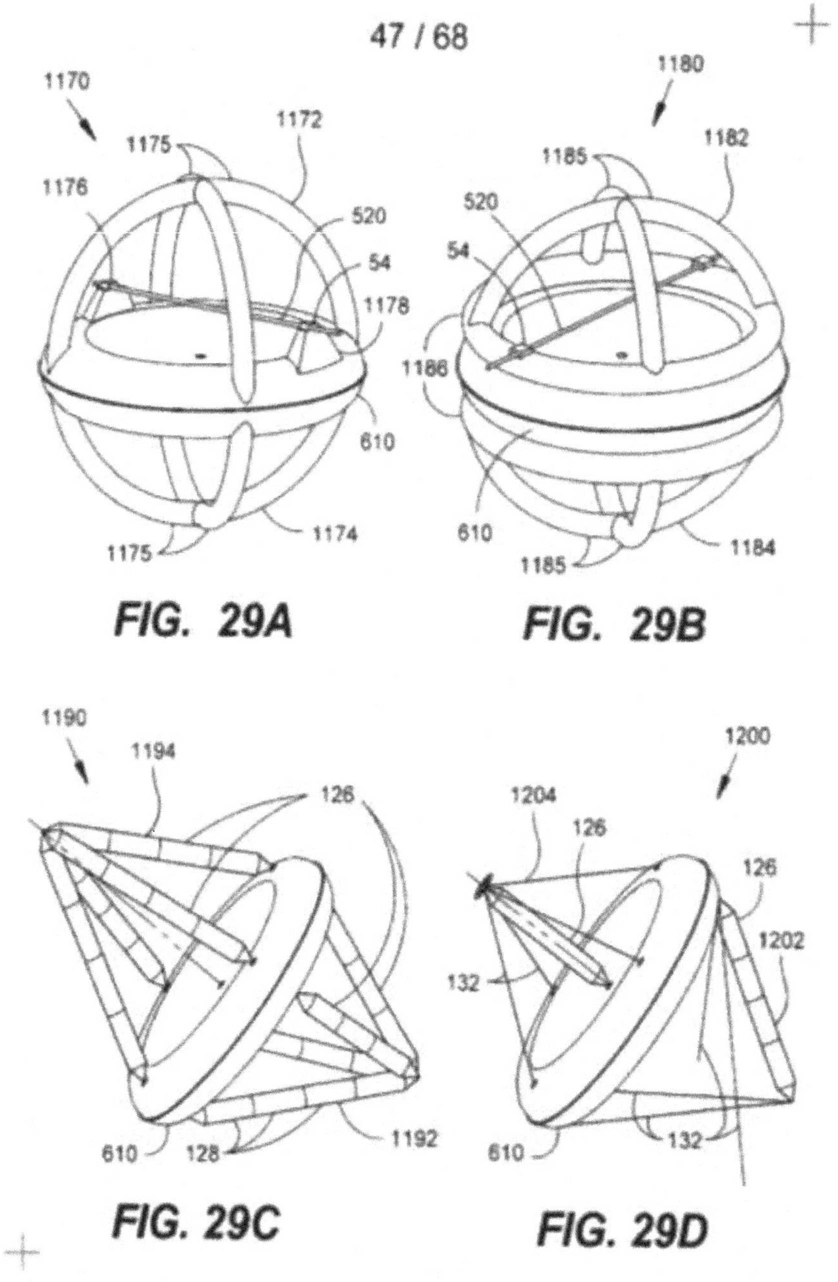

Fig. 29 A–D. An excerpt from the US Patent Application No. 11/254,023 of John R. Essig and James M. Essig illustrating four distinct mechanisms for holding objects at the focal points solar concentrators.

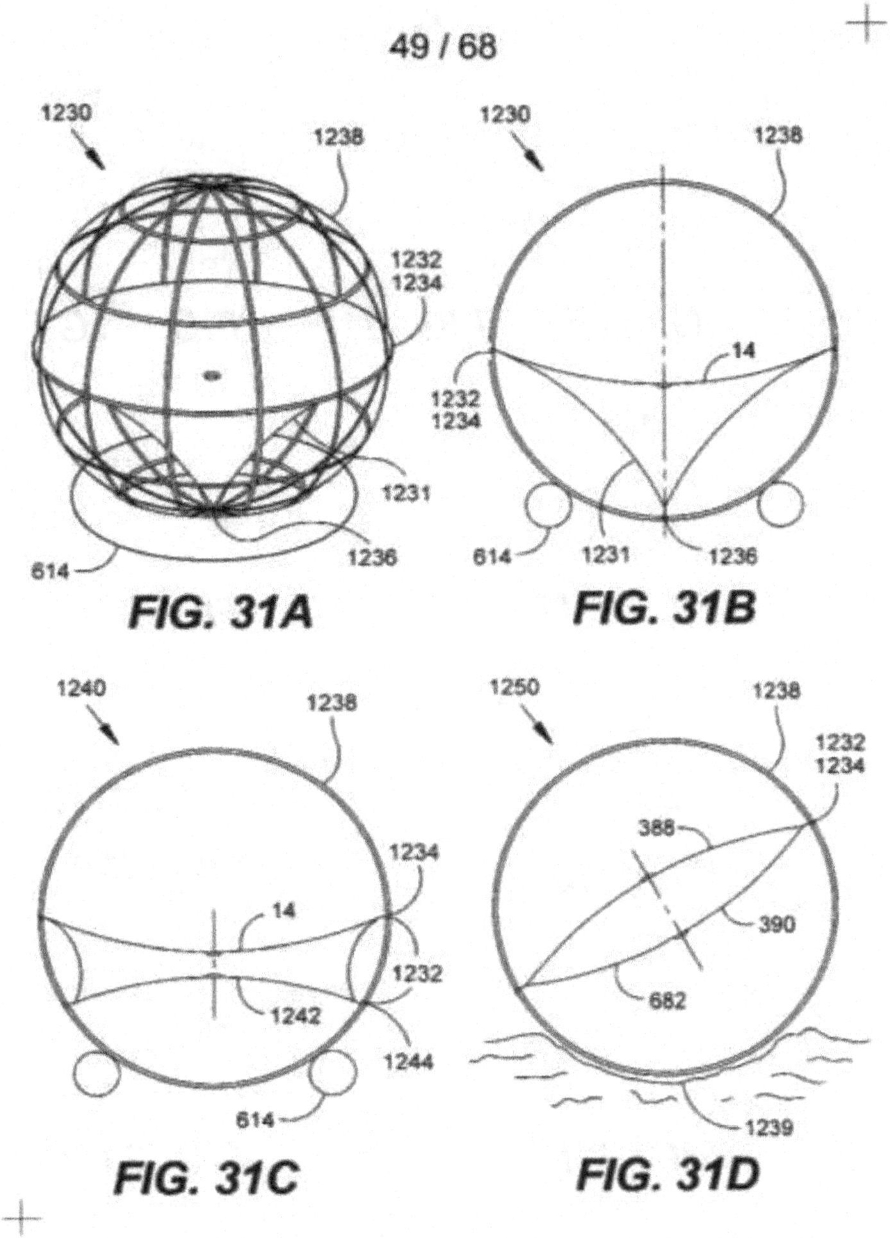

Fig. 31A–D. An excerpt from the US Patent Application No. 11/254,023 of John R. Essig and James M. Essig illustrating four differing mechanisms for deploying electrostatically and/or magnetostatically contoured membranous reflectors.

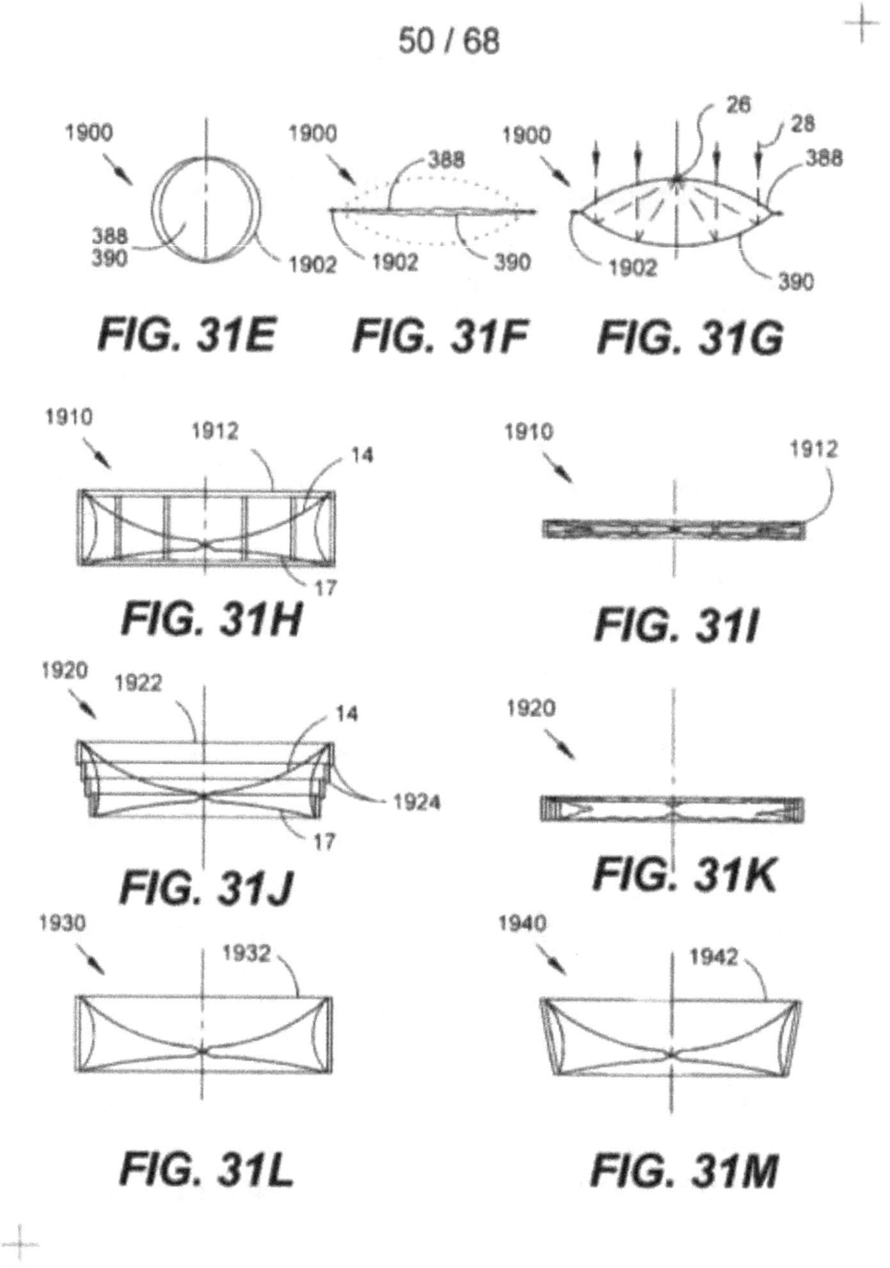

Fig. 31E–M. An excerpt from the US Patent Application No. 11/254,023 of John R. Essig and James M. Essig illustrating a wide variety of combinations of solar concentrating membranes and mechanisms of membrane deployments.

51 / 68

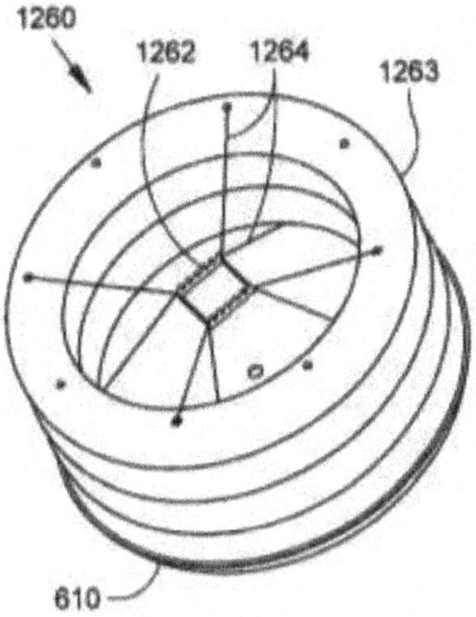

FIG. 32A

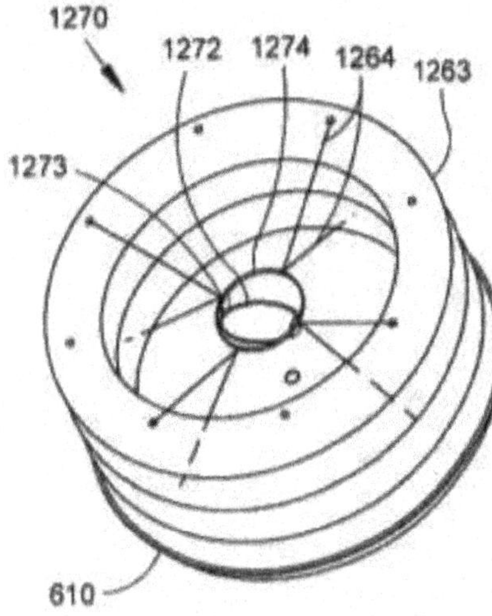

FIG. 32B

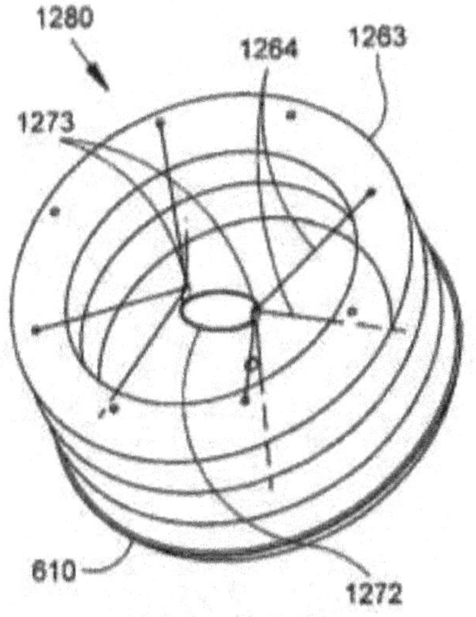

FIG. 32C

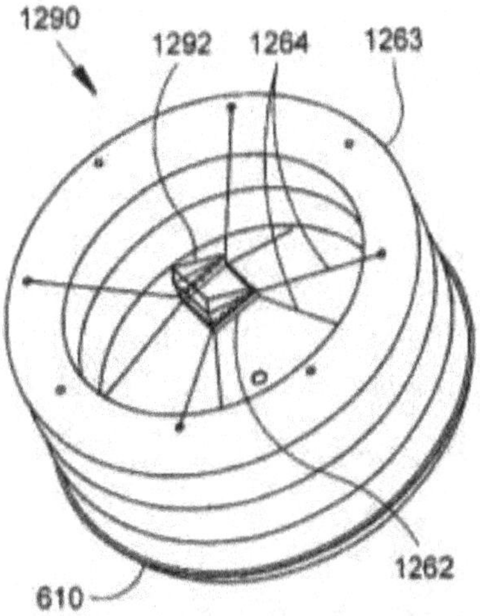

FIG. 32D

Fig. 32A–D. An excerpt from the US Patent Application No. 11/254,023 of John R. Essig and James M. Essig illustrating four distinct types of apparatus for holding object at the focal point of a solar concentrator. Here, the toroidal rings are stacked and deployed by superambient gas pressure.

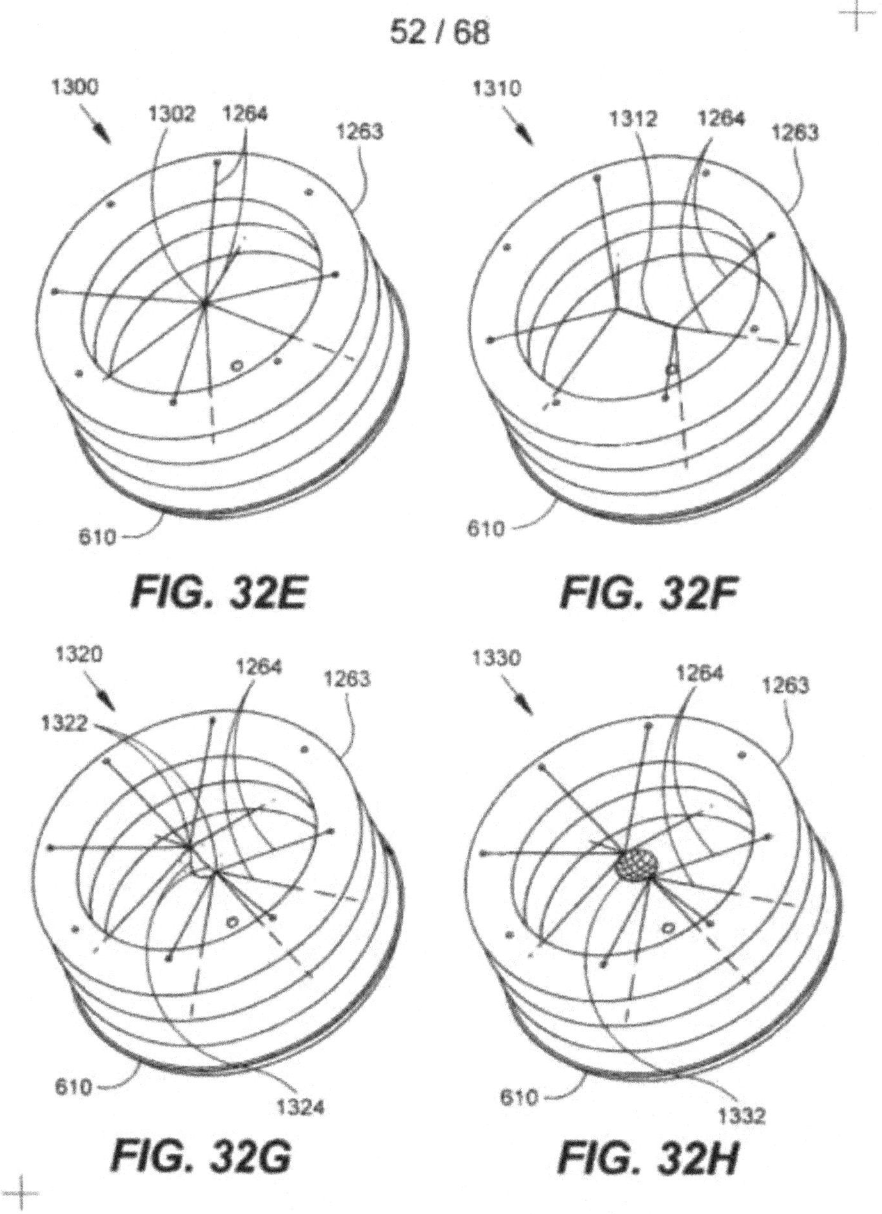

Fig. 32E–H. An excerpt from the US Patent Application No. 11/254,023 of John R. Essig and James M. Essig illustrating four additional distinct types of apparatus for holding object at the focal point of a solar concentrator. Here, the toroidal rings are stacked and deployed by super-ambient gas pressure.

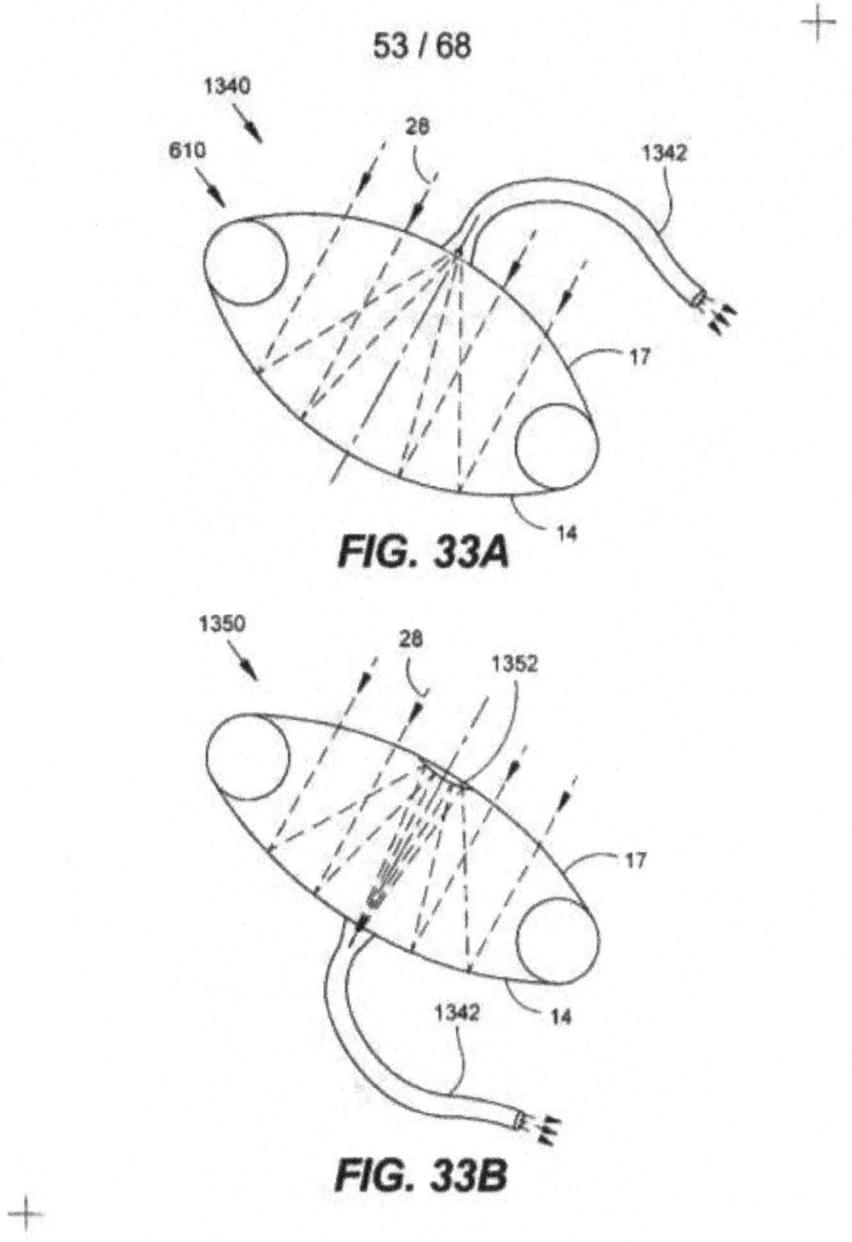

Fig. 33A–B. An excerpt from the US Patent Application No. 11/254,023 of John R. Essig and James M. Essig illustrating two distinct types of apparatus for directing concentrated sunlight down a light-conducting conduit.

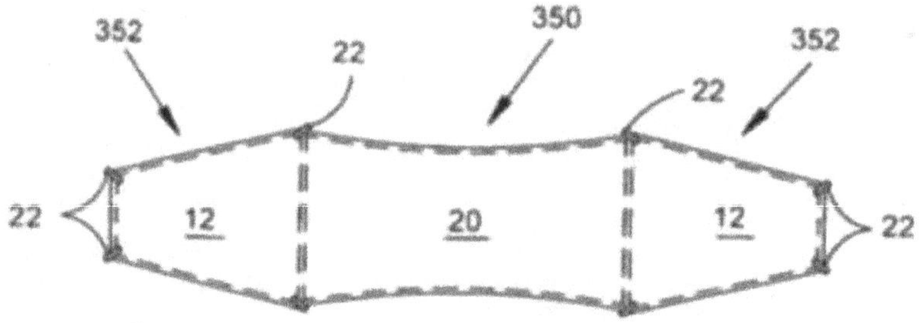

FIG. 54D

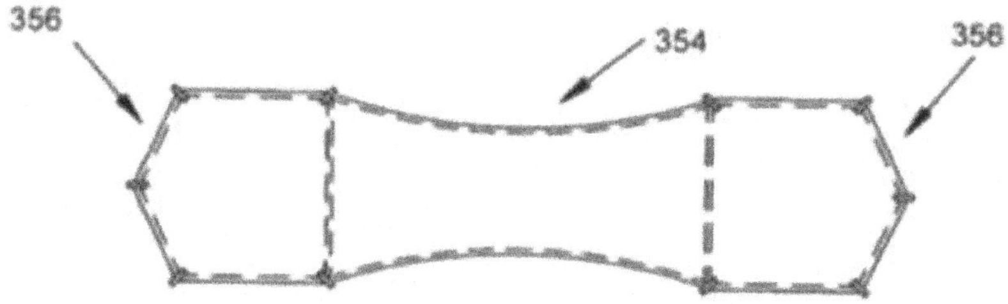

FIG. 54E

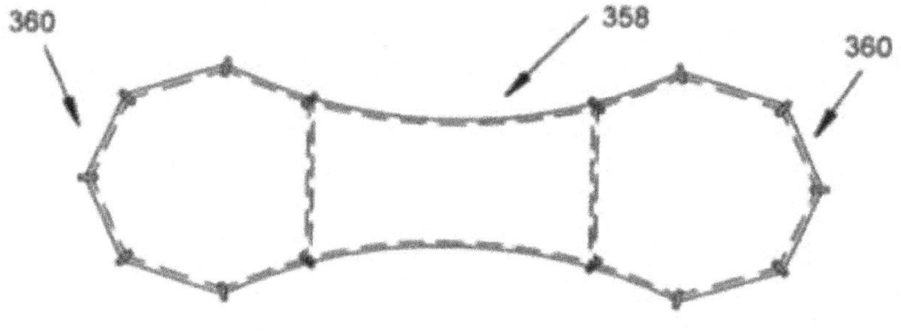

FIG. 54F

Fig. 54D–F. An excerpt from the US Patent No. 6,897,832 show respective 6, 7, and 9 sheet three-dimensional manufacturing patterns for inflatable toroids.

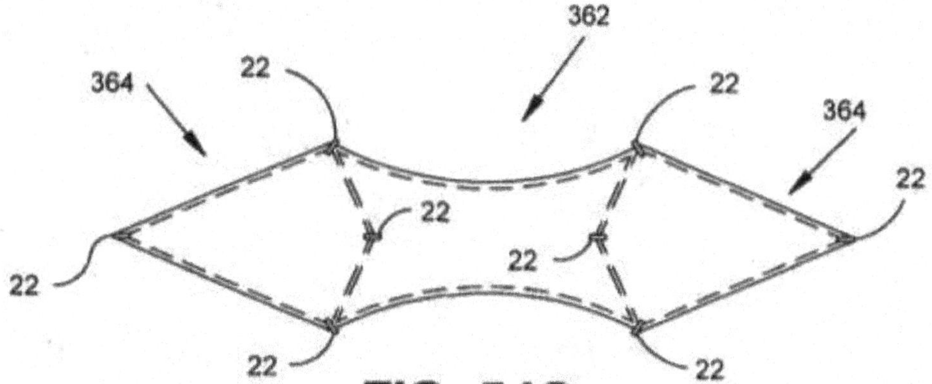

FIG. 54G

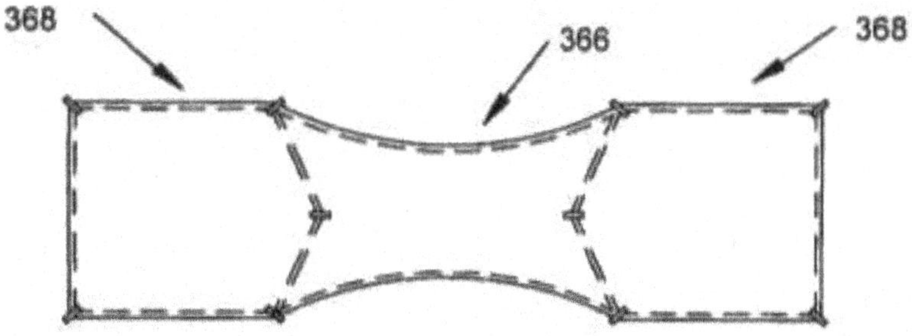

FIG. 54H

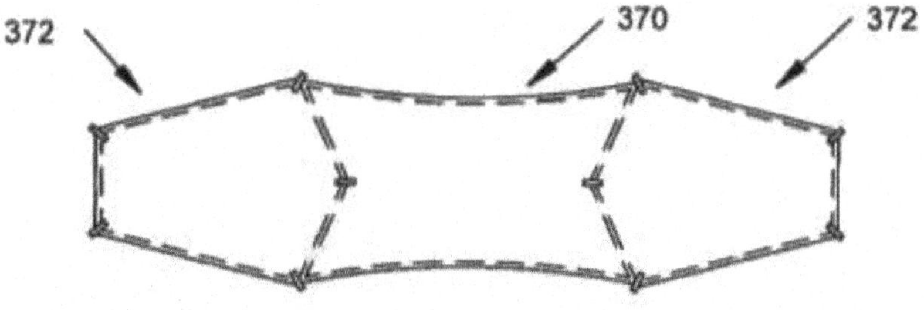

FIG. 54I

Fig. 54G–I. An excerpt from the US Patent No. 6,897,832 show respective 6, 7, and 7 sheet three-dimensional manufacturing patterns for inflatable toroids.

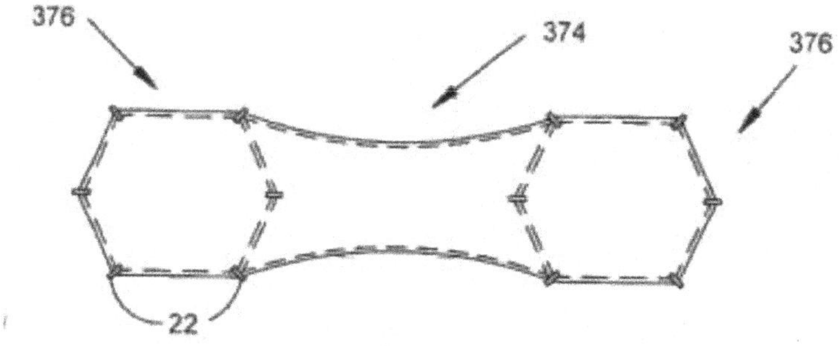

FIG. 54J

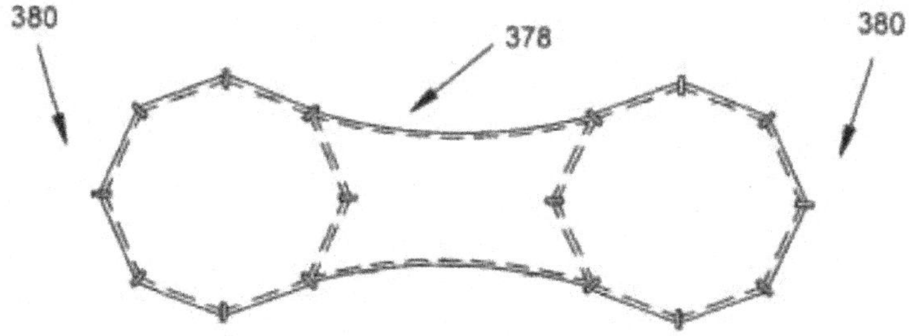

FIG. 54K

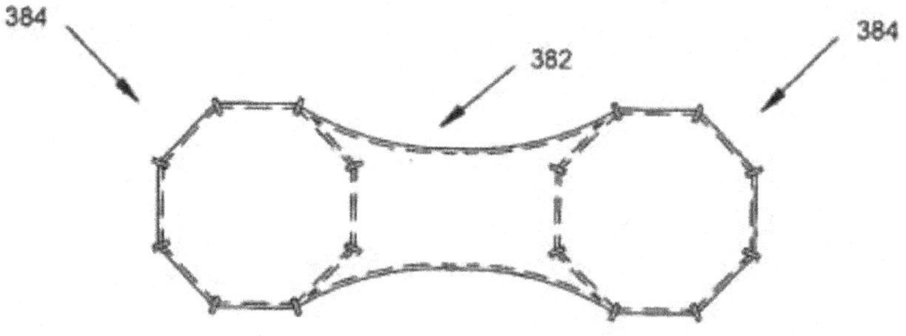

FIG. 54L

Fig. 54J–L. An excerpt from the US Patent No. 6,897,832 show respective 8, 10, and 10 sheet three-dimensional manufacturing patterns for inflatable toroids.

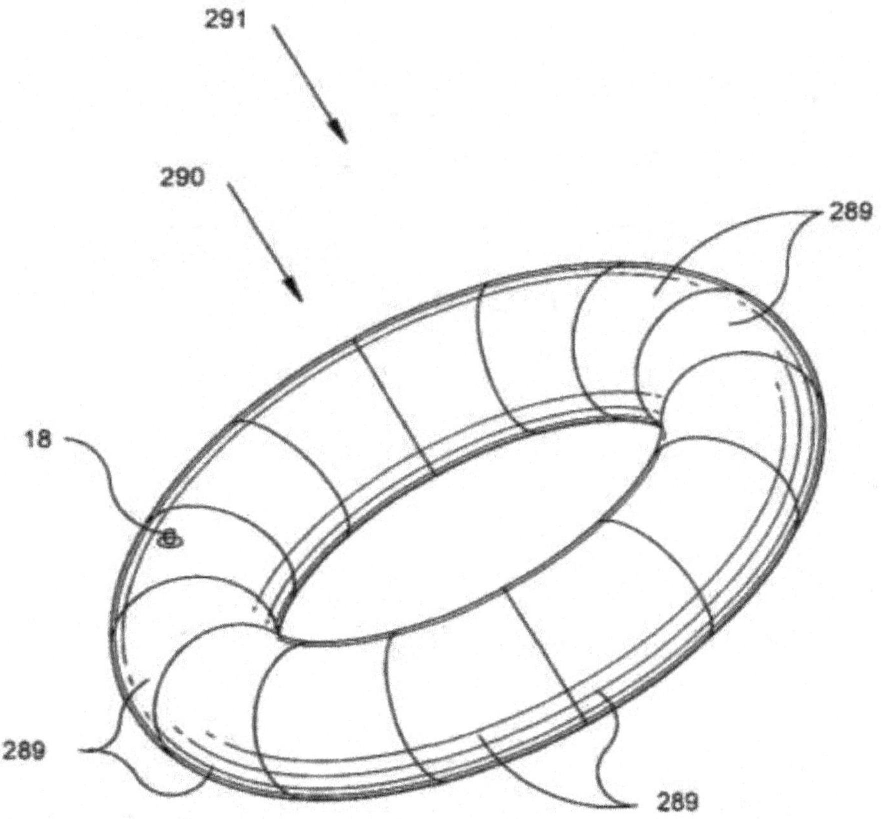

FIG. 55

Fig. 55. An excerpt from the US Patent No. 6,897,832 shows inflatable segmented toroid.

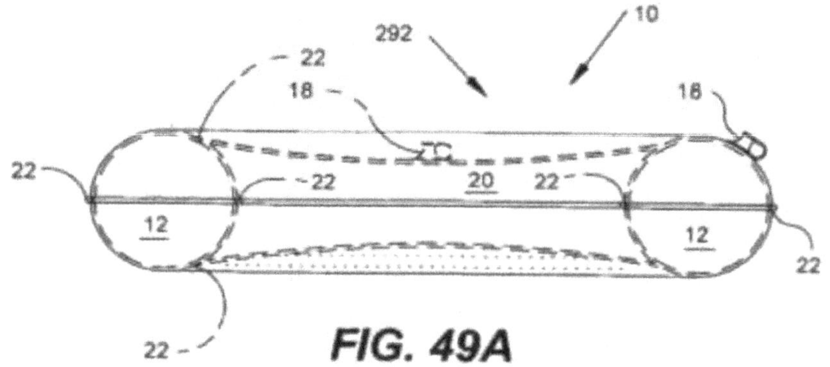

FIG. 49A

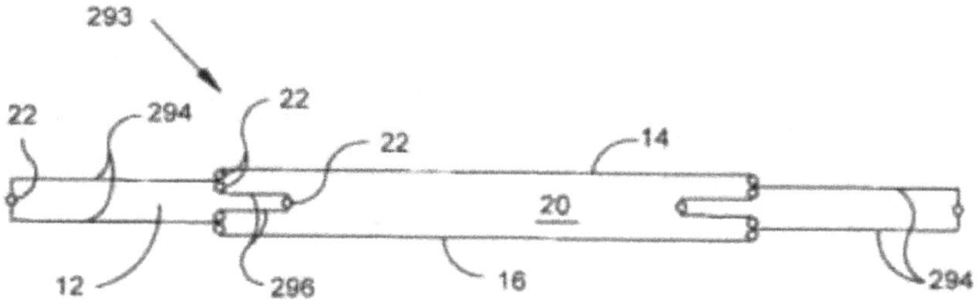

FIG. 49B

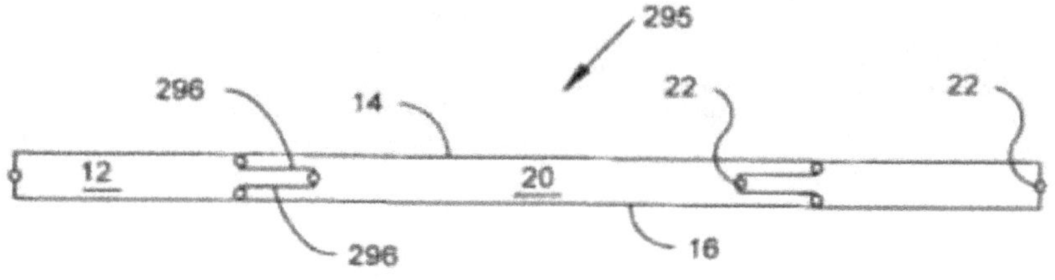

FIG. 49C

Fig. 49A–C. An excerpt from the US Patent No. 6,897,832 show 6 sheet flat manufacturing patterns for inflatable toroids.

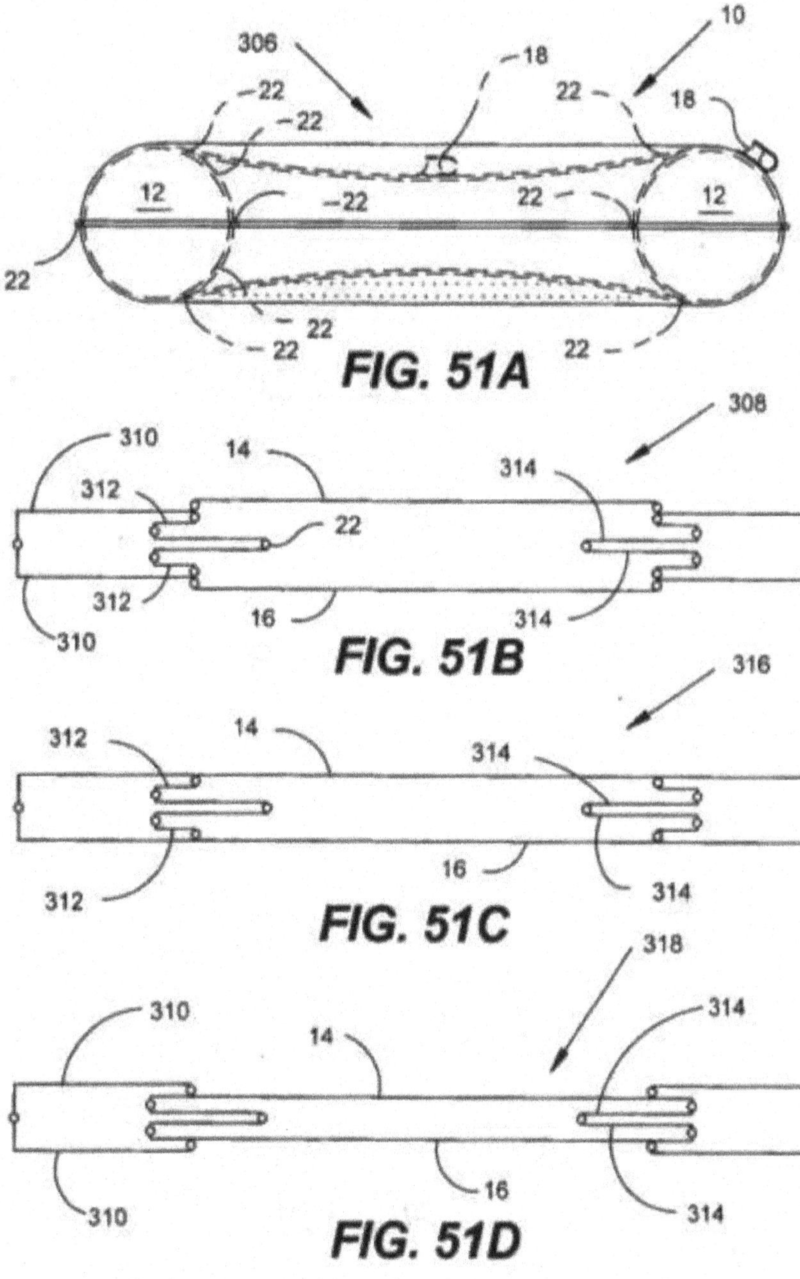

Fig. 51A–D. An excerpt from the US Patent No. 6,897,832 show 8 and 6 sheet flat manufacturing patterns for inflatable toroids.

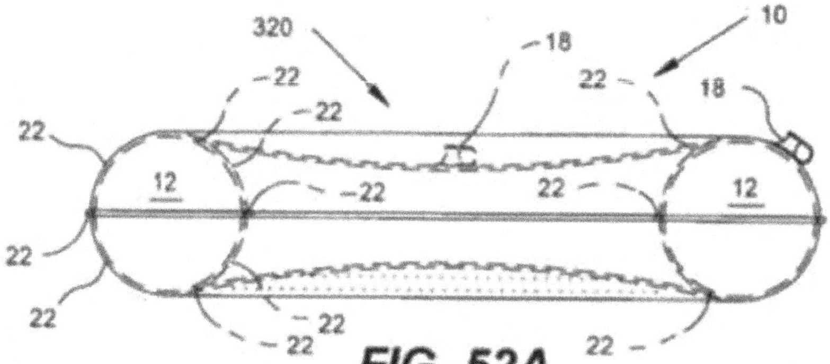

FIG. 52A

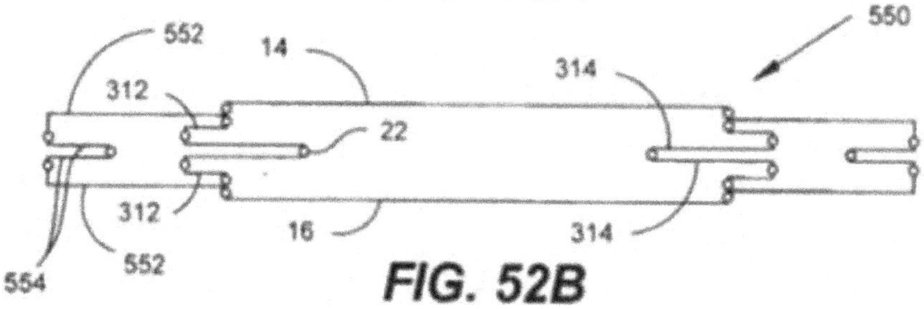

FIG. 52B

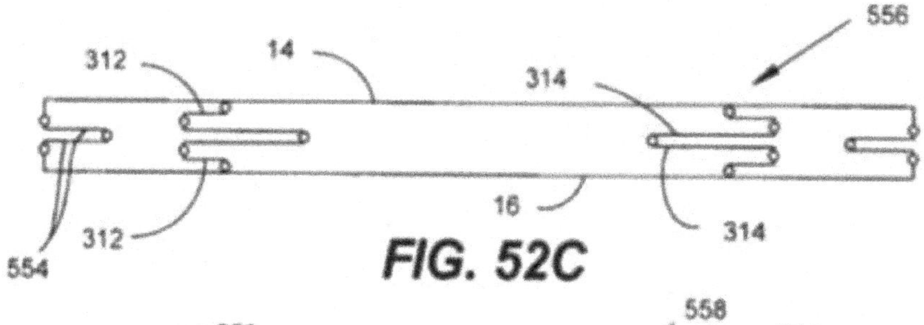

FIG. 52C

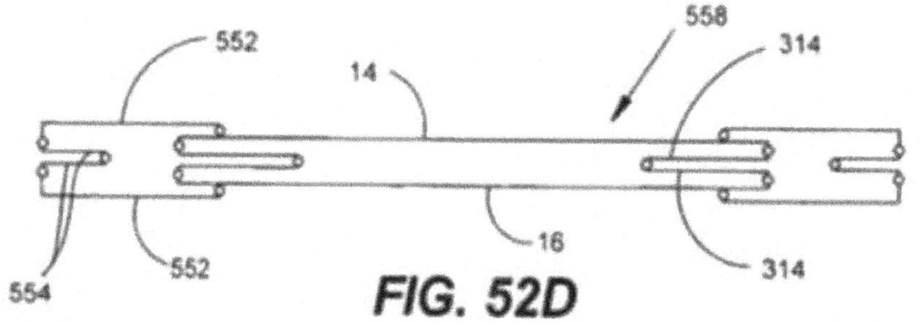

FIG. 52D

Fig. 52A–D. An excerpt from the US Patent No. 6,897,832 show 10 and 6 sheet flat manufacturing patterns for inflatable toroids.

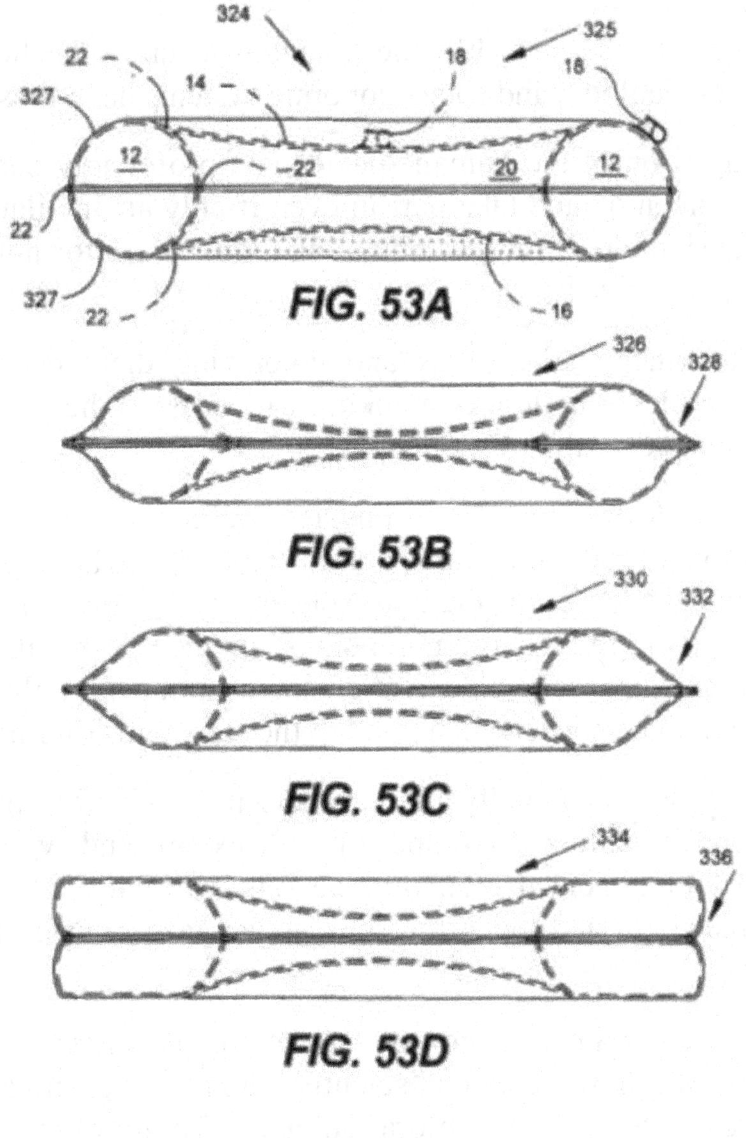

Fig. 53A–D. An excerpt from the US Patent No. 6,897,832 show biased sheet manufacturing patterns for inflatable toroids.

Some scenarios of manned interstellar flight posit that we will need to first achieve manned flight within our planetary solar system and then outward to the Kuiper Belt, and then on to the Oort cloud. We will refer to technologies that enable us to accomplish such flights as pre-interstellar techniques.

Mankind's journey to the stars has the potential to unite the human race where religions, political ideology, and socioeconomic systems have thus far failed.

The outward migration of the human species is a profoundly charitable endeavor that all persons can enjoy and take part in even if only an ancillary support role in many cases. I relish the potential for obtaining living space for innumerable human persons yet to be conceived.

Traveling out to other star systems and colonizing them can help ensure the survival of our species. The longer mankind can survive, the greater the potential for degrees of higher evolution for our species.

Even those with no formal training in physics and engineering can appreciate the outward migration of our species. Certain romantics who prefer to focus on interpersonal relations in lieu of the hard sciences and engineering disciplines can delight in anticipation of the untold numbers of unique personalities and genotypes that will come into existence to an extent commensurate with unlimited numbers of human persons yet to be conceived. Travel to the stars will offer us that potential.

We humans have a love of novelty and a wanderlust to travel to the top of the next proverbial hill. The unlimited scope of the extent and variety of physical phenomenon yet to be discovered and also that of any intelligent life-forms beckons us to reach out and touch the stars as we gaze in awe at the beauty of a starlit midnight sky.

Yes, we can truly build a civilization of love that is of *universal* scope. I feel that we must do so. We now have the scientific and mathematical principles and methods to eventually permit exploration and adventure into the wide open cosmos. Will we have the courage to set sail? I think the answer will prove to be a resounding, "Yes!"

Chapter 5

Going Beyond Chemical Rockets

This chapter presents innovative rocket propulsion technologies other than chemical rockets as well as other forms of propulsion.

A) Ion Thrusters

Two highly capable ion engines summarized at the following link:

http://www.boeing.com/defense-space/space/bss/factsheets/xips/xips.html **(Boeing Company 1995)**

are the Boeing 601HP Thruster (13 centimeters in diameter, has a specific impulse of 2,568 seconds, and produces 18 mN of thrust) and the Boeing 702 Thruster (25 centimeters in diameter, has a specific impulse of 3,800 seconds, and produces 165 mN of thrust).

The Ad Astra Rocket Company's VX-200 VASIMR engine can produce a specific impulse of 5,000 seconds at about 1.12 lb. thrust is described at http://www.adastrarocket.com/aarc/Technology, **(Ad Astra Rocket Company, 2009).**

We can surmise that using a planar array of ion thrusters involving 100 units of either of the first two engines would produce 1.8 newtons and 16.5 newtons, respectively. For the 601HP engine, the array would be only 1.3 meters by 1.3 meters. For the 702 model, the array would be only 2.5 meters by 2.5 meters. Larger arrays incorporating 10,000 units would produce 180 newtons and 1,650 newtons, respectively.

The specific impulse offered by chemical rocket fuels, which can be interpreted as a measure of the amount of momentum transferred to a rocket craft per unit mass of a fuel, is theoretically limited to about 450 seconds. Thus, the higher the I_{sp} value of a rocket fuel, the higher the velocity that a rocket of a given mass can obtain per unit of initial mass of rocket fuel.

Using the rocket equation, we can compute the terminal velocity of a manned spacecraft powered by such planar arrays as follows:

The rocket equation

$$v = v_e \, ln \, (M_0/M_1) \text{ with an } M_0/M_1 \text{ value of } 5$$

For the 601HP thruster,

$$v = (2{,}568 \text{ m/s})(9.81 \text{ m/s}^2)[\ln(5)] = 40 \text{ km/second}$$

For the 702 model

$$v = (3{,}800 \text{ m/s})(9.81 \text{ m/s}^2)[\ln(5)] = 59.997 \text{ km/second}$$

For an M_0/M_1 value of 10, the respective values for V are 58 km/second and 85.835 km/second.

For an M_0/M_1 value of 20, the respective values are 75.47 km/second and 111.67 km/second.

Squaring these M_0/M_1 values would permit the craft to slow down to a stop at its destination using reverse rocket thrust. For extreme velocities, perhaps huge, carbon fiber fabric bags, or other vessels, could be used to contain liquefied fuel where the ratio of fuel mass to bag mass may be as great as 1,000 or more. We are dealing with very low accelerations here, so the strength requirements of the bag need not be as extreme as one might intuitively expect.

Note that in the above rocket equation, v is the terminal velocity after final fuel burn-through, v_e is the exhaust velocity with respect to the spacecraft, M_0 is the fully fueled mass of the entire spacecraft before launch, and M_1 is the final payload of the vehicle or spacecraft dry weight. The ratio M_0/M_1, referred to as the mass ratio, is the initial fueled mass of the spacecraft divided by the mass of the final payload and is, therefore, dimensionless.

The VASIMR (variable specific impulse magnetoplasma rocket) engine technology is described by Ad Astra as being able to permit human travel to Mars in only thirty-nine days. The VASIMR engine is scheduled to be tested onboard the International Space Station (ISS) sometime within the next few years.

VASIMRs use radiofrequency electromagnetic waves to ionize a propellant, which is then heated by magnetic fields in order to produce thrust. The VASIMR rocket under development by the Ad Astra Rocket Company fills the middle ground between the extremely high thrust levels and the relatively low specific impulse of chemical rockets and the very low thrust levels and extremely high specific impulse of ion rockets. The variable specific impulse of the VASIMR engine permits the engine to operate near both ends of the above spectrum.

For the VASIMR engine and its I_{sp} of 5,000 seconds, the Tsiolkovsky rocket equation for a mass ratio of 10 yields a terminal velocity of 112.94 km/second. Squaring the mass ratio to 100 would enable a spacecraft powered by VASIMR technology to slow down to the point where it could refuel after reaching the outer limits of the planetary solar system. Even a mass ratio of 5, or 25, including extra fuel for slowing the craft to a halt, yields a terminal velocity of 78.94 km/second.

Because of limitations in the specific impulse of chemical rocket fuels, current heavy lift chemical rocket boosters require multiple stages to obtain Earth orbit, where two or more stages are sequentially dropped away after each stage's fuel supply is spent.

Thus, there is evidence to support my supposition that planar arrays of electrical thrusters could be practically incorporated into the design of manned spacecraft capable of reaching the outer limits of the planetary solar system.

If engines such as the electric rockets mentioned above could be developed to provide the same specific impulse, but instead use helium or hydrogen ions as a reaction mass, the craft could refuel on Pluto or on other planetoids within the Kuiper Belt, provided some sort of nuclear reactor–powered ice melter or other form of ice melter was brought along.

The power for the ion thrusters could be obtained from improved space-based reactors that operate on similar principles as the Russian Topaz space reactors that the former USSR experimented with in the 1970s and 1980s.

Note that the magnitude of thrust force for state-of-the-art electrical propulsion units, current ion rockets in service, as well as the VASIMR rocket, is small. However, the above engines can produce thrust over much longer time periods than can large, high-power-output chemical rocket boosters. The former electrical propulsion systems can achieve much higher terminal velocities, and with much less fuel than that required by presently available low Earth orbit–capable chemical combustion rockets.

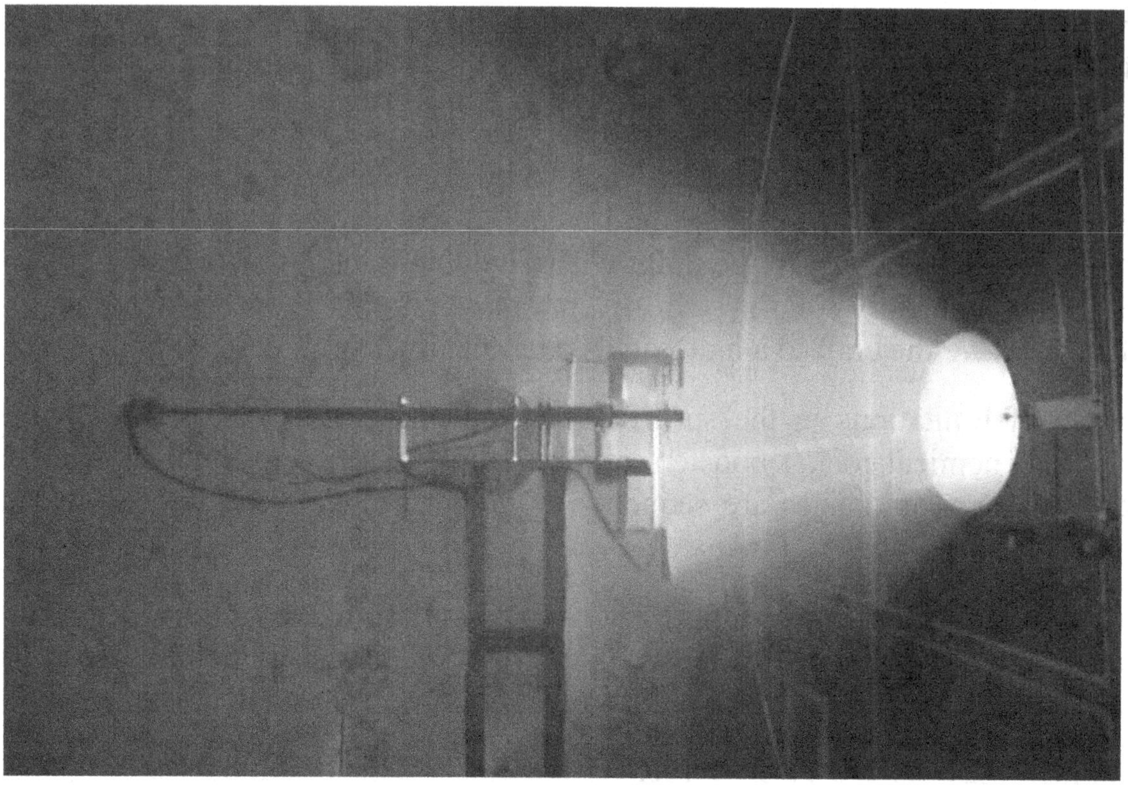

Fig 3. Photograph of the VX-200 Plasma at full power both stages and full magnetic field courtesy of the Ad Astra Rocket Company at: http://www.adastrarocket.com/HiResImagesForPublicRelease/VX-200-FullPowerBothStagesHiRes.jpg

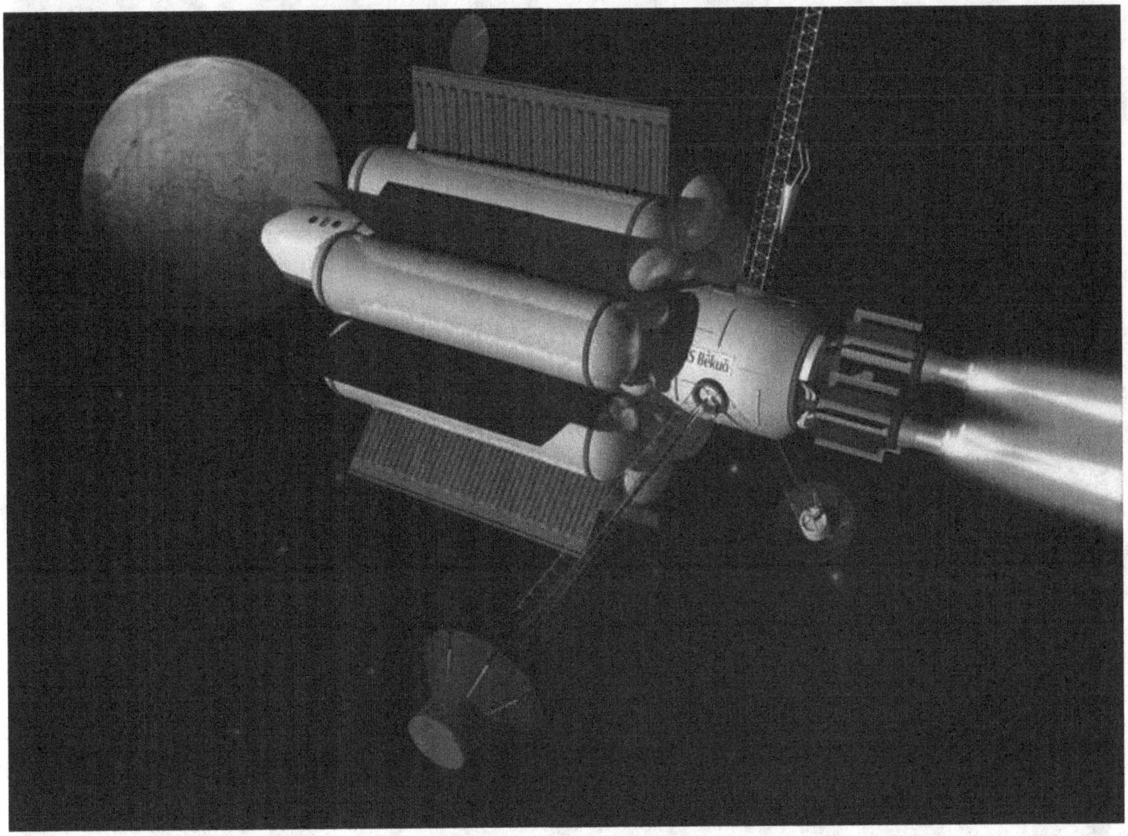

Fig. 4. Credit Bekuo. Courtesy of Ad Astra Rocket Company at http://www.adastrarocket.com/HiResImagesForPublicRelease/BekuoHiRes.jpg Human Mission to Mars with 10 MW nuclear-powered VASIMR Copyright Ad Astra Rocket Company © all right reserved.

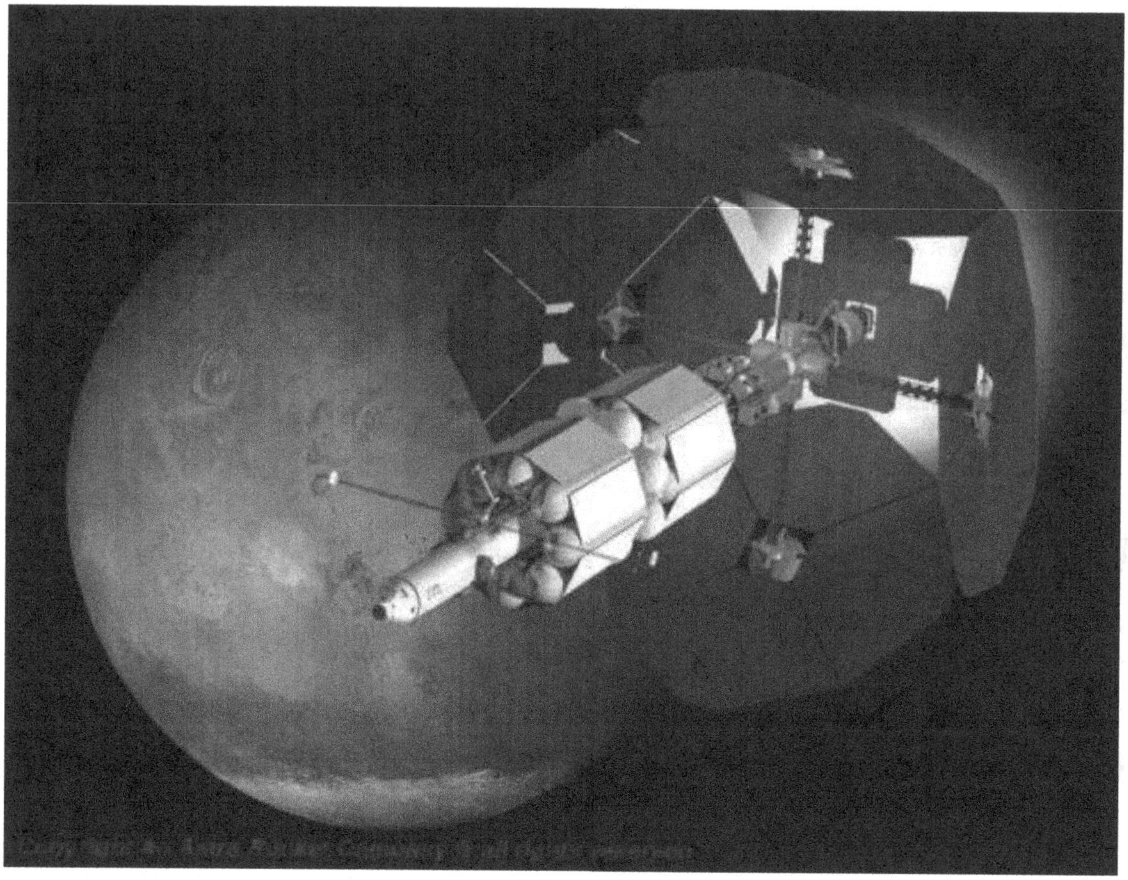

Fig 5. Human Mission to Mars with 200 MW nuclear-powered VASIMR®. Credit Bekuo. Courtesy of Ad Astra Rocket Company at http://www.adastrarocket.com/HiResImagesForPublicRelease/Bekuo200MW-1-MidRes.jpg Copyright Ad Astra Rocket Company © all right reserved.

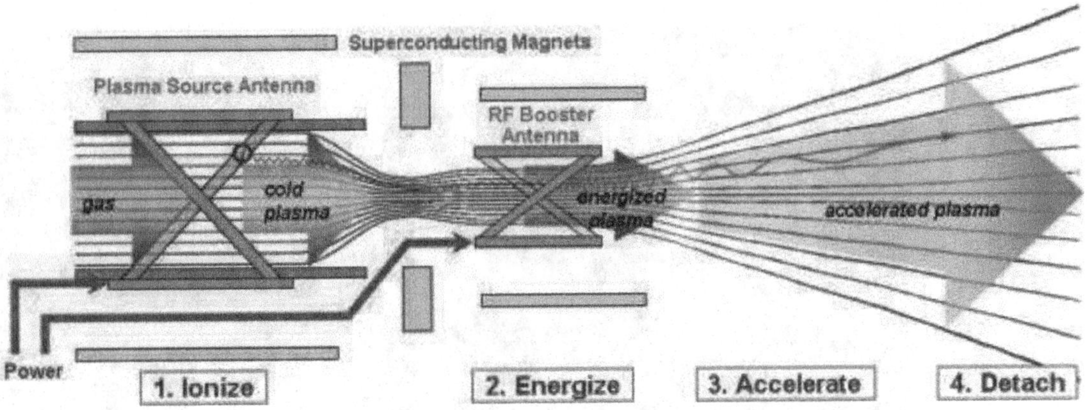

Fig 6. VASIMR® Operating Principles. Courtesy of Ad Astra Rocket Company http://www.adastrarocket.com/HiResImagesForPublicRelease/VASIMR_operating_principles.jpg Copyright Ad Astra Rocket Company © all right reserved.

Below is a photograph of Dr. Franklin R. Chang Díaz, who serves as company president and CEO.

Fig. 7. ASPL Director Franklin Chang Díaz, 2005
http://www.adastrarocket.com/HiResImagesForPublicRelease/Franklin-ASPLHiRes.jpg

B) Electrical Ion Rockets

Another rocket technology involves an ion rocket designed mainly by Canadian scientists. Some claim it will enable manned travel to and from Mars in less than thirty-nine days when scaled up using nuclear power to energize the ion rocket. We might very well be booking flights on Planetary Cruise Lines' vessels bound for Mars and beyond sometime later this century. Anyone willing to book a few weeks at the Mars Hyatt Regency? If you can afford the cost, and I am still alive when such voyages become commonplace, please take me with you.

Note that electrical rocket engines are just what the name implies. These engines generally rely on the establishment of an electrodynamic accelerating potential to accelerate ions. Some examples of electrical propulsion systems are described below.

In gridded electrostatic ion thrusters, electrons are used to bombard xenon gas in order to ionize the gas. The positive xenon ions are then directed to a multigrid system where they are accelerated by the electrical potential set up between a first grid and a second grid. A cathode is placed near the engine that emits a beam of electrons into the xenon ion beam. This is necessary in order to prevent the xenon ions from falling back into the second grid, which tends to attract the ions. The electrical potential between the first and second grid is typically about 1 to 2 kilovolts, thus resulting in the xenon ions exiting with a kinetic energy of between 1 and 2 kilo-electron volts KeV (Wikipedia, 2006).

Hall Effect thrusters rely on the acceleration of xenon or bismuth ions down a cylindrical anode to a cathode composed of a plasma. The electrical charge due to the xenon or bismuth ions is neutralized as these ions pass into the cathode due to the cancellation of the opposite charges. This results in a kinetic energy stream in the form of the accelerated reaction mass in such a manner that the exhaust does not get pulled back into the exhaust nozzle. A massive particle stream is directed away from the rocket, thereby resulting in a net momentum gain, and thus a net velocity gain by the spacecraft.

Hall Effect thrusters contain a member that sets up a radial magnetic field between the member and the surrounding acceleration tube. Due to the relatively large mass of the ions, the ions are affected only slightly by the magnetic field. The electrons produced near the end of the element are affected far more by the magnetic field and are held in place, thus resulting in the creation of the cathode. The Hall Effect current is caused by a portion of the electrons spiraling down toward the anode where these electrons collide with the xenon or bismuth in order to produce the ions (Wikipedia, 2007).

Field emission electrical propulsion systems rely on the acceleration of metal atoms with a low ionization potential and a high atomic mass. These systems consist of a reservoir of liquid metal, a small slit or opening through which the liquid metal can flow, followed by an accelerator ring. Upon reaching the slit in the emitter, an electrical field is applied between the emitter and the acceleration ring. This field causes the liquid metal atoms to become ionized. The positive ions are then accelerated by the electric field produced by the emitter and the acceleration ring. An external source of electrons neutralizes the positive ions, thereby preventing undue electrical charge build up on the spacecraft hull.

Pulsed inductive thrusters (PIT) involve the application of orthogonal electric and magnetic fields to accelerate a reaction mass stream. A small amount of gas is released to spread across an induction coil. The coil is briefly magnetized by a bank of capacitors, thus inducing a circular electric field in the gas that ionizes it. The ions revolve in the opposite direction as the original pulse current and are ejected out of the engine. The PIT engine requires no electrodes, and the power output of PITs can be easily scaled up to megawatts by increasing the number of pulses per unit of time.

Electrodeless plasma thrusters utilize electromagnetic fields to ionize a gas. The ionized gas is then transferred to a chamber where it is accelerated by an oscillating electric and magnetic field. The electrodeless plasma thruster has separated ionization and acceleration stages and permits not only the thrust to be varied, but also the specific impulse associated with the reaction mass.

Other forms of electrical propulsion include the resistojet, arcjet, microwave electrothermal thrusters (MET), and the helicon double-layer thruster.

Given the large number of different forms of electrical propulsion systems, many other designs yet to be dreamed up will most likely be developed in the future. The invariant mass-specific impulse of electrical propulsion systems is ultimately limited only by the exhaust velocity of the systems and therefore the momentum to invariant mass ratio of the emitted exhaust species. The mathematical limit of the invariant mass specific impulse of electrical propulsion systems is infinity for portions of the ejected exhaust streams. This is so because the relativistic momentum of ever higher energy, particle-accelerator-like, electrical propulsion systems can be designed to be arbitrarily high with respect to the invariant mass of the ejected species. How to power such systems remains a different topic altogether.

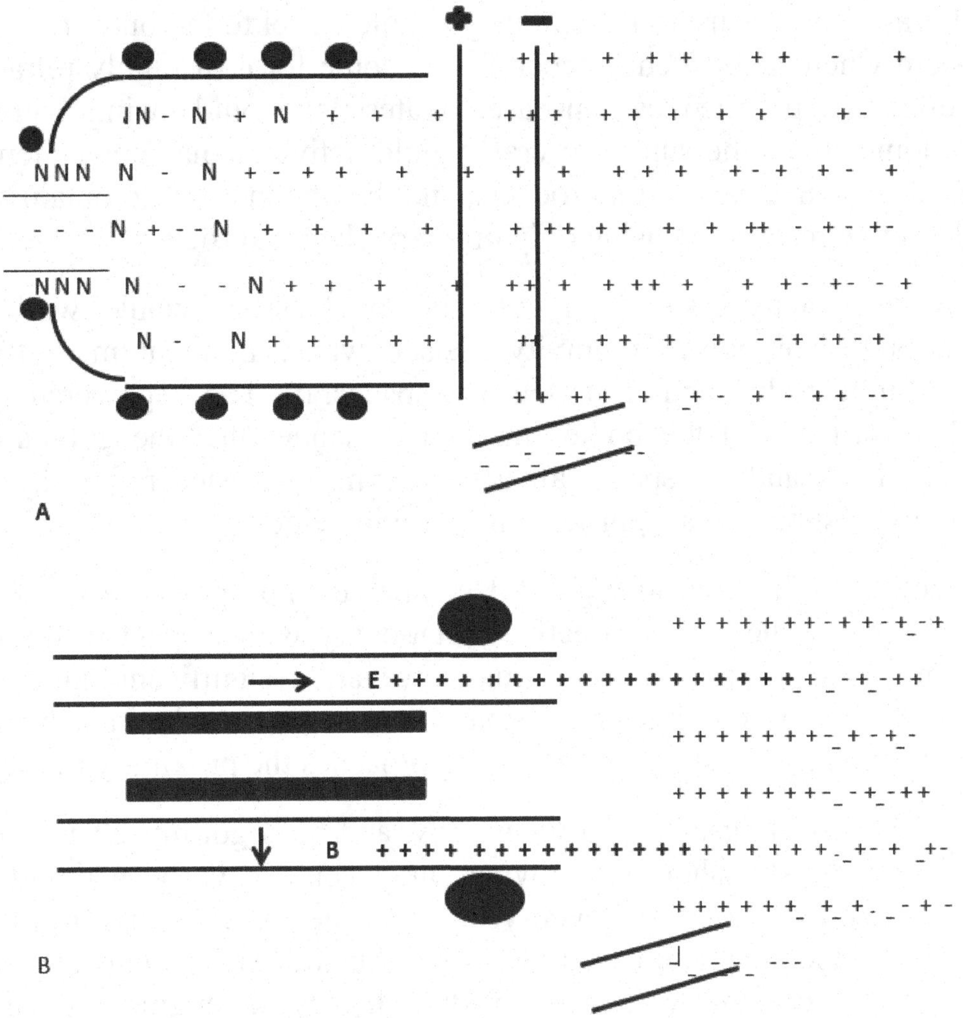

Fig. 8A–B: Figure 8A depicts a gridded electrostatic thruster. The circles indicate magnets. The Ns, plus signs, and minus signs indicate neutral atoms, positive ions, and electrons. The two vertical lines on the right indicate the positive and negative grid. Note the neutralizing stream of electrons. Figure 8 B depicts a Hall Effect thruster. The vectors E and B represent the radial magnetic field and the electric field, respectively. A positive ion exhaust is directed to the right out the conduit containing the letters *E* and *B*. The elongated rectangles represent the inner magnetic coil. The ellipses near the centered portion of the image indicate the outer magnetic coil. Note that both figures depict cross-sectional aspects of the thrusters. The neutralizing electron stream is indicated by the minus signs within the obliquely oriented conduit.

C) Powered Gravity Assists

Powered gravity assists using the sun may enable travel to the outer reaches of the solar system where a shielded spacecraft (via some form of highly refractory and highly reflective, polished carbon-based materials) would swing within a few million kilometers of the sun. The craft would activate solar power ion rockets, chemical rockets, nuclear thermal rockets, and the like in order to obtain a velocity that could not be reached by using only ordinary chemical rockets.

Solar-powered gravity assists and other powered assists would work with an effectiveness of non-powered gravity assists, where momentum is transferred between orbiting bodies and an unpowered spacecraft. The unpowered spacecraft is pulled toward one of the bodies in such a manner that the gravitational tug between the body and the spacecraft acts to sling the spacecraft away from the orbiting body, resulting in a significant net gain in velocity.

The sun could not be used in a gainful manner for non-powered gravity assists. This is so because the sun is effectively stationary with respect to the center of momentum frame of the solar system. However, a significant increase in the efficiency of the spacecraft's propulsion systems can be obtained by judicious application of rocket thrust as a spacecraft approaches the proximity of the sun.

I used to play with slingshots as a young boy, and even got myself in trouble one time for firing the slingshot at a moving car. I could have been reported to the police, but since I was a naive eleven-year-old at the time, the driver only showed up at our front door step and asked me to reimburse him for the cost of a broken side rearview mirror. However, even we adult physics and engineering geeks often have a childlike fascination with the practical aspects of large-scale slingshots in the form of gravitational assists, whether powered or not.

D) Solar Dive and Fry Maneuvers

Solar dive and fry maneuvers might permit small but significant fractions of the speed of light to be reached where a manned spacecraft would dive in toward the sun approaching the sun at a minimum distance of a few million kilometers. The craft would then rapidly deploy an ultra-high mass specific area solar sail, thereby providing the spacecraft with the desired boost in velocity. The sail could take the form of an ultra-thin, metalized, carbon fiber-based membrane and the like materials. Some scenarios anticipate some form of metalized monolithic carbonaceous membrane with a thickness as small as a few nanometers or less.

I have performed some simple calculations that suggest that a single pass Sundiver mission could obtain a maximum terminal velocity well beyond 25 percent the speed of light with materials that might plausibly be developed. Such materials might include netlike weaves of carbon graphene materials where the graphene would somehow be metalized with extremely refractive and highly reflective metals or metallic compounds. Note that graphene is a one-carbon atom thick membranous form of carbon and has been measured to be two hundred times stronger than typical construction-grade steel.

Another potential sail material consists of highly reflective metalized carbon nanotubes woven into a membrane that has a thickness of about one nanometer or more.

Yet another potential sail material consists of nano-scale diamond threads that have a metallic deposition on them. Certain forms of artificial diamond may have much higher thermal conductivity and hardness than perfect natural diamond. These artificial forms of diamond have a dissimilar blend of carbon isotopes than do natural diamonds.

Beta carbon nitride (β-C_3N_4) material might be fashioned into ultra-strong threads by virtue of the extreme atomic bond strength between the constituent carbon and nitrogen atoms. Carbon nitride may in some forms be harder than diamond and thus may have stronger atomic bonds.

The above carbon-based sails are currently science fiction even though they are, in theory, possible to construct. The chief limiting factor for the production of such sails in the near term is the extreme cost per unit of mass of the conjectured materials of composition. Perhaps some sort of high-strength, temperature-resistant, metalized polymer would need to be used in the meantime involving greater minimum distances of approach to the sun.

Instead of "sailing the ocean blue," to use the parlance of the New World explorers and pioneers who did so in centuries past, we may be "sailing the ocean black" in the coming decades. Thus, we may realize, in some sense, Albert Einstein's youthful thought experiment or daydream of riding along a beam of light. The glory days of sun sailing and, eventually, sailing on artificially generated, immensely powerful beams of visible and infrared light, microwaves, and radiofrequency radiation may one day become reality. Such mechanisms may usher in a bold new renaissance of pioneering and may ultimately open up the entire Milky Way Galaxy and realms beyond to human exploration and colonization.

Solar sails and beam sails would take advantage of the momentum possessed by photons or particles of light. These photons or electromagnetic energy waves would either be absorbed or reflected from a large-area sail with very low surface-area density. The impinging photon stream would push on the sail in a manner similar to that of the wind pushing on the wind sails of ships in centuries past and on the sails of modern pleasure craft and racing yachts.

Photons are massless—or, at the very least, must be limited to utterly minuscule but finite masses due to sensitivities of certain theoretical paradigms. However, photons do have non-zero momentum. Energy and mass are interchangeable, and even though photons have no mass, they do in fact carry momentum.

The momentum of photons has long been observed and is defined in the following simple equation.

$$P = \hbar k = h\nu/c = h/\lambda$$

Here, P is the photon's momentum, ħ is the reduced Planck constant, k is the photon's wave vector, h is the Planck constant, ν is the frequency of the photon, λ is the photon wave-length, and c is the speed of light.

Various experimental methods have determined the upper limits of photon mass. Three such results are as follows:

$$m \lesssim 10^{-14}\ \mathrm{eV}/c^2,\ m < 3\times10^{-27},\ \mathrm{eV}/c^2 \text{ and } m = 10^{-18}\ \mathrm{eV}/c^2$$

Technically, photons should have no mass but might have a minuscule mass as may be allowed by sensitivities of certain relevant theories (Wikipedia).

The reduced Planck constant is equal to

$$1.054571628(53) \times 10^{-34}\ \mathrm{J{\cdot}S} \text{ or } 6.58211899(10) \times 10^{-16}\ \mathrm{eV{\cdot}s}$$

In empty space, the energy of a photon is $E = Pc$. Therefore, $P = E/c$.

Now imagine a photon having the energy equivalence of 1 kilogram of matter. The energy of the photon would be $MC^2 = 9 \times 10^{16}$ joules.

As a result, the associated momentum would be $[9 \times 10^{16}\ \mathrm{J}]/[3 \times 10^{8}\ \mathrm{m/s}] = 3 \times 10^{8}$ kg m/s.

The formula for relativistic momentum is $Mv\gamma = Mv\{1/\{1 - [(v/C)^2]\}^{1/2}\}$. So a kilogram invariant mass traveling at a Lorentz factor of 2 or at 0.8667 C has a momentum of (1 kg)(260,010,000 m/s)(2) = 5.2002×10^{8} kg m/s. The Lorentz factor is properly referred to as the Lorentz Transformation Factor and is equal to $\{1/\{1-[(v/C)^2]\}^{1/2}\}$.

Now, consider a photon with an energy of 9×10^{19} joules. The momentum of our mythical photon would be 3×10^{11} kg m/s. Also consider that a kilogram invariant mass traveling at a Lorentz factor of 1,000 or at a velocity of 0.999998 C has a momentum of (1kg)(300,000,000 m/s)(0.999998)(1,000) = 2.999994×10^{11} kg m/s.

You can see that as the percentage of total energy of a relativistic mass in the form of relativistic kinetic energy approaches 100%, the momentum of the relativistic mass is exactly equal to that of its photonic energy equivalence.

Although not a rigorous argument for non-zero photonic momentum, the case for the plausibility of non-zero momentum based on the interconvertability of mass and energy has been roughly established in the above argument.

Particle physics, solar sails, and even the US Air Force–funded research and development involving highly accurate, airborne, high-powered lasers, work because of the known relation of non-zero photon momentum to photon energy. The details are somewhat more complicated, but I think you can get a general idea of the respective concepts.

The momentum of a photon, as well as its energy, is directly proportional to its frequency and inversely proportional to its wavelength. For example, an optical light photon with an energy of 1 electron volt will have 1 million times the wavelength of a photon with an energy of 1 million electron volts, which is an example of a gamma ray photon.

Space light sailing is a wonderful concept that does not require the spacecraft to carry fuel to accelerate or decelerate. In theory, velocities virtually equal to C can be obtained by a light-sail-driven spacecraft; even spacecraft large enough to carry human crew members.

Light-sail-powered craft lie somewhere in the exotic, between nuclear rockets and the as-yet highly speculative concepts of warp drive, antigravity, wormhole travel, and the like. The good news is that light sail spacecraft concepts are very well grounded in theory.

E) Nuclear Thermal Rockets

Nuclear thermal rockets were the subject of active research during the early 1960s through the early 1970s in the US under a project known by the acronym NERVA, or Nuclear Engine for Rocket Vehicle Application. NTRs are viewed by some astronautical engineers as potential propulsion systems to enable manned travel throughout the solar system. Such systems generally anticipate using a nuclear reactor to heat a working gas. The thermally expanded gas would then be ejected out of the back end of the rocket chamber to provide specific impulses in the range of 1,000 seconds (David Darling, "Internet Encyclopedia of Science").

One of the main hurdles to overcome in fielding any high-end specific impulse nuclear thermal rocket is the need to insulate the rocket fuel heating and exhaust chambers from the extremely hot gas. Mechanisms using cooler-intake gas drawn from the gas reservoir to buffer the critical contact points of the rocket from the extremely hot gas have been studied. Various critical surfaces can be made extremely reflective to black body radiation emitted by the hot exhaust gas, which can reach temperatures approaching 3,000 K or more. Mirror equipment might be used to focus the heat from the hot nuclear reactor components, where gas temperatures higher than that of the nuclear reactor heating elements might be obtained.

Note, once again, that the specific impulse can be interpreted as a measure of the amount of momentum transferred to a rocket craft per unit of fuel mass. For practical chemical rockets, I_{sp} is theoretically limited to about 450 seconds. Research into chemical fuels that offer significantly higher specific impulses has been conducted, but in general, these fuels have so far proven very unstable or can only be produced in small quantities at present.

F) Magnetic Plasma Bottle Sail

A Magnetic Plasma Bottle Sail, or Magsail, has been proposed as a propulsion mechanism where a spacecraft would be surrounded by an extended plasma that is electrodynamically coupled to the spacecraft. The solar wind would push on the magnetic plasma bottle. Due to the bottle's large capture area, the spacecraft could reach a terminal velocity of around 900 kilometers per second or more. Such a velocity would enable manned missions in huge colony ships to Alpha Centauri in about 1,300 years and to Barnard's Star in about 1,800 years. Missions could be sent out in droves to stars as far as 100 light-years away with travel times of only about 33,000 years, both Earth and ship time. The entire Milky Way Galaxy could be colonized by humanity in about 24 million years without the need to achieve relativistic velocities and the associated need to shield the craft from impact with relativistic dust grains, gas atoms, and molecules.

The Magsail for practicality is limited to roughly the velocity of the solar wind which has a maximum velocity on the order of 1,000 kilometers per second. A Magsail can in theory be pushed faster than the driving plasma stream; however, practical kinematical relations between the maximum solar wind speed and an outgoing Magsail craft effectively limit such craft to about the speed of the solar wind. The total mass of the plasma comprising the electrodynamically reactive bottle can, in theory, be made very small.

More practical interstellar space-faring technologies can likely be developed that will enable human-crewed spacecraft to reach highly relativistic velocities and thus enable travel with ease between the stars by taking advantage of relativistic time dilation. However, some academics feel that our first manned colonization missions to other star systems will likely take the form of slow multigenerational space arks that travel for thousands of years between stars.

Even if we were somehow limited to such slow technology, would we have the courage to embark on such a noble agenda and set sail to new worlds? Some of these worlds may be teeming with life-forms that could be considered every bit exotic, if not more so than the creatures that ply the depths of our oceans or roam our forests or live underground.

Future grade-school biology textbooks might be adorned with elegant and beautiful photographs of creatures we have yet to envision. Some researchers feel that in our own solar system, there might exist plant or animal life-forms in locations such as the aquatic environments below the icy surface of the moon Europa, or perhaps within similar environments on the moons of Ganymede, Enceladus, Titan, as well as others.

We live in a unique era. We are in the precursor planning stages of venturing out into interplanetary space and then to worlds beyond. What will we discover when we finally do so?

H) Fission Fragment Drive

Another technique that could be used to produce terminal velocities from between 0.01 C to 0.05 C is the so-called fission fragment drive. The higher end of the above velocity range can be obtained with large mass ratios and discardable FFD stages, as will be explained in the chapter on fission fragment drives found near the end of this book. The FFD would rely on open-ended nuclear fission reactors that would exhaust neutrons, gamma rays, and other fission fragments to the vacuum of space from behind the spacecraft. The fission fragments that travel toward the spacecraft would be absorbed by the reactor and its forward located shielding. The net momentum delivered to the spacecraft by the combined effects of the immediately backwardly ejected fission products and any forwardly impinging then backwardly reflected fission fragments can result in very significant spacecraft velocities.

The heat generated by the absorbed ionizing radiation from the fission process can be used to drive a turboelectric generator that can power: ion, electron, photon, MHPD, EHPD, and electro-magneto-hydrodynamic-plasma drive systems. These electrical propulsion systems can augment the chief fission fragment drive feature.

At a conservative 0.01 C, the fission fragment drive could enable travel to Alpha Centauri in only about 400 years and to Barnard's Star in only about 580 years. The more extreme value of 0.05 C could reduce these two travel times to 80 years and 116 years, respectively. The Milky Way could be colonized by huge space arks in only 1.4 million years, which is on the order of the time period that our hominid ancestors had been roaming the surface of the earth. Some researchers feel that velocities as high as 0.1 C are, in principle, attainable by fission fragment drive technology. Expressed in units of C or of the speed of light, nuclear fission rocket fuels can provide a maximum specific impulse of about 0.04 C, which corresponds to an exhaust velocity of 0.04 C.

Chapter 6

Current Nuclear Fission Reactor Technologies

This chapter provides a brief overview of the more prominent types of nuclear reactors already in existence. Such a summary provides context for further discussion of potential nuclear reactor applications for external and internal power sources for long-mission-duration interstellar space-arks and colony ships. The review of current reactor types provides a context in which to bound and ground the proposed reactor applications in real-world existent and plausible yet to be developed fission-powered systems. So a large range of nominally performing reactor types are considered in order to ensure a more thorough analysis of potential suitability for various classes of interstellar vehicles, some of which may require reactor types dissimilar to those required for other classes of spacecraft. These reactors may also provide power for solar system–based colonies and outposts.

Nuclear fission reactors generally referred to simply as nuclear reactors, or reactors, come in a variety of varieties and sizes.

Some of the newer designs are as follows: advanced boiling water reactor, molten salt-cooled reactor, liquid metal-cooled reactor, and high temperature helium gas-based Brayton cycle supercritical CO_2 reactor.

Reactors that operate at high temperatures are generally more efficient in converting heat from fission reactors to electrical energy or other useful process heat.

High-temperature reactors for which the generated heat is rapidly carried away can operate at higher-power outputs.

Reactor concepts also include systems that would burn long-life high-activity isotopes. Such systems have the benefit of providing additional energy by burning much of the fission waste products in addition to reducing the amount of potentially dangerous radioactive waste.

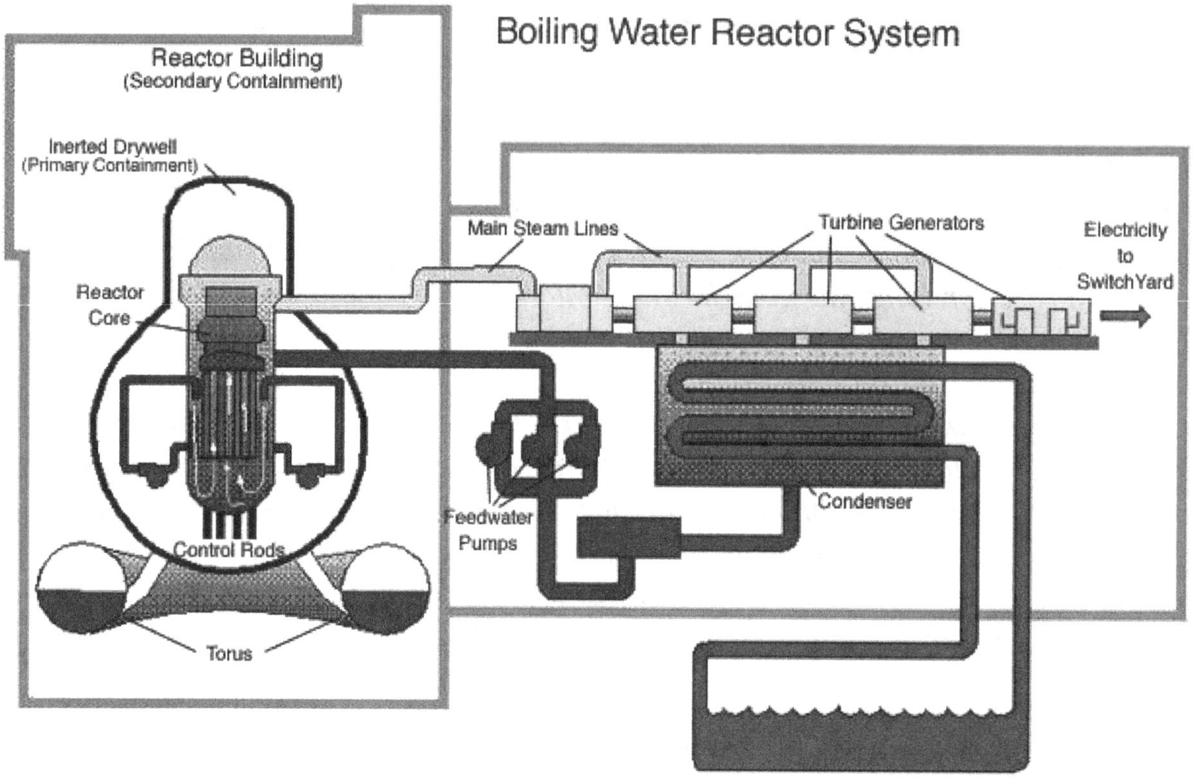

Fig. 9. Illustration of a boiling water reactor system. Such a system could conceivably be applied on a colony-type interstellar spacecraft. For such applications, the colony would be set rotating before the boiling water reactor would be activated. Since the colony would lack sufficient gravitational mass to simulate one Earth g, some other means of simulating gravity would be necessary. For spacecraft undergoing small accelerations, rotating can simulate gravity to enable the reactor system to work just as it would on Earth. (Image was borrowed from public domain online sources.)

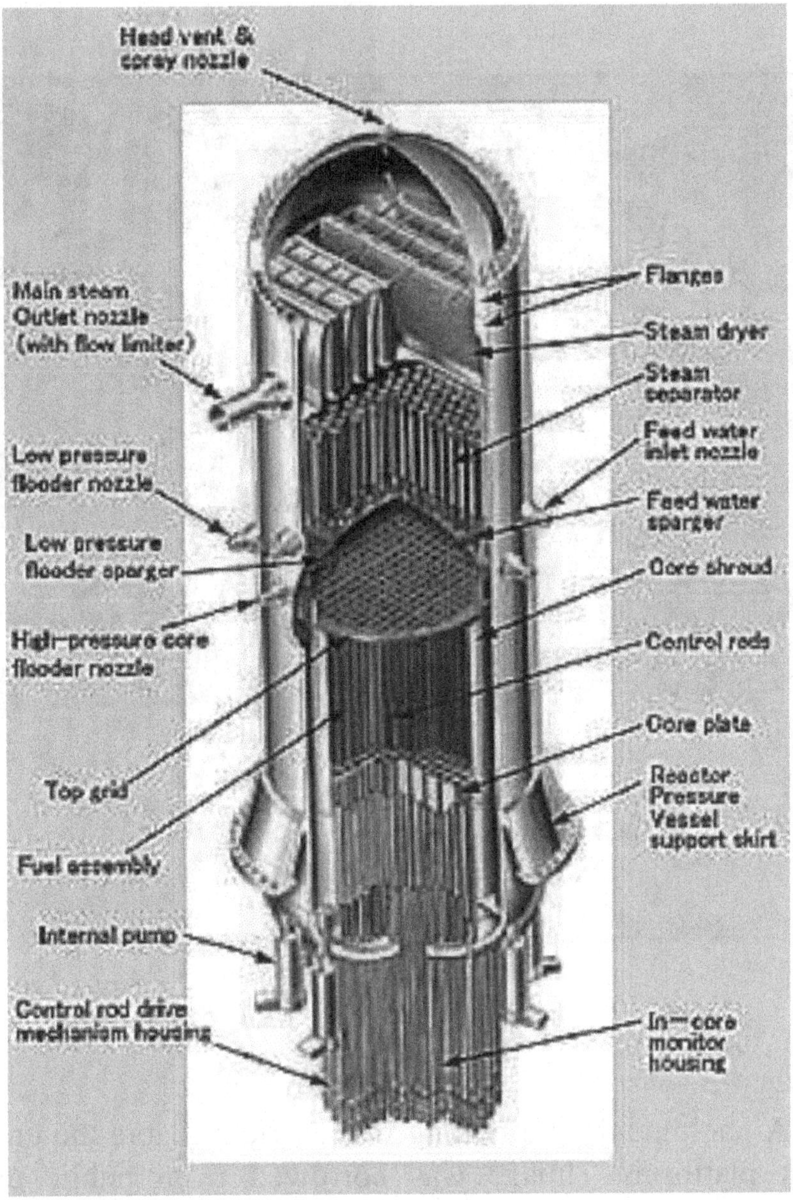

Fig. 10. The core of a boiling water reactor system. As can be intuited from the diagram, boiling water reactors are complex mechanisms, having several effluent carrying features. (Image was borrowed from public domain online sources.)

Fig. 11. A photograph of a boiling water reactor. Note the unit dwarfing of the work platforms. (Image was borrowed from public domain online sources.)

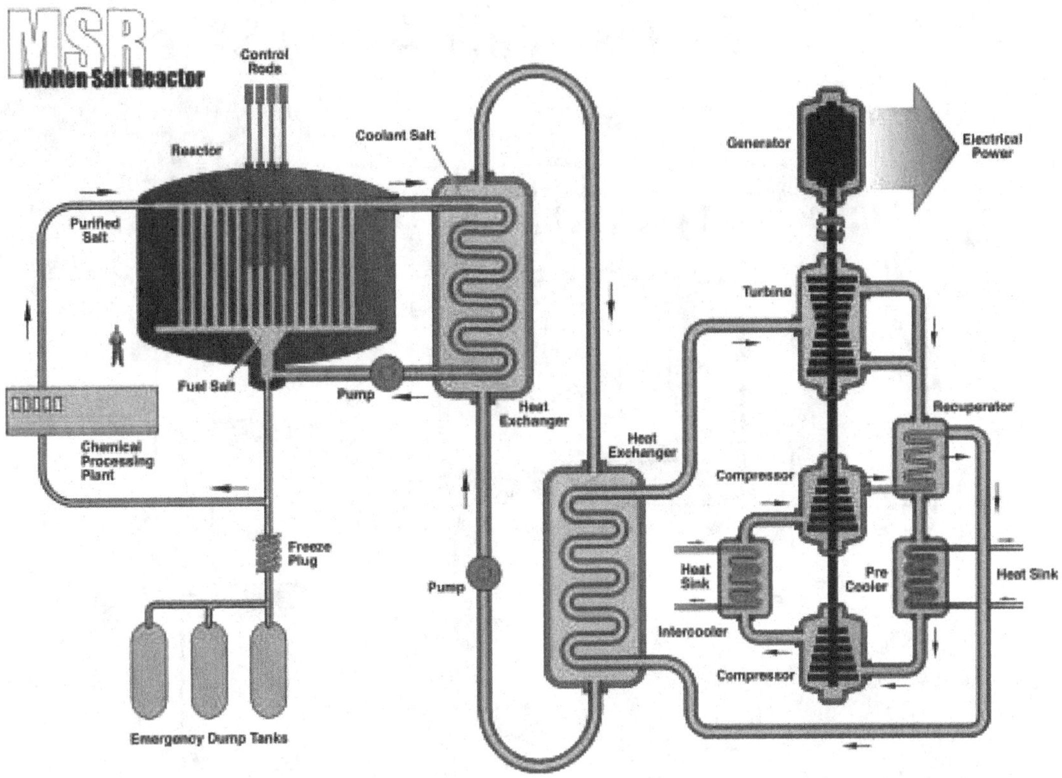

Fig. 12. Illustration of a molten salt-cooled reactor. Note the use of salt as a cooling effluent. In cases where the emergency dump tanks would require simulated gravity feeds, a means for providing artificial gravity would be required such as can be enabled by a rotating colony ship. (Image was borrowed from public domain online sources.)

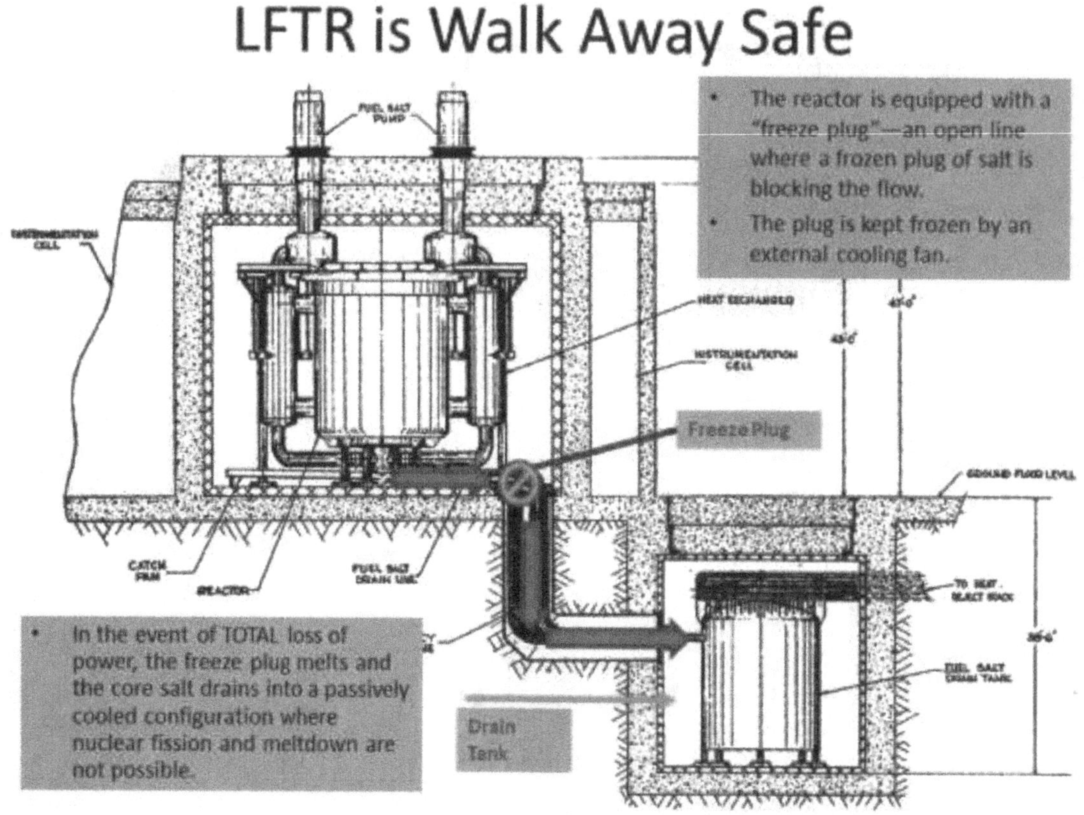

Fig. 13. A worst-case scenario for failure of a liquid salt reactor plant. As indicated in the diagram, a mechanism exists to prevent core meltdown in the event of a total power failure. (Image was borrowed from public domain online sources.)

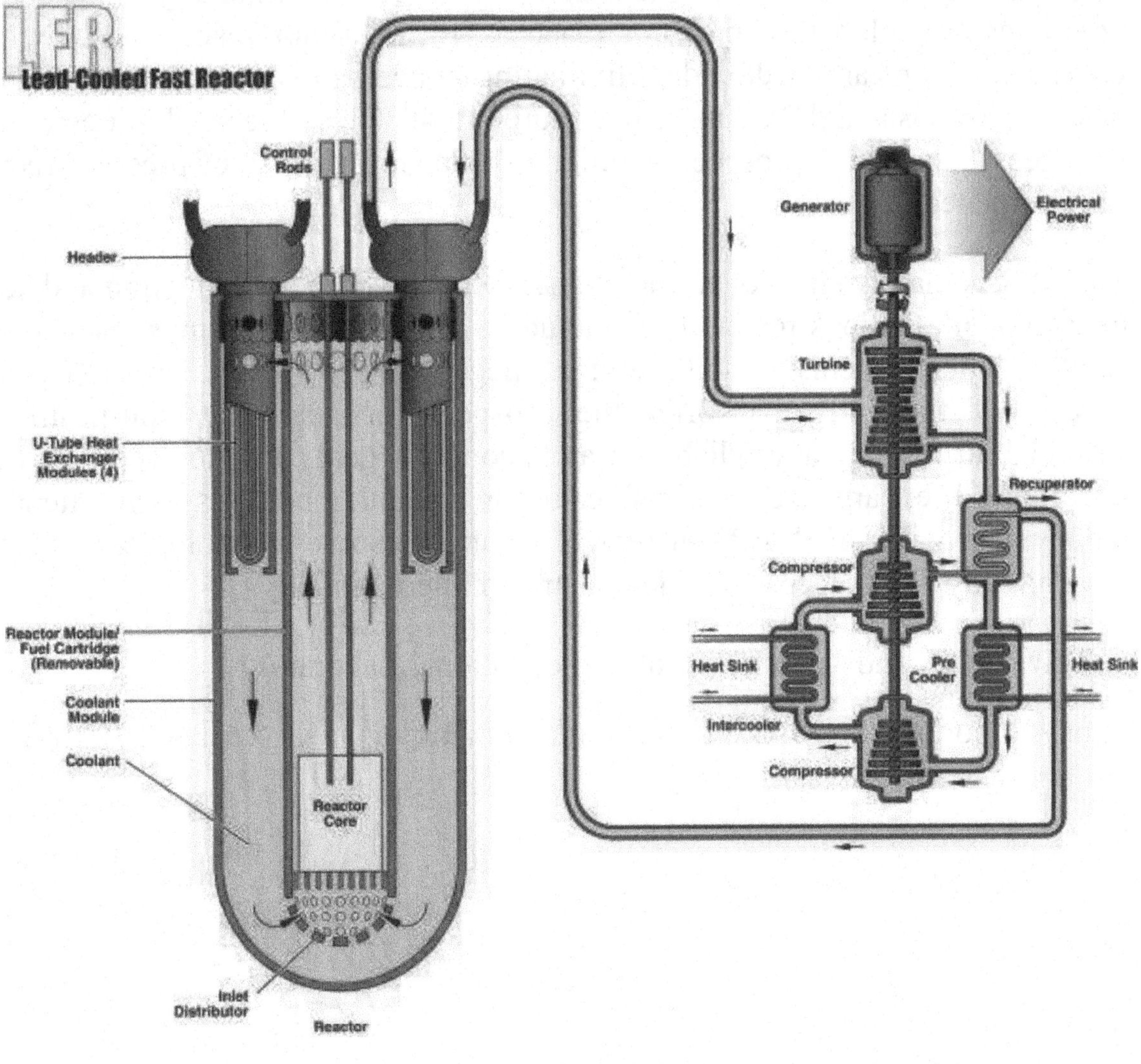

Fig. 14. The lead-cooled fast reactor (LFR) system features a fast-spectrum lead or lead/bismuth eutectic liquid metal-cooled reactor and a closed fuel cycle for efficient conversion of fertile uranium and management of actinides.

The lead (Pb) coolant exhibits very low parasitic absorption of fast neutrons, and this enables the sustainability and fuel cycle benefits traditionally associated with liquid metal-cooled fast-spectrum reactors. Pb does not react readily with air, water/steam, or carbon dioxide, eliminating concerns about vigorous exothermic reactions. It has a high boiling temperature (1,740°C), so the need to operate under high pressure and the prospect of boiling or flashing in case of pressure reduction are eliminated.

The LFR is mainly envisioned for electricity and hydrogen production and actinide management. Options for the LFR include a range of plant ratings and sizes from small modular systems to multi-hundred-megawatt-sized plants. Two key technical aspects of the LFR that offer the prospect for achieving non-proliferation, sustainability, safety and reliability, and economic goals are the use of Pb coolant and a long-life, cartridge-core architecture in a small, modular system intended for deployment with small grids or remote locations. Some technologies for the LFR have already been successfully demonstrated internationally.

Credit: Image and caption care of Idaho National Laboratory

https://inlportal.inl.gov/portal/server.pt/community/nuclear_energy/277/lfr/2254

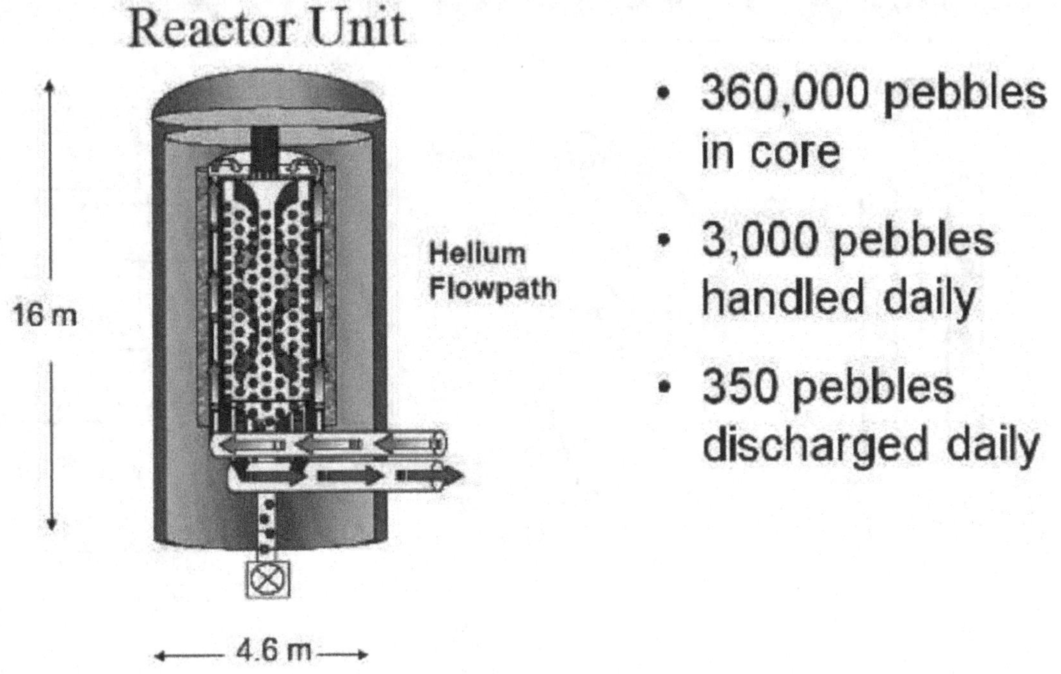

Fig.15. Illustration of the pebble bed reactor system. Note the indicated rates for pebble intake and discharge. (Image was borrowed from public domain online sources.)

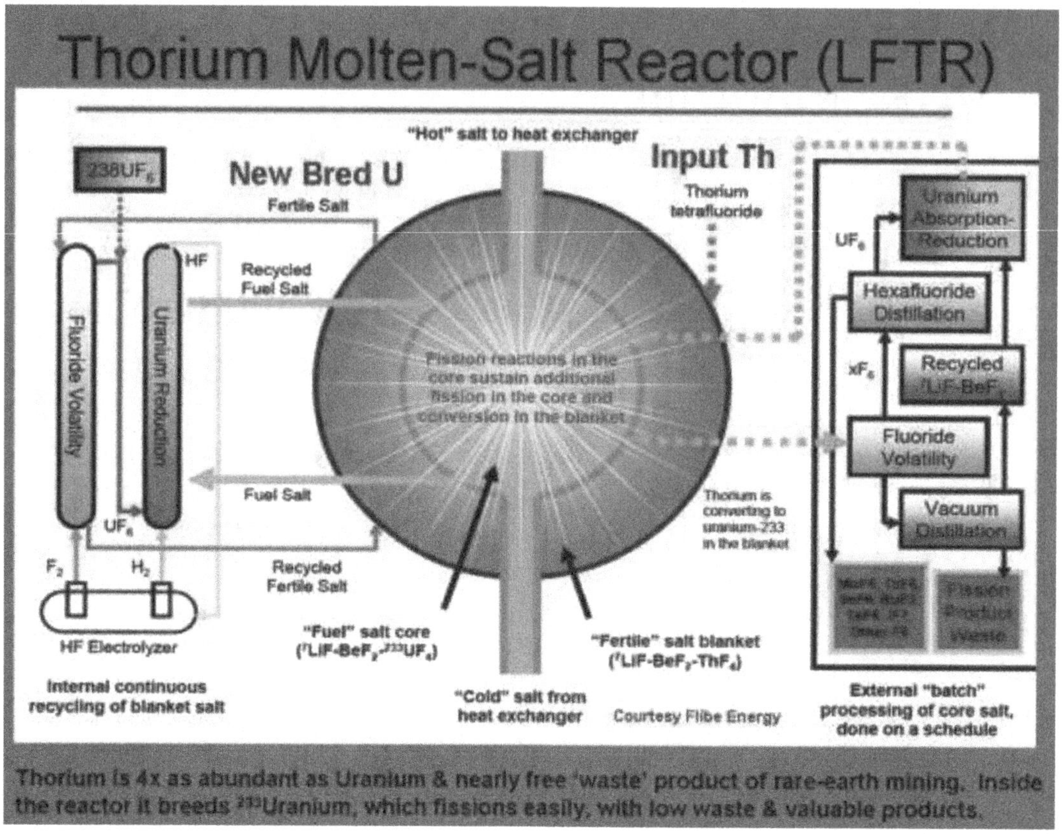

Fig. 16. Illustration of the breeding cycle of U-233 from thorium tetraflouride. Diagram (Courtesy of Flibe Energy) and caption adopted from https://www.google.com/search?q=Liquid+Fluoride+Thorium+Reactor+public+domain+images&tbm=isch&tbo=u&source=univ&sa=X&ei=hW0kU-OVDLKl0gGYp4EI&ved=0CCoQsAQ&biw=1365&bih=792&dpr=0.75#facrc=_&imgdii=_&imgrc=us54zGwvzoBhrM%253A%3BDJnnpYMfZ4QOpM%3Bhttp%253A%252F%252Fi.imgur.com%252FlObCP.jpg%3Bhttp%253A%252F%252Fclimatecolab.org%252Fplans%252F-%252Fplans%252FcontestId%252F4%252FplanId%252F15102%3B1363%3B1033

Reactor design is a multi-aspect process.

For example, the operating temperature is a major design consideration. The operating temperature of the core of a nuclear reactor is a significant contributor to the mass-specific power output of the reactor. Since some reactor designs operate at temperatures as high as 1,600°C, such reactors could conceivably offer small but nontrivial spacecraft thrust from thermal cooling and associated radiative emissions to space.

A second aspect is cooling. A reactor, whether of high or lower temperatures of operations, can provide more turboelectric power for cases where reactor core heat extraction rates are maximized to facilitate rapid heat extraction such as for driving thermal-mechanical dynamos.

Accordingly, multicycle heat thermomechanical steam-driven systems can provide increased utilization of core-source heat. The generated electrical energy may be used to power electrical propulsion systems such as ion or electron rockets, photon rockets, electrodynamic-hydrodynamic-plasma-drives, interstellar ramjets, and magnetic plasma bottle sails and magnetic field-effect sails.

The inclusion of progressively lower phase change and operating temperature of steam or other gas working effluents can significantly increase the overall efficiency in converting core heat to electrical energy.

Regenerative mechanisms can also be employed to more efficiently capture core-source heat for operating dynamos.

A primary cycle such as a helium gas–cooled high-temperature module may provide primary driving of a thermo-mechanical dynamo. A secondary water steam–driven system may offload heat from the high-temperature helium gas for which the water steam would drive thermomechanical dynamos. A tertiary steam system might use a lower-phase-change-working effluent such as an alcohol or other low boiling temperature volatile.

Methanol and ethanol are candidates for lower-temperature steam effluents. Another useful effluent is ammonia (NH_3). It boils at −33.34 °C (−28.012 °F) at a pressure of 1 atmosphere. The following table includes four alcoholic compounds with boiling points lower than water and thus may be of use in an after cycle for turboelectric energy generation that would follow a water steam–driven process.

Compound	Ordinary Name	Melting Point (°C)	Boiling Point (°C)	Solubility in H_2O at 23°C
CH_3OH	Methyl alcohol	-97.8	65.0	Infinite
CH_3CH_2OH	Ethyl alcohol	-114.7	78.5	Infinite
CH_3CH_2Cl	Ethyl chloride	-136.4	12.3	0.447 g/100 mL
$CH_3CH_2CH_2OH$	Propyl alcohol	-126.5	97.4	Infinite

Still additional effluents may include common refrigerants.

Refrigerant Boiling Points

Low-Pressure Boiling Points

R-11	Trichlorofluoromethane	CCl3F	74.5°F
R-113	Trichlorotrifluoroethane	CCl2FCClF2	117.6°F
R-123	Dichlorotrifluoroethane	CHCl2CF3	82.2°F

High-Pressure Boiling Points

R-12	Dichlorodifluoromethane	CCl2F2	-21.6°F
R-22	Chlorodifluoromethane	CHCl F2	-41.5°F
R-114	Dichlorotetrafluoroethane	CClF2CClF2	38.6°F
R-134a	Tetrafluoroethane	CF3CH2F	-15.1°F
R-500	Azeotrope	R-12/152a	-28.3°F
R-502	Azeotrope	R-22/115	-49.8°F
R-404A	Azeotrope	R-125/143a/134a	-52.1°F
R-409A	Azeotrope	R-22/124/142b	-31.8°F

R-410A Azeotrope	R-32/125	-61.0°F
R-416A Azeotrope	R-134a/124/600	-10.0°F
R-507 Azeotrope	R-125/143a	-52.8°F
Boiling Points		
R-13 Chlorotrifluoromethane	CCl F3	-114.6°F
R-503 Azeotrope	R-23/13	-126.1°F
R-23 Trifluoromethane	CHF3	-115.6°F

So given the above brief examples of progressively lower temperature effluents, we can see of several, five in this case, steam cycles can be cooperated to extract additional energy from core-source heat.

Thermoelectric cells and/or infrared operative photovoltaic or photoelectric cells can convert waste heat leftover from the thermomechanical systems to provide additional power.

For example, common so-called thermo-photovoltaic materials operate on light from emitter sources at 900°C to about 1300°C. However, as long as the source temperature is higher than the TPV units, electrical energy can be generated from the incident radiant flux.

Common materials used in TPV cells include silicon, germanium, gallium antimonide, indium gallium arsenide antimonide, indium gallium arsenide, and indium phosphide arsenide antimonide.

The efficiency of a thermoelectric device is given by, η, as

$$\eta = (\text{energy generated})/(\text{heat absorbed}) = [(T_H - T_C)/T_H]\{\{[1 + (ZT)]^{1/2}\} - 1\}/\{\{[1 + (ZT)]^{1/2}\} + (T_C/T_H)\}$$

Here, $ZT = [\sigma(S^2)T]/\lambda$.

S is the Seebeck coefficient, λ is the thermal conductivity, σ is the electrical conductivity, and T is the temperature.

In an actual thermoelectric device, two materials are used. The maximum efficiency is then given by

$$\eta_{max} = \text{(energy generated)/(heat absorbed)} = [(T_H - T_C)/T_H]\{\{[1 + (Z\check{T})]^{1/2}\} - 1\}/\{\{[1 + (Z\check{T})]^{1/2}\} + (T_C/T_H)\}$$

Here, T_H is the hot junction temperature, T_C is the temperature of the surface being cooled, $Z\check{T}$ is the modified dimensionless figure of merit.

$$Z\check{T} = [[(S_p - S_n)^2]\check{T}]/\{[(\rho_n\kappa_n)^{1/2} + (\rho_p\kappa_p)^{1/2}]^2\}$$

Here, ρ is the electrical resistivity, $\check{T}$ is the average temperature between the hot and cold surfaces and n and p denote properties related to the n- and p-type semiconducting thermoelectric materials, respectively.

Note that thermoelectric materials can be used to cool, which is a process commensurate with the ability of the thermoelectric material to generate electrical power from heat. Since the heat must come from somewhere, the heat conversion is expressed as a heat loss term.

Materials used or under consideration for thermoelectric device applications include the following:

Bismuth chalcogenides
Lead telluride
Inorganic clathrates,
Magnesium group IV compounds
Silicides
Skutterudite thermoelectrics
Oxide thermoelectrics
Half Heusler alloys
Electrically conducting organic materials
Silicon germanium
Sodium cobaltate
Functionally graded materials
Nanomaterials
Tin selenide

Tin selenide (SnSe) has a ZT of 2.6 along the *b* axis of the unit cell. This is the highest value known so far.

Thermoelectric cells can be applied in a layered arrangement with or without a Dewar insulating mechanism, which may also be multilayered. Moreover, the use of more than one species or class of thermoelectric materials may be used to convert thermal energy to electrical power.

A third consideration in nuclear reactor design is radiation resistance of structural components of the reactor core and other portions exposed to high-ionization energy fluxes. Materials used for such reactor elements need to be highly radiation resistant and/or have self-healing properties.

For example, the lattice or other micron and submicron scale structures of metals, alloys, and other materials used in high-radiation environments can experience significant degradation over the design life or operating life cycle of a nuclear reactor.

The radiative degradation can include as nonlimiting examples the dislocation of atomic and molecular bonds within the materials and the chemical transmutation of the materials.

Such degradation can adversely affect the mechanical strength of the irradiated materials, the refractory properties of the materials, and even the materials' resistance to further radiation-induced damage.

A fourth major design aspect includes methods and materials for reducing radiative oxidation. Radiative oxidation can result from oxidation reactions due to ionization of atoms and molecules by gamma and neutron radiation and other high-energy species. Such ionization can promote chemical oxidation.

A fifth design aspect is the need for structural and containment materials that are compatible when interfaced with each other under the thermal stress of the high operating temperature of the nuclear reactors, especially in locations near or within the core.

For example, high operating temperatures can cause material intermixing near interfaces. The high radiation intensities can accelerate this process and potentially lead to failure of the critical components of the reactor.

A sixth design aspect is the requirement for high-pressure effluent conduits that can operate in radiation intense environments. In some designs, resistance to high-temperature, chemically corrosive environments is necessary.

At the Idaho National Laboratory, the US Department of Energy's premier nuclear reactor technology research and innovation facility, the Advanced Test Reactor (ATR) is used to conduct tests on fuel burnup rates, radiation resistance of materials, and other aspects of the elements and materials used in nuclear reactor design and construction.

Accordingly, several experiments from the same organization or research group, or from several distinct organizations and groups can be conducted simultaneously and/or in a temporally overlapping sequence.

Mechanisms for temperature control, radiation flux control, and corrosion control are provided in the ATR. Other variables can be examined in a highly controlled manner.

The individual compartments of the ATR are compact, thus providing a means for several experiments to be conducted simultaneously, and then so in a manner that provided useful data for scaled-up designs and the future implementation of the tested technologies.

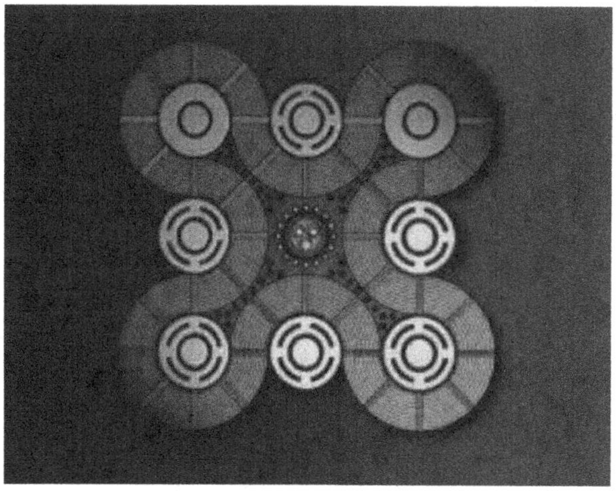

Fig. 17. The unique serpentine design of the ATR reactor core provides nine experimental locations. Each location has its own test conditions, and experiments can be run simultaneously. (Credit: Image and caption National Scientific User Facility at http://www.inl.gov/featurestories/2008-01-21.shtml)

Fig. 18. Actual photograph of the Advanced Test Reactor operating at the Idaho National Laboratory. Credit: National Scientific User Facility at http://www.global13.org/wp-content/uploads/2011/08/atr.png

Fig. 19. Below is a three-dimensional perspective view of the ATR.

FIGURE 20 The INEEL's Advanced Test Reactor is the most powerful test reactor operating in the United States. Credit: Idaho National Laboratory.

FIGURE 21 Test Reactor Area west of Idaho Falls is home to the Advanced Test Reactor. Credit: Idaho National Laboratory.

Mixed oxide fuels (MOF) are blends of enriched plutonium and uranium dioxide (UO_2). Such fuels are a consideration for utilizing weapons-grade plutonium derived from dismantled nuclear weapons stocks.

New Reactor and Fuel Technologies

Among the research efforts to develop better nuclear reactor and fuel technologies, Lightbridge is among the more prominent companies mentioned in open literature.

Lightbridge is in the process of designing and testing metallic fuels for use in currently operating pressurized water reactors and in upgraded designs of PWR.

Uranium dioxide fuel is a common mainstay for the nuclear electric utility industry. However, only about 6 percent of the uranium atoms in UO_2 fuel burn-up or fission. In metallic fuels such as UZr_2, the percentage of burn-up can be as high as 21 percent.

Several issues arise within the fuel rod assemblies of both UO_2 and UZr_2 fuel types.

For example, fission gas release in UO_2 fuels can cause gas bubbles to form and lead to gas propagation away from the location in the fuel where the gas atoms originally form as fission fragment products. These gas products (typically xenon and krypton) can cause the fuel portions to swell leading to damage to the fuel rods via cladding failure. Gas bubble formation can also occur in metallic fuels investigated by Lightbridge; however, these gas atoms tend to remain where they form and, at worst, form microscopic bubbles. Thus, the result of gas bubble based swelling in the metallic fuels is strongly mitigated than otherwise.

Crystalline dislocation and lattice structure change in reactor fuels, thereby resulting in swelling of the fuel portions.

The fuel rod assemblies being studied by Lightbridge have a four lobe cross-sectional shape in a helical pattern that enhanced thermal conductivity and heat transfer out of the fuel. The fuel rods are cladded; however, the cladding is thicker at the ends of the lobes to provide additional resistance from contact stresses that can occur at these contact locations.

The geometry of the metallic fuel rods allows for minimal diametric rod expansion while still permitting the rod to expand as utilization toward burn-up progresses.

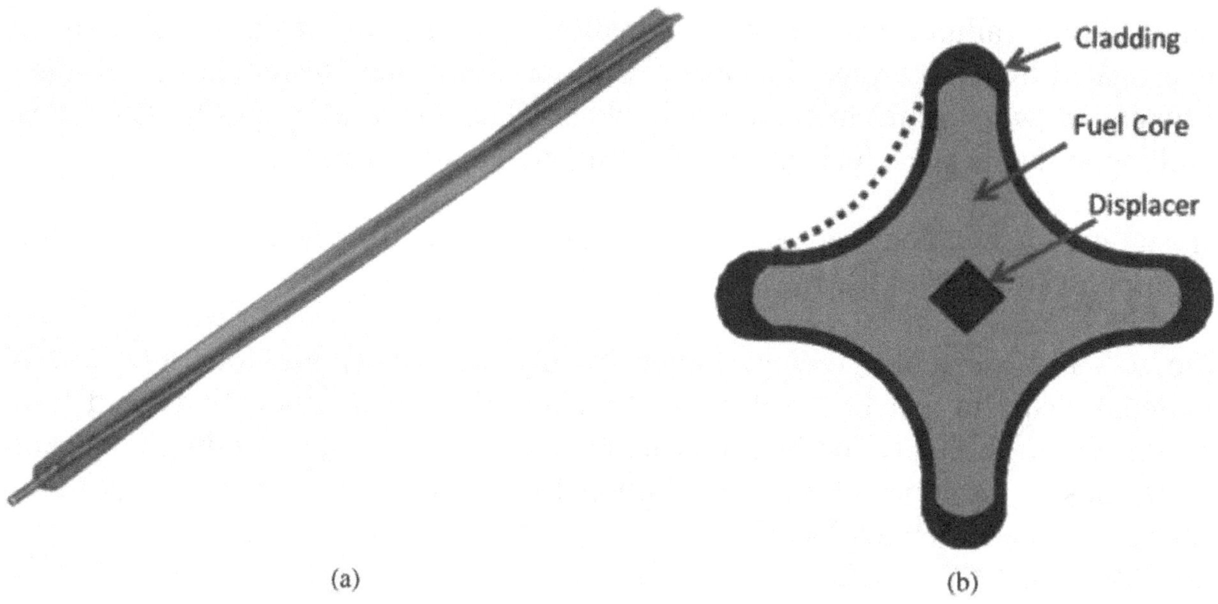

Fig. 22. The schematic of Lightbridge's metallic fuel rod for PWRs: (a) segment of rod showing helical twist and (b) cross section showing the fuel core, central displacer, and cladding. The dashed line identifies the typical change in fuel rod profile at end of life due to swelling deformation (not to scale). (Credit: Lightbridge.)

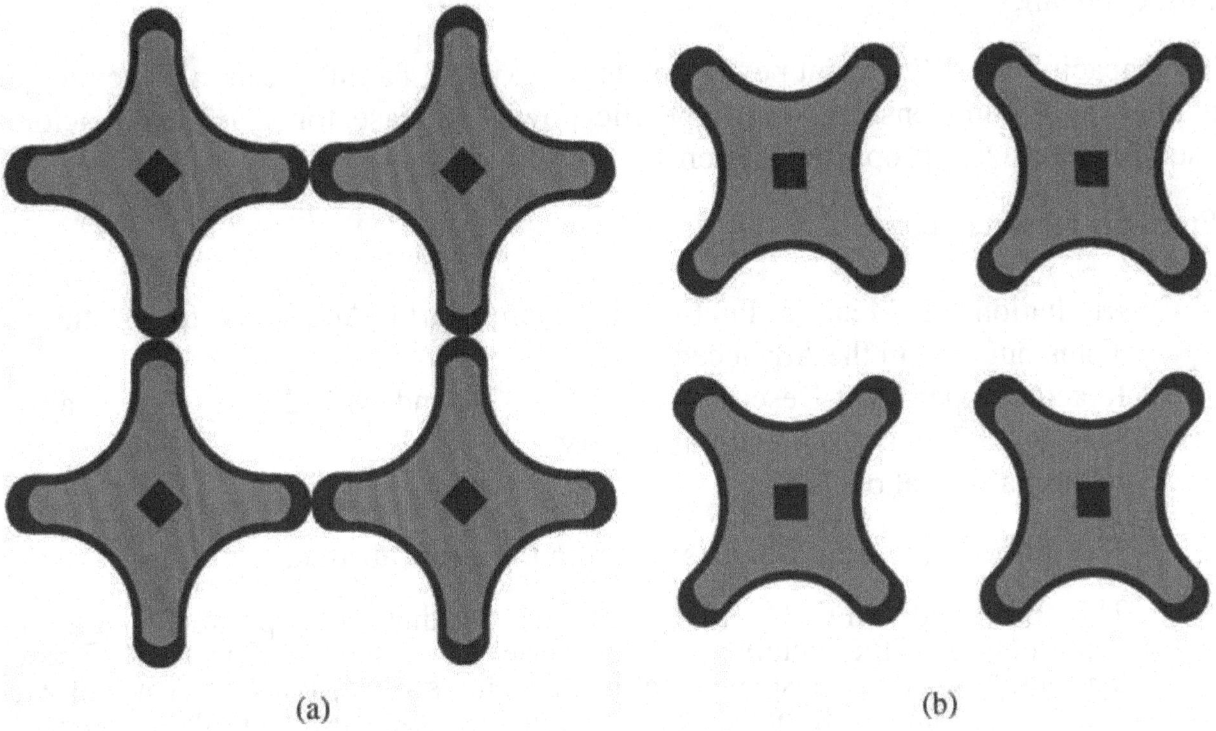

Fig. 23. A schematic cross section of the metallic fuel rod aligned in a square lattice array (a) showing the self-spacing plane wherein rod-to-rod contact eliminates the need for spacer grids and (b) axially halfway between self-spacing planes. Credit: Lightbridge.

The efforts of Lightbridge are aimed, at least in part, to provide advanced reactor fuels to enable power uprates as high as 17 percent in existing PWR and as high as 30 percent in new-build PWRs.

Applications of these exemplar metallic fuels for fission reactor–powered starships, such as space-arks, can enable higher effective specific impulses where only one burn cycle is used. For scenarios where inflight fuel reprocessing and refinement is accomplished, the specific impulses for the fuel are significantly further enhanced.

The capacity for a 17 percent power uprate for existing facilities can in principle be applied as a commensurate mass-specific power increase for space-ark reactors used for energizing propulsion systems.

From the research report:

> Irradiation of Metallic Fuels With Rare Earth Additions for Actinide Transmutation in the Advanced Test
> Reactor Experiment Description for AFC-2A and AFC-2B, by S. L. Hayes
> March 2007 Idaho National Laboratory
> Idaho Falls, Idaho 83415;
>
> The following excerpt is borrowed from the paper abstract:
>
> The proposed AFC-2A and AFC-2B irradiation experiments are a continuation of the metallic fuel test series in progress in the ATR. These experiments will consist of metallic fuel alloys of U, Pu, Np, Am and Zr, some with minor additions of rare earth elements meant to simulate expected fission product carry-over from pyro-metallurgical reprocessing, to be irradiated to
> Burn-up levels of ≥ 10 and ≥ 25 at.-% burn-up.
>
> The fuels and cladding are being studied for swelling effects and diffusion effects between the fuel and the cladding. Rare Earth lanthanide elements are added to the fuel in small amounts to simulate introduction of the lanthanides in the fuel reprocessing stage.
>
> During fuel burning, irradiation, thermochemical, and thermomechanical processes can cause the fuel cladding to mechanically fail. Moreover, the fuel can diffuse into the cladding and vice versa. The diffusion regions mechanical, thermal, and chemical properties can significantly change, and then so in localized unpredictable manners thus affecting the performance of the fuel, thermal conductivity of the fuel and cladding, and chemical activity in or near the fuel and cladding diffusion zone. In some reactor types, such multi-aspect changes may conceivably result in reactor failure. Reactor system failures can become quickly catastrophic for internal spacecraft life-

support and living infrastructure at worst, although at the very least, such mishaps can disable the spacecraft propulsion systems thereby reducing or eliminating propulsive power output. So, the authors deemed the digression herein on fuel integrity an important topic to address.

Since fast fission powered spacecraft ($v \geq 0.1$ C) will require large mass-ratios, such spacecraft will need to utilize fuel and reactor structural elements that can function at high burn-up percentages, and long duration irradiation.

Note that some commercially designed nuclear power plants have a design and/or certified operating live for 60 years or longer. Thus, knowledge of fuel and structural performance within the radiation intensive environments of nuclear reactors is necessary for mission involving long thrusting intervals and can be obtained from study of current and decommissioned reactors. Since practical field data is mainly available with certainty only from nuclear power plants on Navy ships such as submarines and aircraft carriers, and well established commercial electric utility plants, a presentation on well-established reactor technologies herein is warranted. If nothing else, at least studies of the various existing and decommissioned reactors can provide a minimal bounding condition with which to study reactor applications for slow and fast fission powered space-arks and colony ships.

As of march 2007, the following tabulated data describe some of the experimental fuel configurations.

AFC-2A & AFC-2B Fuel Test Matrix.
Rodlet Metallic Fuel Alloy† U-235 Enrichment
1 U-20Pu-3Am-2Np-15Zr 93%
2 U-20Pu-3Am-2Np-1.0RE*-15Zr 55%
3 U-20Pu-3Am-2Np-1.5RE*-15Zr 45%
4 U-30Pu-5Am-3Np-1.5RE*-20Zr 55%
5 U-30Pu-5Am-3Np-1.0RE*-20Zr 65%
6 U-30Pu-5Am-3Np-20Zr 93%
†Alloy composition expressed in weight percent.
*RE designates rare earth alloy (6% La, 16% Pr, 25% Ce, 53% Nd).

As can be seen in the above reformatted table (Credit INL), the concentration of Uranium-235 is much higher than in Uranium Dioxide fuels as indicated by the level of U-235 enrichment quoted above.

Another table from said paper is presented below.

AFC-2A and AFC-2B Rodlet Constituent Masses.

Fuel Column Constituent Masses (g)

Density Total Bond

Rodlet 1	13.98	0.152	4.570	4.251	0.320	1.523	0.001	1.257	0.251	0.009	0.005	0.229	0.000	1.143	0.426
Rodlet 2	13.98	0.152	4.510	2.480	2.029	1.523	0.001	1.257	0.251	0.009	0.005	0.229	0.076	1.143	0.426
Rodlet 3	13.98	0.152	4.456	2.005	2.451	1.523	0.001	1.257	0.251	0.009	0.005	0.229	0.114	1.143	0.426
Rodlet 4	12.87	0.210	2.840	1.562	1.278	2.104	0.001	1.736	0.347	0.012	0.007	0.351	0.105	1.403	0.426
Rodlet 5	12.87	0.210	2.889	1.878	1.011	2.104	0.001	1.736	0.347	0.012	0.007	0.351	0.070	1.403	0.426
Rodlet 6	12.87	0.210	2.945	2.739	0.206	2.104	0.001	1.736	0.347	0.012	0.007	0.351	0.000	1.403	0.426
Total 2.558		1.088	22.211	14.915	7.296	10.882	0.005	8.982	1.795	0.062	0.037	1.737	0.366	7.635	

Note: RE designates rare earth alloy (6% La, 16% Pr, 25% Ce, 53% Nd).

Note the large predominance of the total weight contribution of the actinides. For the high burnup rates theoretically practical with the above fuels and subsequent reprocessing during a mission, it is easy to see how very fast Keplerian velocities could be attained for large fission-powered starships. Further considering an accelerator based neutron source and/or a low-level fusion-derived neutron source, it becomes a smaller mental stretch to suggest that most if not virtually all of the actinides could be burned without the need for reprocessing. The final inert waste products could be exhausted as an ion stream of optimal velocity with respect to the spacecraft frame.

The Advanced Test Reactor and other facilities at the Idaho National Laboratory can likely play a major role in the study and selection of fuel types, cladding, and other reactor core elements for manned fission-powered interstellar spacecraft. If funding is made available to undergo a major interstellar space travel program at the national and international level, we can expect the ATR to figure prominently in the reactor and fuel design process.

Fuel cladding can be intermediated by a fuel liner that contains the fuel to prevent or mitigate fuel-cladding diffusion. Such lining is a research agenda for companies and governmental programs desiring to overcome the issues of fuel and cladding diffusion.

Metallic fuels are generally well suited for fast neutron reactors of which most in current operation are sodium cooled reactors. Additionally, metallic fuels are easy to fabricate and reprocess compared to metal-oxide fuels. Metal-oxide fuels have the benefit of being more difficult to reprocess and refine in terms of burned-up states so such fuels do have some merit regarding nuclear weapons proliferation issues. However, processing difficulties make the oxide fuels more likely to be discarded in longtime storage facilities and can pose a potential radioactive hazard for hundreds, and even many thousands, of years until the long-lived fission products have largely decayed.

Because metallic fuels are highly suitable for reprocessing and display high burnup levels in practical test reactors, these fuels are well suited for space-ark and colony ships for which a large mass ratio is required along with efficient use of fissile elements and isotopes. Even a large mass ratio along with an inability to burn most of the fissile isotopes and the inclusion of much inert composition of the reactor fuel associated with oxide fuels can drastically reduce the theoretical terminal velocity performance of a fission-powered spacecraft.

Another form of metallic fuels is also being investigated, which are configured as uranium fuel foils sandwiched between two plates, say, ones made of aluminum. These fuels include, among other alloys, uranium molybdenum blends generally of low levels of enrichment, which are typically considered to be less than 20 percent by weight. These monolithic plates, as currently being fabricated or having been fabricated for testing, are on the order of about 1 to 2 millimeters thick.

A special friction welding process has been developed to efficiently and precisely weld the layered components together. Accordingly, a portion of a layer of material is heated by heavy friction, after which two portions of material are placed in contact and thermally weld or bonded. A modification of this process was specifically developed to enable more than two-layered surfaces to be bonded simultaneously.

The bonding process enables excellent quality bonds between similar and dissimilar materials and enables materials that would otherwise be difficult to weld to be welded. Aluminum, titanium, and other hard-to-weld materials can thus be easily welded. The strength and homogeneity of the bonds so produced tends to be as good as or superior to those produced by more traditional welding processes. As of 2009, test plates on the order of from one square meter to several square meters have been fabricated.

Friction welding has safety advantages as well. No high-power intensity high-voltage arcs are needed, and no high temperature combustion torches are needed. This reduces the otherwise-associated additional degrees of freedom for fuel-reprocessing accidents, as well as accidental electrocution of arc welders and fires, which can occur in the presence of very hot flames and pressurized combustants and oxidizers. Since long-duration missions will require a stable and adequate supply of oxygen with a margin of excess availability for contingency purposes, it may be important, or at least prudent, to limit the possibility of runaway fires and the resulting smoke and noxious fumes and oxygen depletion that can otherwise result.

The present interest of the nuclear power industry in the monolithic plate fuels is in the ability to fabricate such fuels as low-enrichment forms in support of the agenda of National Nuclear Security Agency (NNSA) and other organizations within the United States and abroad to limit or reduce the availability of weapons-grade or highly enriched uranium. Such is an important objective in the case where space-ark fuel would be manufactured on Earth or on any other planetary body and launched into Earth or planetary orbit.

Issues of weapons-grade uranium production can conceivably arise even under the condition where the fuel production would be done on other planetary bodies, within the asteroid belt or in proximity to free-ranging asteroids, or in interplanetary solar-orbiting manufacturing plants.

One major benefit to using the low-enrichment plate forms of metallic fuels is the ease of reprocessing to concentrate the unfissioned fissile atoms while at the same time reducing the risk of nuclear weapons proliferations.

Atomic recoil from fission fragment impacts can significantly displace fuel and fuel cladding atoms to distances on the order of microns to tens of microns. This process can not only cause the thermal, chemical, and mechanical failure of fuel/cladding units, but it can be potentially very dangerous were it to occur in high-power environments such as commercial nuclear power plants and fission-powered starships.

Some good news is that the thermal deposition of layers of alloyed material between the fuel and the cladding can greatly mitigate such recoil-based displacement and mitigate the risk of fuel element failure.

Usually, aluminum, silicon, and zirconium are considered for such research.

However, we are interested in the theoretical performance of carbonaceous supermaterials such as carbon nanotubes, graphene, graphene oxide paper, boron-nitride nanotubes, vapor-deposited diamond, carbon nitride, and the like has on the inter-fuel and cladding medium. Carbon can be highly reactive when heated; however, carbon supermaterials have really tight and energetic bonds. So the atomic recoil of carbon atoms might be greatly reduced in terms of how far the carbon atom would travel in the carbon spacer.

A recoiling spacing atom can knock other atoms loose to form a dispersive jet of atoms of decreasing kinetic energy. So a single collision can, in principle, effect the displacement of numerous atoms.

Other supermaterials that might be of use include super-refractories such as tantalum hafnium carbide which has a melting point of 4215°C or (7,619°F). Although many of these materials are superhard but brittle, the high melting point may compensate for this, thus enabling reduced atomic bond breakage and atomic dislocation.

Another fascinating prospect involves nanotechnology spacer manufacture. Such techniques may be able to produce yet to be discovered carbon phases or phases of other solid atomic and molecular compounds. Perhaps elemental or molecular compounds not commonly considered for supermaterial production can be coaxed into such configurations of supermaterial constituents...

White dwarf density and strength supermaterials might be ideal but have never been directly observed, produced, or measured. Such materials may have a density of 1 metric ton per cubic centimeter, so a layer 1 micron thick having a surface area of 10 square meters need to have a mass of only one metric ton. A layer 0.1 microns thick having an area of 100 square meters would have a mass of 1 metric ton, and a layer 0.01 microns thick having an area of 1,000 square meters would have a mass of 1 metric ton.

Candidates for such stabilized white dwarf degenerate matter may include carbon, especially, but also any other elemental or molecular compositions.

The white-dwarf degenerate matter spacers need not be melted or otherwise disintegrated, but they may, in theory, be reused. However, eventually, these spaces would need disintegration, where the resulting mixture could be purified and remanufactured.

One hundredth of a nanometer scale technology, or 0.01 nanotechnology, can, in principle, be used to manufacture and remanufacture the degenerate-matter materials.

Nanoscale self-assembly methods may at some future time enable reactor fuel and cladding to have much lower displacements with or without a spacing layer. Such nanoscale self-assembly may even enable spent fuel to be self-reprocessing for the rejection of inert components and reassembly of new fuel. The reassembled fuel may also contain alloying or oxidation components for reinstallation into nuclear reactor cores.

Self-assembly may conceivably also be used for reactor structural members aside from the fuel cladding. Such elements may include fuel rod or plate-support structures, steam effluent pipes, primary reactor shielding and core confinement structures, reactor fuel in core arrangement mechanisms, and other portions subject to long-duration high-radiation flux, which can include gamma rays, neutrons, and fission fragments.

Neutrons, being electrically neutral, would be able to easily penetrate the degenerate spacers, and perhaps even

Other factors are significant players in monolithic fuel plate research and development. The chemical erosion of the cladding and expansion of the fuel plate can occur as a result of the formation of fission fragments, which are less dense than the fuel.

Loss of thermal conductivity of local portions of the plates can lead to localized heating. The loss of thermal conductivity is a function of multiple processes and is sensitive to chemical changes in the fuel and cladding phases as well as temperature. Once non-trivial local heating starts to occur, the change process can spin out of control. Such scenarios can reduce the effectiveness of the nuclear reactor control systems at best. At worst, it can lead to catastrophic accidents.

However, the biggest contributor to fuel plate swelling is the development of gas pockets mostly from xenon and krypton, which are major fission products.

Perhaps the gas-based swelling could be mitigated by specifically microstructured reactor fuels and cladding.

For example, the fuel and cladding might be designed with a network of circular cross-sectioned veins, or arteries, having the same cross-sectional area. The width of the arteries could be sufficiently small, and the distance between arteries could be sufficient to enable gas bubbles to easily grow to the point at which the bubbles would burst into the arteries, where the gas could then be vented or vented and collected for other purposes.

Alternatively, a mammalian-like circulatory system of branching arteries and capillaries of a variety of cross-sectional areas may be used. The smallest capillaries may be only one nanometer to ten nanometers in diameter and may be as close as ten nanometers to, say, thirty nanometers to adjacent capillaries. Thus, the gas bubbles on average could grow to only a limiting value before encroachment on one or more capillaries.

The size, cross-sectional shape, and the degree of tapering along the length of one or more arteries, or capillaries, may be uniform or variable from one artery or capillary with respect to another, and/or may be uniform or varying along the length of a given capillary.

The system of arteries and/or capillaries can include a precisely regular three-dimensional matrix of conduits orthogonally oriented—oriented mainly in parallel, obliquely oriented, or in other orientations such as somewhat random fractal patterns commonly found in animal life-forms.

The conjectured arteries and capillaries may be simply borehole-shaped pores or vacancies within the fuel cladding itself and/or be lined with a material that is more refractive than the fuel cladding itself. However, using linings of higher strength than the fuel can be problematic because the gas bubbles formed within the fuel may preferentially deform the fuel instead of rupturing into the gas conduits.

Nanotechnology self-assembly would be a very handy tool for permitting direct gas ejection from the fuel and/or cladding, and/or for maintaining the integrity of the conduits. The nanoscale process could be mediated by large molecules, or perhaps by micrometer-scale machines with mechanically and molecularly distinct components.

One or more pumps may be included within the fuel/cladding and structural supports to provide a pressure gradient sufficient to extract the fission fragment gases produced.

It is conceivable that photolithography, ion beam lithography, or ion beam deposition could be used to assemble the conducted fuel and cladded. Nanoscale or microscale three-dimensional fuel and cladding printing techniques are also plausible.

Another mechanism to prevent localized overheating of the fuel may include thermal diodes having extreme conductivity. The diodes would have variable conductivity as a function of working temperature. Thus, when a local portion of fuel and/or cladding became excessively hot, the conductivity of locally affixed thermal diodes would increase, thereby strongly moderating the temperature fluctuation of the fuel from optimal operating temperatures.

The thermal diodes may be arranged by length, uniform or varying cross-sectional size and/or shape, and matrix pattern of disposition within the fuel and cladded in the same or similar manners as the gas conduits. As one, but not necessary, option, the conduits may contain an inner, an outer, or an interior lining of thermal diodic materials. As another, but not necessary, option, the conduits themselves may be comprised of thermal diodic material.

More than one class or species of material composition can be used for the conduits. The same applies for the thermal diodes. Additionally, conduit materials of compositions as well as those for the diodes, may have the same or dissimilar refractory, mechanical, and chemical properties, as well as thermal, conductivities.

A fascinating prospect for wicking away excess heat from a reactor fuel and cladding, or more generally, from the larger core of a reactor is the use of diamond. Diamond is currently very expensive and available from industrial sources in relatively small quantities. However, nanotechnology self-assembly mechanisms may at some future time enable the mass production of ordinary diamond or isotopically optimized diamond.

Isotopically optimized diamond has much higher heat conductivity than ordinary diamond, which also has extremely high thermal conductivity. However, the operating temperatures of such reactor cores would need maintenance below the decay temperature of diamond. Temperatures that would leave refractories unscathed, and those unable to melt high-temperature steels and alloys can be sufficient to cause diamond to transition to graphite. The reaction can be explosive, as well as adding chemical volatility within or near the reactor core.

Light-Water Reactors

Light-water reactors are a major research-and-development agenda. Although it is true that LWRs are a well-established technology for commercial electrical power generation within the United States, many, if not most, of these reactors are being considered for operating lifetime enhancements. With many of these reactors reaching their currently certified design lives during the 2030s, new research into life-cycle enhancements is a major agenda of the Department of Energy and the nuclear electric utility industry.

Many of the commercial reactors currently operating have been certified to operate for sixty years. Life-cycle enhancements can increase the operating times to eighty years or more.

Consider a scenario where a space-ark reactor for propulsion is based more or less on current-design life capacities and that 50 percent of the energy generated within the reactor core is converted to spacecraft kinetic energy. Further assume that the spacecraft, including the reactor, has a mass of 1,000 metric tons. Still further, assume a reactor power output of one gigawatt. The Newtonian velocity of the spacecraft will be 1,760,681 meters per second, or about 1,760 kilometers per second. This is equal to 0.0058689 C. Such a space-ark could reach the Centauri system in about 682 years or Barnard's Star in 1,295 years. These results assume constant spacecraft invariant mass. Increasing the spacecraft mass and power output linearly and proportionally will enable the same terminal velocity during the same time interval.

Reducing the above 1,000-metric-ton spacecraft power generation to 100 megawatts while holding the other design parameters constant results in the spacecraft attainment of the same velocity after one thousand years of thrust. Increasing the spacecraft mass to 10,000 metric tons while assuming a power generation of one gigawatt and thrusting time of one thousand years will enable the same terminal velocity. Increasing the spacecraft mass and power output linearly and proportionally will enable the same terminal velocities during the same time intervals.

These Newtonian scenarios are plausible using more or less current levels of light-water reactor technology.

Such space-arks may conceivably be greatly scaled up in mass so as to comfortably support live and awake crew members and passengers. For journeys lasting on the rough order of one thousand to ten thousand years, a very large spacecraft is likely required to·provide comfortable living space, the mass requirement for redundant and replaceable system and structural components, as well as food, oxygen, and water and/or the ability to regenerate these essential metabolites.

Other factors are significant players in monolithic fuel plate research and development. The chemical erosion of the cladding and expansion of the fuel plate can occur as a result of the formation of fission fragments that are less dense than the fuel.

The loss of thermal conductivity of local portions of the plates can lead to localized heating. The loss of thermal conductivity is a function of multiple processes and is sensitive to chemical changes in the fuel and cladding phases as well as temperature. Once non-trivial local heating starts to occur, the change process can spin out of control. Such scenarios can reduce the effectiveness of the nuclear reactor control systems at best, and at worst, they can lead to catastrophic accidents.

However, the biggest contributor to fuel plate swelling is the development of gas pockets mostly from zenon and krypton, which are major fission products.

Perhaps the gas-based swelling could be mitigated by specifically microstructured reactor fuels and cladding.

For example, the fuel and cladding might be designed with a network of circular cross-sectioned veins or arteries having the same cross-sectional area. The width of the arteries could be sufficiently small, and the distance between arteries could be sufficient to enable gas bubbles to easily grow to the point at which the bubbles would burst into the arteries where the gas could then be vented or vented and collected for other purposes.

Alternatively, a mammalian-like circulatory system of branching arteries and capillaries of a variety of cross-sectional areas may be used. The smallest capillaries may be only one nanometer to ten nanometers in diameter and may be as close as ten nanometers to, say, thirty nanometers to adjacent capillaries. Thus, the gas bubbles, on average, could grow to only a limiting value before encroachment on one or more capillaries.

The size, cross-sectional shape, and degree of tapering along the length of one or more arteries or capillaries may be uniform or variable from one artery or capillary with respect to another, and/or may be uniform or varying along the length of a given capillary.

The system of arteries and/or capillaries can include a precisely regular three-dimensional matrix of conduits orthogonally oriented, oriented mainly in parallel, obliquely oriented, or in other orientations such as somewhat random fractal patterns commonly found in animal life-forms.

The conjectured arteries and capillaries may be simply borehole-shaped pores or vacancies within the fuel cladding itself, and/or be lined with a material that is more refractive than the fuel cladding itself. However, using linings of higher strength than the fuel can be problematic because the gas bubbles formed within the fuel may preferentially deform the fuel instead of rupturing into the gas conduits.

Nanotechnology self-assembly would be a very handy tool for permitting direct gas ejection from the fuel and/or cladding, and/or for maintaining the integrity of the conduits. The nanoscale process could be mediated by large molecules, or perhaps by micrometer-scale machines with mechanically and molecularly distinct components.

One or more pumps may be included within the fuel/cladding and structural supports to provide a pressure gradient sufficient to extract the fission fragment gases produced.

It is conceivable that photolithography, ion beam lithography, or ion beam deposition could be used to assemble the conducted fuel and cladded. Nanoscale or microscale three-dimensional fuel and cladding printing techniques are also plausible.

Another mechanism to prevent localized overheating of the fuel may include thermal diodes having extreme conductivity. The diodes would have variable conductivity as a function of working temperature. Thus, when a local portion of fuel and/or cladding becomes excessively hot, the conductivity of locally affixed thermal diodes would increase, thereby strongly moderating the temperature fluctuation of the fuel from optimal operating temperatures.

The thermal diodes may be arranged by length, uniform or varying cross-sectional size and/or shape, and matrix pattern of disposition within the fuel and cladded in the same or similar manners as the gas conduits. As one but not necessary option, the conduits may contain an inner, an outer, or an interior lining of thermal diodic materials. As another, but not necessary option, the conduits themselves may be comprised of thermal diodic material.

More than one class or species of material composition can be used for the conduits. The same applies for the thermal diodes. Additionally, conduit materials of compositions, as well as those for the diodes, may have the same or dissimilar refractory, mechanical, and chemical properties, as well as thermal conductivities.

A fascinating prospect for wicking away excess heat from a reactor fuel and cladding, or, more generally, from the larger core of a reactor is the use of diamond. Diamond is currently very expensive and available from industrial sources in relatively small quantities. However, nanotechnology self-assembly mechanisms may at some future time enable mass production of ordinary diamond or isotopically optimized diamond. Isotopically optimized diamond has a much higher heat conductivity than ordinary diamond, which also has extremely high thermal conductivity. However, the operating temperatures of such reactor cores would need maintenance below the decay temperature of diamond. Temperatures that would leave refractories unscathed, and those unable to melt high temperature steels and alloys can be sufficient to cause diamond to transition to graphite. The reaction can be explosive, as well as adding chemical volatility within or near the reactor core.

Chapter 7

Interplanetary and Interstellar Barges Powered by Radio-Isotopic Thermal Generators and Solar Sails

This chapter includes descriptions of interstellar travel technologies involving electrical propulsion systems based on power sources that are currently well understood but which are herein contemplated on scales far more extensive than any previous applications of the technologies. The primary power plants covered in this chapter have been routinely used in unmanned spacecraft for decades, but only for systems capable of ordinary Keplerian velocities. In this chapter, scaled-up applications for slow interstellar barge like missions are contemplated.

Several scenarios exist for potential use of various isotopes to power slow interstellar freighters on several thousand-year missions between adjacent star systems at velocities conceivably close to 0.005 C while permitting higher-energy density fuels to be used for high-Lorentz-factor transport of persons and time-critical commodities such as food, medical equipment, and others.

Barges utilizing ^{42}Ar, ^{137}Cs, ^{90}Sr, ^{14}C, and/or ^{60}Co may be of use for transporting non-time-critical supplies, or perhaps for powering huge multigeneration spacecraft or world Zonds. Descriptions of energy budgets for these spacecraft are included in this chapter.

Note that higher-energy fuels involve nuclear-fission fuels, nuclear-fusion fuels, and matter-antimatter fuels. Radioactive decay is not normally considered nuclear fission although it does involve the release of nuclear energy. Nuclear fission generally involves two orders of magnitude more energy release per unit mass of fuel than do non-fissile RTG fuels. The above classes of higher-energy fuels are covered in the subsequent chapters. The RTG radio-nuclide fuels can be reserved for slow barge travel, thereby permitting higher-energy fuels to be used for time-critical missions such as the transport of key commodities and personnel.

The aspect ratios of RTGs can be configured in flight so as to prevent excess heat buildup and to maximize the conversion of heat to electrical energy as the mass-specific power levels of the RTGs drop due to time-dependent decreases in radioactive-fuel nuclear decays. The shape and size of the RTGs can be chosen for specific mission criteria. Such additional morphological degrees of freedom can enable more appropriate interaction of the RTGs and the spacecraft with the interstellar and intergalactic medium including, but not limited to, natural variations in: (1) plasma distribution by density, charge, and species; (2) neutral gas distribution by density and species; (3) ambient stellar and quasar light distributions according to energy spectrum and/or power flux density; (4) and interstellar and intergalactic magnetic-field intensity and vector field orientation. Such considerations can be important for drag reduction, and, for certain conditions, intentional increases in drag such as for rerouting and deceleration. Such natural variations can enhance or degrade spacecraft performances including, but not limited to, propulsion-system efficiency.

Barges Utilizing Argon-42

Argon-42 undergoes beta decay with an energy release of 0.60 MeV. Argon-42 has an atomic mass of about 41.96 amu, or about 42 times the mass of the proton at about (42)(938) MeV/[C^2]. Thus, the fractional mass specific energy yield of this decay is about (0.6)/[(42)(938)] = 0.00001523. The half-life of ^{42}Ar is 33 years.

Now assume that the mass of the ^{42}Ar is 90 percent of the total mass of the spacecraft system and that the rest of the mass is taken up by shielding, crew quarters, and highly efficient RTG-powered ion or electron rockets. Further assume that the decay energy is captured and converted into spacecraft kinetic energy with near–100 percent efficiency. The relativistic Lorentz factor of the spacecraft after nearly all of the ^{42}Ar has decayed is 1 + [(0.00001523)(0.90)] = 1.000013707, which corresponds to a velocity of 0.005235 C.

We can contemplate mass-driver-launched, or beam-sail-launched, ^{42}Ar fuel pellets that would be used to power the spacecraft upon rendezvous with and capture by the spacecraft. In such scenarios, arbitrarily high Lorentz factors are achievable, commensurate with velocities of translational travel through space of C – e, where *e* is positive but can be made vanishingly small to the extent that the supply time and distance of pellet stream projection is unlimited. However, it is hard to see how such a system could work, given the modest half-life of the ^{42}Ar and its limited mass specific yield compared to nuclear fission, nuclear fusion, and matter-antimatter reactions. Assuming the radioactivity could somehow be suppressed, such a system might be feasible provided that the decay products of ^{42}Ar could be jettisoned so that the inert leftover mass could be discarded. The process of capture and discardment could happen repeatedly. A similar justification can be made for each of the additional scenarios described below where other radionuclides are considered as the primary RTG fuel source.

Alternatively, a linear series of mass-driver stations positioned along the travel route could be utilized to accelerate the fuel pellets. Optionally, a circulinear or helical type of pellet runway distribution pattern could be set up in relatively close proximity to the sun or other stellar locations, where the craft could continue to accelerate and jettison the spent decay product. Perhaps the decay product could be used as a reaction mass in an ion rocket propulsion system. Similar fuel configurations can be applied in cases of arbitrary RTG fuel usage, including, but not limited to, the additional scenarios provided below.

The craft could utilize a Stellar Cycler type of mechanism where an electrically charged tow line or other charged member attached to the craft would react via the Lorentz turning force. Such interaction would permit the craft to travel in a circular iterative motion as the craft accelerated through either the background interstellar magnetic field(s) or the solar or stellar magnetic field(s).

The fuel pellets might optionally be guided by a Stellar Cycler type of charged towline so that they can gently catch up with the spacecraft.

Jettisoned spent stages could be rerouted back to a processing station where the inert mass could be recycled or repurposed. A Stellar Cycler type of mechanism could be used to redirect the spent stages.

Note that Karl Schroeder was the originator of the Stellar Cycler concept and has, no doubt, already thought of similar applications before I was even aware of the general concept. I first read of his concept about four years ago.

Given the possibilities described in the previous few paragraphs, extremely high relativistic velocities for the craft are conceivable. Similar Stellar Cycler configurations can be applied for cases of arbitrary RTG fuel usage including but not limited to the additional scenarios provided below.

Another mechanism might entail the use of an electrodynamic funnel such as a magnetic scoop, which would collect and collate interstellar ions, which would then be directed into RTG units. The collision energy of the interstellar ions with the RTG fuel mass would be recycled via the thermal energization of turbo-electric systems involving multicycle steam turbines. Alternatively, perhaps some form of Stirling generator could be utilized.

Another option would entail the heat generated by the collisions being processed directly by the RTG mechanism. Any enhanced radioactivity of the RTG fuel mass could then be harnessed to power ultra-efficient ion or electron rockets, magneto-hydrodynamic-plasma-drive, electro-hydrodynamic-plasma-drive, or electromagneto-hydrodynamic-plasma-drive systems that react against the interstellar-charged medium. Yet once again, a similar justification can be made for each of the additional scenarios described below.

Note that a Stirling engine operates through a cyclical compression and expansion of air or other gas, which is the working fluid, at different temperature levels so that there is a net conversion of heat energy to mechanical work. A Stirling generator would simply convert the resulting mechanical energy into electrical energy such as through a linear induction–based mechanism or by a traditional rotating dynamo type of mechanism.

Refueling can be achieved in flight by the provision of mass-driver-launched fuel pellets that catch up to the spacecraft. The relative velocities of the pellets can be adjusted so that the relative kinetic energies of the pellets with respect to the spacecraft are at the low-end range of Keplerian velocities. Small relative velocities at pellet intake provide a measure of safety since even high relative Keplerian velocities could prove disastrous in the event of a collision. Perhaps the best way to fuel the spacecraft by pellet streams is to launch the pellets before the spacecraft is launched so that the spacecraft can catch up to the pellet or by launching the pellets closely behind the spacecraft as the craft passes the pellet launch facilities. Another option would be to employ the Stellar Cycler concept, where highly charged pellets would make circuitous routes that intercept the spacecraft in cases where the initial velocity vector of the pellets is oblique or orthogonal with respect to the velocity vector of the spacecraft. Such operational configurations also apply for the additional scenarios described below.

Now 0.005235 C is fast enough to travel 4 light-years in 764.088 years. This is fast enough to arrive at Alpha Centauri in only about 780 years, or to arrive at Barnard's Star in only about 1,100 years. Reverse thrust can be achieved via linear induction magnetic braking using large superconducting cables, reverse mini-magnetospheric plasma propulsion systems (M2P2 systems), or reverse thrust interstellar ramjet types of systems.

Now the results in the previous paragraph are ideal and assume that the spacecraft can be brought up to speed in a small portion of the total transit time for single-batch fueling. However, multi-batch refueling might be used for enhanced travel velocities and reductions in travel time.

Even for a system that had an efficiency of only 10 percent, a Lorentz factor of 1.0000013707 could be obtained, which corresponds to a velocity of 0.001656 C. This can get us to Alpha Centauri in only about 2,500 years and to Barnard's Star in about 3,700 years.

Translational Force

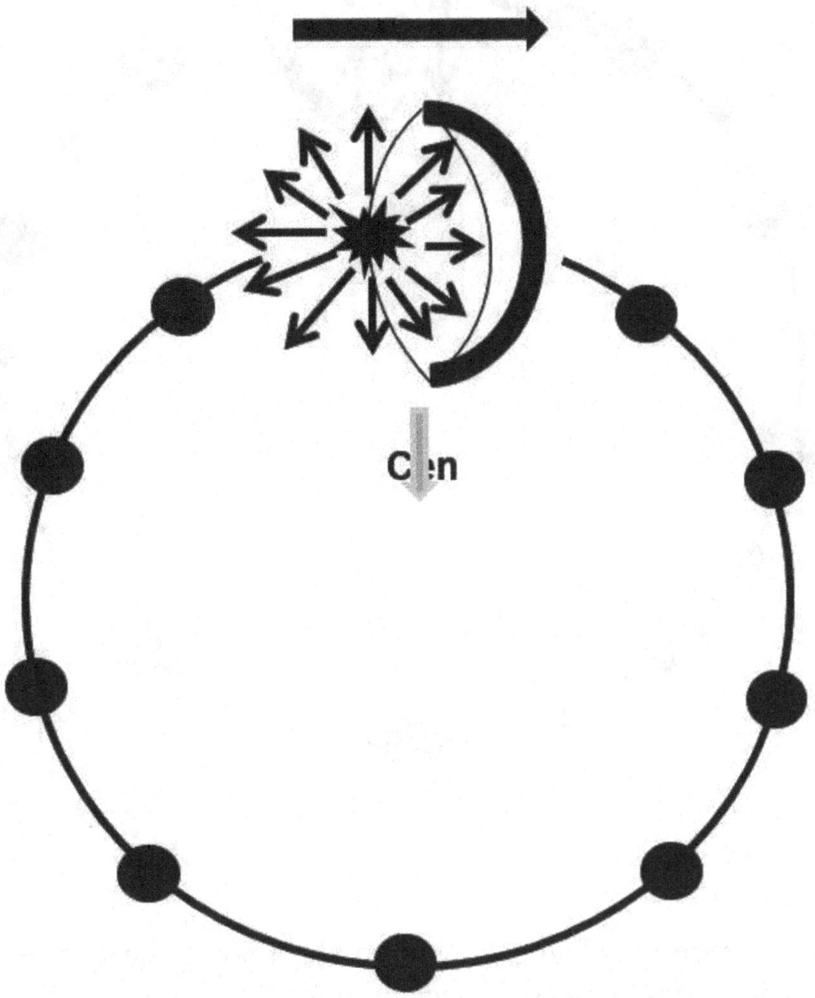

Fig. 24A. A stationary pellet stream fuel stream. The downward-pointing gray arrow indicates centripetal acceleration.

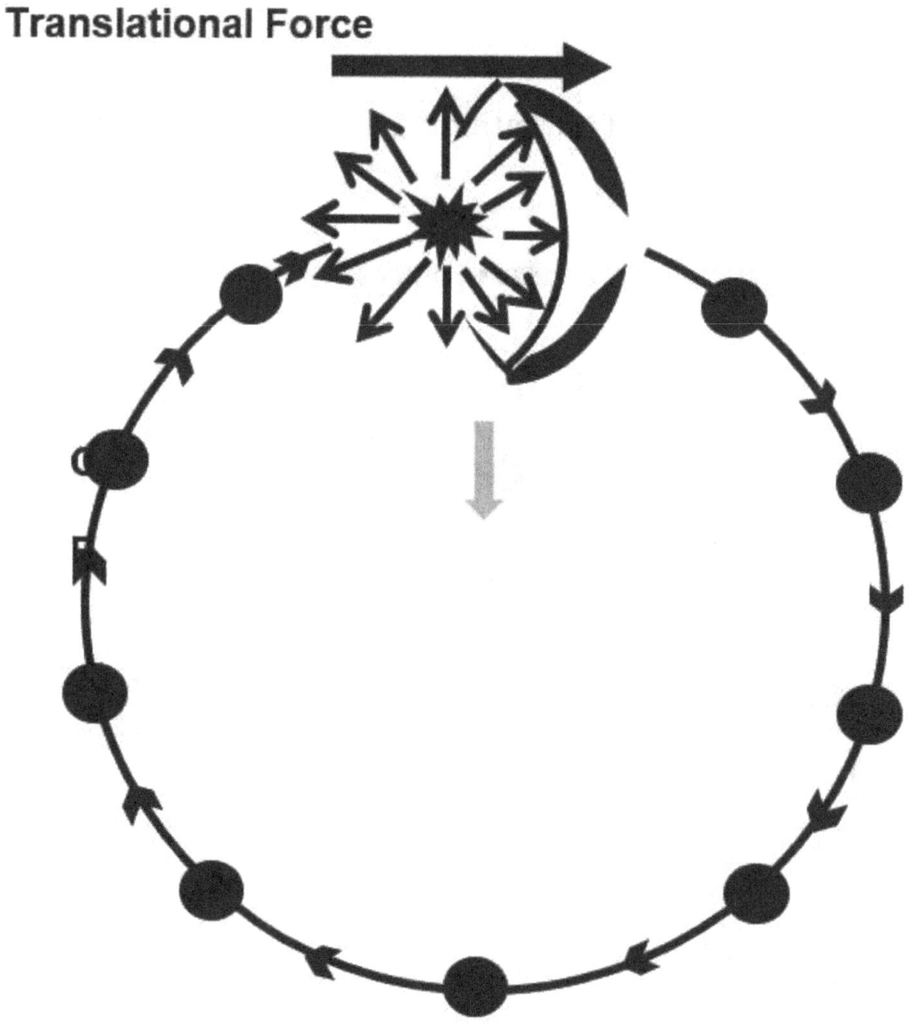

Fig. 24B. A revolving pellet stream fuel stream. The downward-pointing gray arrow indicates centripetal acceleration.

Barges Using Cesium-137

Cesium-137 undergoes beta and gamma decay into ^{137}Ba. The energy released in the decay process is 1.174 MeV. The half-life of ^{137}Cs is 30.07 years, and it decays into the metastable isotope of ^{137}Ba-m, which is a short-lived isomer with a half-life of 2.55 minutes. The atomic mass of the cesium-137 atom is 136.907 amu or (931.494)(136.907) MeV/[C^2] = 125,528 MeV/[C^2]. Thus, the fractional mass specific decay yield of this decay sequence is about 1.174/125,528 = 0.0000093525.

Now assume that the mass of the ^{137}Cs is 90 percent of the total mass of the spacecraft system and that the rest of the mass is taken up by shielding, crew quarters, and highly efficient RTG-powered ion or electron rockets. Further assume that the decay energy is captured and converted into spacecraft kinetic energy with near 100 percent efficiency. The relativistic Lorentz factor of the spacecraft after nearly all of the ^{137}Cs is converted into ^{137}Ba is 1 + [(0.0000093525)(0.90)] = 1.00000841725, which corresponds to a velocity of 0.0041 C.

Now 0.0041 C is fast enough to travel 4 light-years in 975.6 years. This is fast enough to arrive at Alpha Centauri in only about 1,000 years or to arrive at Barnard's Star in only about 1,500 years. Reverse thrust can be achieved via linear induction magnetic braking using large superconducting cables, reverse mini-magnetospheric plasma propulsion systems (M2P2 systems), or reverse-thrust interstellar ramjet systems.

Even for a system that was only 10 percent efficient, a Lorentz factor of 1.000000841725 could be obtained, which corresponds to a velocity of 0.0013 C. This can get us to Alpha Centauri in only about 3,160 years and to Barnard's Star in about 4,550 years.

Barges Using Strontium-90

Using ^{90}Sr-fueled RTG apparatus will enable manned missions to Alpha Centauri in 520 years, to Barnard's Star in 750 years, and throughout our Galaxy supercluster over the next few billion to several billion years.

Strontium-90 undergoes beta-decay with an energy release of 0.546 MeV into an electron and ^{90}Y (which has a half-life of 64 hours), which then undergoes beta decay into an electron and the stable isotope of ^{90}Zr. The total energy released by this strontium-90 decay sequence is 2.826 MeV. Strontium-90 has an atomic mass of about 90 amu, or about ninety times the mass of the proton, or about (90)(938) MeV/[C^2]. Thus, the fractional mass specific yield of this decay sequence is about (2.826)/[(90)(938)] = 0.000033475. The half-life of ^{90}Sr is 28.8 years.

Now assume that the mass of strontium-90 is 90 percent of the total mass of the spacecraft system and that the rest of the mass is taken up by shielding, crew quarters, and highly efficient RTG-powered ion or electron rockets. Further assume that the decay energy is captured and converted into spacecraft kinetic energy with near 100 percent efficiency. The relativistic Lorentz factor of the spacecraft after nearly all of the ^{90}Sr is converted into ^{90}Zr is 1 + [(0.000033475)(0.90)] = 1.00003, which corresponds to a velocity of 0.00786 C.

Now, 0.00786 C is fast enough to travel 4 light-years in 508.9 years. This is fast enough to arrive at Alpha Centauri in only about 520 years, or fast enough to arrive at Barnard's Star in only about 750 years. Reverse thrust can be achieved via linear induction magnetic braking using large superconducting cables, reverse mini-magnetospheric plasma propulsion systems (M2P2 systems), or reverse-thrust interstellar ramjet types of systems.

Even for a system that has an efficiency of only 10 percent, a Lorentz factor of 1.000003 could be obtained, which corresponds to a velocity of 0.00245 C. This can get us to Alpha Centauri in only about 1,670 years and to Barnard's Star in about 2,400 years.

Barges Utilizing Carbon-14

A ^{14}C-fueled RTG apparatus could theoretically be used to enable manned missions to stars within 100 light-years of Earth in 20,408 years with a single fueling while permitting higher-energy density fuels to be used for the high-Lorentz-factor transport of persons and time critical commodities.

Carbon-14 undergoes beta decay into the stable isotope of ^{14}N and yields 0.156476 MeV. After one half-life period of ^{14}C of 5,730, plus or minus 40 years, an initial one-kilogram sample will contain only 1/2 of a kilogram of ^{14}C atoms. The mass of the ^{14}C atom is 14.003241 amu.

Thus, in the decay process, in 5,730 years, a fraction of about (1/2)(0.156476)/[(931.494 028)(14.003241)] = 0.000005998 of the mass of a sample of pure ^{14}C is converted into energy.

Now assume that the mass of the ^{14}C is 99.9 percent of the total mass of the spacecraft system and that the rest of the mass is taken up by shielding, crew quarters, and highly efficient RTG-powered ion or electron rockets. We will assume that the ^{14}C decay energy is captured and converted into spacecraft kinetic energy with near–100 percent efficiency. The relativistic Lorentz factor of the spacecraft after half of the ^{14}C decays is equal to 1.000006, which corresponds to a velocity of 0.00345 C. After nearly 100 percent fuel burnup, all else being the same, the relativistic Lorentz factor of the spacecraft will be 1.000012, which corresponds to a velocity of 0.0049 C.

Now, 0.00345 C is fast enough to travel 100 light-years in 28,980 years and 0.0049 C is fast enough to cover 100 light-years in 20,408 years. Reverse thrust can be achieved via linear induction-based magnetic braking using large superconducting cables, reverse mini-magnetospheric plasma propulsion systems (M2P2 systems), or reverse-thrust interstellar ramjet types of systems. Reverse thrust can be achieved via linear magnetic induction braking using large superconducting cables, reverse mini-magnetospheric plasma propulsion systems (M2P2 systems), or reverse-thrust interstellar ramjet systems.

Even for a system that was only 10 percent efficient, for a 50 percent fuel burnup, a Lorentz factor of 1.0000006 could be obtained, which corresponds to a velocity of 0.0011 C. In 90,909 years, the ship could cover 100 light-years.

Barges Utilizing Cobalt-60

A ^{60}Co-fueled RTG apparatus will enable manned missions to Alpha Centauri in 620 years, to Barnard's Star in 890 years, and throughout our local Galaxy supercluster over the next few billion to several billion years.

Cobalt-60 undergoes beta decay into the stable isotope of ^{60}Ni, which emits two gamma rays with energies of 1.17 MeV and 1.33 MeV, respectively. The radiative power produced by 1 gram of cobalt-60 is about 20 watts per gram, or 20 kilowatts per kilogram. This results in about 600 billion joules per kilogram being emitted over a one-year period. After a half-life period of 5.27 years, an initial one-kilogram sample will have a power output of 10 kilowatts.

The ^{60}Co would need to be distributed in a planar dish–like configuration. A dish- or platelike RTG arrangement could provide a large surface area. Such a large area is necessary so that the ^{60}Co would not melt through its housing or otherwise damage or destroy the RTG units.

The mass difference between a ^{60}Co atom and a nickel-60 atom is about 0.003 amu and the mass of a ^{60}Co atom is 59.9338222 amu. Thus, through the decay process, in 5.27 years, a fraction of about (1/2)(0.003)/(60) = 0.000025 of the mass of an initial sample of pure cobalt-60 is converted into energy.

Now assume that the mass of the ^{60}Co is 40 percent of the total mass of the spacecraft system and that the rest of the mass is taken up by shielding, crew quarters, and highly efficient RTG-powered ion or electron rockets. We will also assume that the conversion of the ^{60}Co decay energy is captured and converted into spacecraft kinetic energy with near 100 percent efficiency. The relativistic Lorentz factor of the spacecraft after half of the ^{60}Co decays is equal to 1.00001, which corresponds to a velocity of 0.0045 C. After a nearly 100 percent fuel burnup, all else being the same, the relativistic Lorentz factor of the spacecraft will be 1.00002, which corresponds to a velocity of 0.006475 C.

A speed of 0.006475 C is fast enough to travel 4 light-years in 617.76 years. This is fast enough to arrive at Alpha Centauri in only about 620 years or to arrive at Barnard's Star in only about 890 years. Reverse thrust can be achieved via linear magnetic-induction braking using large superconducting cables, reverse mini-magnetospheric plasma propulsion systems (M2P2 systems), or reverse-thrust interstellar ramjet systems.

Even for a system that is only 10 percent efficient, a Lorentz factor of 1.000002 could be obtained, which corresponds to a velocity of 0.002 C. This can get us to Alpha Centauri in only 2,000 years and to Barnard's Star in about 2,900 years.

Conclusion

The good news is that we know how to build RTGs and have used them in our interplanetary space probes. Consider the possibility that our civilization would decide to ramp up the nuclear fission reactor infrastructure given the extreme situations such as global warming and ecological disasters such as the huge oil spill that occurred in the Gulf of Mexico. In principle, much safer nuclear fission reactors can be built compared to the reactors that have recently failed in Japan. Among the new reactor concepts are the pebble bed reactor, proton catalyzed fission reactor, and fusion-initiated nuclear fission reactors and others. The latter two reactor types are currently not reduced to practice although research and development activities are currently underway. There are also improved designs for traditional fuel rod types of reactors. There must then be some good use for all of that nuclear waste. Why not at least research powering starships with it?

Augmenting the human life expectancy to 1,000 years, 10,000 years, and even longer durations through nanotechnology-based DNA and tissue repair and regeneration might put all the galaxies within our local group, and perhaps even within our local supercluster, available for human space missions and colonization over the next 10 billion years. Such increased human life expectancies would fuel the need to colonize the galaxy and beyond.

We might power such very long RTG-enabled voyages with a combination of short half-life and long half-life radionuclides within very large RTG apparatuses. Such an application would power a starship's daily living infrastructure for periods of billions of years' transit time for missions to galaxies dozens of millions of light-years distant.

Energy recycling and IR through RF electromagnetic frequency insulation to mitigate energy loss to space could provide an excellent means to sustain daily living in huge world Zonds. Such huge spacecraft could enable humanity to populate our local supercluster over the next few billion to several billions of years.

We can contemplate using such RTG propulsion systems to power ships that can be induced to enter any hyper spatial dimensions such as might provide shortcuts to very cosmically distant regions in space and in time within our universe. However, that is another story to be addressed in a book currently in the planning stages.

Erich von Däniken has proposed that extraterrestrial civilizations have made contact with human civilization several millennia ago and perhaps at some public level, interacted with the ancient human civilizations including those of Ancient Egypt, Greece, and perhaps the Mayans. Accordingly, the extraterrestrials may even have had one or more colonies on Earth such as the mythical civilization of Atlantis that seems to be at least loosely referred to in some ancient writings (Erich von Däniken, 2005).

Depictions in ancient cave arts resembling rocket-thrust plumes, electromagnetic antennas, and jet or rocket planes have been interpreted in some highly conjectural speculations by Däniken as representing extraterrestrial technologies.

If his speculations are correct, then perhaps ETI have already interacted with humans in ancient times. We now have electromagnetic antennas, rockets, jet planes, and our technology, including our spaceships, might not be far behind—on a cosmic scale—any technology, science, medicine, and nuclear physics that some previously visiting ETI cultures and civilizations may have possessed.

Check out Erich von Däniken's website at http://www.evdaniken.com/

Note that there is currently no substantiated evidence for any extraterrestrial civilizations and the pseudoscientific notion of ultra-terrestrials. However, the possibility of such life-forms is not beyond the bounds of mature scientific discussion. This possible existence of such life-forms has become more plausible, given the finding of on the order of 1,000 extrasolar planets as of the time of this editing, including some candidates for planets within habitable zones.

Nuclear energy can definitely get us to the stars and beyond. All the kinematical configurations of nuclear energy propulsion systems and refueling and staging methods have not yet been thought of. Consequently, the number of such possible systems yet to be dreamed up is probably very, very large.

We must not fear nuclear energy. Even though the weaponization of nuclear energy has given it a bad name for many citizens of our global civilization, we are now freer to use nuclear energy for our peaceful bidding for the colonization and exploration of our universe and beyond. Can we afford not to take advantage of this wonderful opportunity!

We know of just two nuclear forces, the strong nuclear force which is mediated by eight types of gluons. Gluons are superposition of color states of which there exist only three. There are a total of nine superposition of color states. The color singlet state is

$$(r\bar{r} + b\bar{b} + g\bar{g})/\sqrt{3}.$$

Singlet color states can interact with other singlet color states, but not with color states. If the above state could be measured, there would be equal probabilities of the measurement indicating a red-anti-red, blue-anti-blue, and green-anti-green state.

There are a total of eight color states, where each state corresponds with a gluon. These eight states are linearly independent and are as follows.

$$(r\bar{b}+b\bar{r})/\sqrt{2} \quad -i(r\bar{b}-b\bar{r})/\sqrt{2}$$

$$(r\bar{g}+g\bar{r})/\sqrt{2} \quad -i(r\bar{g}-g\bar{r})/\sqrt{2}$$

$$(b\bar{g}+g\bar{b})/\sqrt{2} \quad -i(b\bar{g}-g\bar{b})/\sqrt{2}$$

$$(r\bar{r}-b\bar{b})/\sqrt{2} \quad (r\bar{r}+b\bar{b}-2g\bar{g})/\sqrt{6}$$

The weak nuclear force is mediated by the W+, W-, and Zo weak force bosons. Might there exist additional nuclear forces and associated bosonic force mediators? The answer may very well be yes. We can dream of the fantastic applications of the two known nuclear forces that are yet to come, and even more so, of exotic applications of any yet-to-be-discovered nuclear forces.

These slow starships might act as interstellar freighters that would gradually carry ores, equipment, and supplies, from one star system to another. However, perhaps faster spacecraft could be used to transport persons and for other time-critical missions among the stars.

Such long travel times and high-end Newtonian velocities for these starships may also be practical via the development of suspended animation, or perhaps near-freezing hibernation states of crew members' bodies. Once again, such states might be regulated by nanotechnological control so that potentially dangerously degrading pathogens, redox reactions, and ionizing cosmic radiation can be dealt with.

At the following URL from Centauri Dreams is an excellent short article on light sails with a reference to the Sundiver concepts (Gilster, 2009): http://www.centauri-dreams.org/?p=10190

The Planetary Society, (Planetary Society, 2011). discusses their light sail concept at: http://www.planetary.org/programs/projects/solar_sailing/

Fig. 25. LightSail-1 artist depiction by Rick Sternbach. © The Planetary Society.

Fig. 26. Lightsail-1 Stowed Configuration. Diagram showing Lightsail-1 in its stowed (launch) configuration: prior to opening of solar panels and the deployment of solar sails. Solar cells are shown in blue. (Credit: The Planetary Society)

Steve Howe's concept of an antimatter sail involves the catalyzation of a uranium infusement within a sail by antiprotons captured by the sail. The exhaust is directed appropriately in order to produce high-end Newtonian, and even mildly relativistic, velocities. Such a craft can be considered a rocket.

A good brief primer on Steve Howe's concept (Gilster, 2004) is available at the following URL:

http://www.centauri-dreams.org/?p=28

Chapter 8

Growing the Beanstalk

Recall the story of Jack and the Beanstalk. You will be familiar with this childhood tale of how Jack found a beanstalk that reached well into the sky. Accordingly, Jack climbed the beanstalk and eventually got in trouble with a giant who lived in a kingdom up in the sky.

Well, the concept might actually have application in surface-to-Earth-orbit travel.

For example, carbon graphene, carbon nanotubes, boron nitride nanotubes, and like materials, may be used to form a long cable that would reach into geosynchronous orbit. At the upper end of the cable, there would exist a counterweight type of space habitat or other infrastructure. The space habitat would include its judicious extension beyond geo-synchronous orbit so that the net pulling force at geo-synchronous orbit would be balanced by the centripetal acceleration of the outwardly extended portion of the space habitat.

The cable may be a mere few square centimeters in cross-sectional area to perhaps as much as a square meter or more in cross-section. A larger cable would permit the holding of a much larger counter-weight. Thick cables could be assembled from orbit downward and from the ground upward via the braiding of strings perhaps as small as one square millimeter in cross-section or less. A zip-line type of mechanism would repeatedly run up and down the cable as a work in progress in order to increase its cross-sectional area.

The zip-line mechanism can be powered by an electric motor in a manner similar to modern high-rise building elevators. One option is to employ a traditional rotor-based electric motor within the zip-line car for which a geared coupling to the cable or other coupling would propel the car up and down the cable. Another option would include a magnetic form of propulsion similar to that of a Maglev train. Either way, the electrical energy used to power the car's motion may be generated internally from within the car through solar panels attached to the car and/or the cable, or perhaps a high-energy-density electric generator within the car. The generator may include a compact nuclear reactor, a fuel cell, or other high-energy-density units. It may one day be possible to use rechargeable galvanic cells or batteries to provide ascent and descent along the cable upon development of suitable energy-storage densities. The batteries may optionally be recharged from solar-energy stations along the cable.

Alternatively, the electrical power used to drive the cars may be supplied from an Earth-bound electrical generation facility and/or a space-based one. Space-based generators may best be photovoltaic in nature so as to take advantage of the enormous solar power output. However, compact space-based nuclear-fission reactors as a power source are also conceivable.

As another kinematic option, the zip car or elevator may be powered by a pulley system attached between the ground and the space-hab. A counterweight may be used that travels in the opposite direction as the ascending or descending elevator to at least partially recycle the system's mechanical energy. The pulley system may be powered by electromotive means and/or stream-driven or sterling types of mechanical power system. Once again, the required electrical power can be derived from solar energy and/or nuclear fission reactor systems. One option would be to use turboelectric systems heated by solar concentrators and/or nuclear fission reactors.

Upon completion of the cable and the space-hab, travel may be routinely accomplished along the cable by any of the previously described motive mechanisms.

The cables would ordinarily need to be attached near the equator. If not attached near the equator, two or more spaced cables would be required in configurations resulting in no drift of the space-hab with respect to Earth's surface.

Habitats each having a mass of 1 billion metric tons could house about 10 million to 100 million persons. Supplies could be routinely ferried between ground and space, and waste may be processed in space in situ or transported back to Earth for processing.

Optionally, farming can be done on the space-habs, thus taking advantage of solar energy for photosynthesis. Livestock can be raised on the habs and fed by the farm grown grains. Electrical power may alternatively be sent from the ground up to the habitats to provide life support and energy needed for farming activities and oxygen and freshwater regeneration.

One-G gravity within the hab can be simulated by rotating personnel and residence quarters. The rotating living quarters can be affixed to the bulk of a hab via a mechanical material coupling, but preferably would instead use a magnetic-bearing type of coupling so that there would be no frictional wear and tear on the coupling joints. The magnetic bearing may be permanent or electromagnet based. Alternatively, a high-temperature superconducting mechanism may provide the attachment via the Meissner effect. The Meissner effect is commonly displayed in high school and college physics classroom labs, where a portion of superconducting material is placed over a permanent magnet. As long as the hovering material is cool enough to remain superconducting, it will continue to hover by itself in place over the magnet even in cases where the superconductor is placed off-center of the magnet vector flux lines.

Alternatively, magnetic clothing can be worn by the residents. Here, the magnetic clothing would be held in an attractive relation to a "floor" through a magnetic coupling. The floor may consist of permanent magnets, and/or a continually generated electromagnet field.

Travel tubes can be used to link adjacent habs to effectively form a planetary ring world.

The installation of 1,000 such habitats may provide living space for 10 billion to 100 billion human persons, where each hab has a mass of 1 billion metric tons.

Large interstellar spacecraft may be assembled in Earth's orbit without the need to carry the required components, materials, and other supplies into space via rocket or space-plane vehicles.

Thus, beanstalks on steroids may open up a golden new era for humanity where the human population is unlimited in numbers in the limits of the depths of eternity.

Now what about beanstalks on other planetary bodies within the solar system?

For one, Earth's moon would be a great candidate. Habitats and way stations can be centered in lunar geostationary orbit just as it can be for Earth.

The moon has lots of titanium and aluminum in its surface composition. Thus the moon would provide an excellent source of materials for the construction of more extensive infrastructure in lunar-geostationary orbit and farther abroad in our solar system.

Lunar regolith may be fashioned into concrete blocks for use in shielding habitats in Earth, lunar, and solar orbit from cosmic rays. The blocks may be manufactured in lunar orbit or on the lunar surface and then transported to lunar orbit by a beanstalk.

Mars is another good candidate for beanstalk construction. Mars, apparently, has some fairly complex minerals in its surface composition as well as water-ice. This complex chemistry is conducive to the manufacture of compact and powerful solid rocket fuels, methane, and other hydrogenous fuels. Liquid oxygen and hydrogen can be manufactured from water and ice within and under the Martian surface. All these materials may be transported to Martian geosynchronous orbit by way of beanstalks.

Since Mars is similar in size to Earth and has nearly the same day length, Martian geosynchronous orbit may support perhaps as much as one thousand habitats, each having a mass of 1 billion metric tons. Each habitat could be powered by similar means as those contemplated previously for Earth. Thus, perhaps a Mars planetary ring world may support upward of 100 billion persons simultaneously.

The gas giant planet moons of Europa, Io, Ganymede, Titan, Callisto, and the like can likewise support beanstalks ring worlds. Some of these moons, such as Europa, have very extensive oceans below its icy crusts and thus would provide an excellent resource for the manufacture of LOX (liquid hydrogen) rocket fuels, as well as plenty of fresh water for ring-world habitats. Oxygen for breathing can be manufactured from such aqueous regions within and on these moons.

Beanstalks constructed around the gas giant planets of Jupiter, Saturn, Uranus, and Neptune can likewise be conduits for the transport of valuable resources. Helium-3, which is an ideal nuclear-fusion fuel, may be extracted from the atmospheres of these planets, where it would then be used for nuclear fusion power interplanetary and interstellar-manned vessels. This fusionable isotope may enable terminal velocities for relativistic rockets that are as high as 0.3 C to 0.4 C with large but reasonable mass ratios.

Beanstalk ring worlds may also be constructed around Pluto and the other minor planets fairly recently discovered and any others yet to be discovered. The same can be generally said for any sizable Kuiper Belt objects and Oort Cloud objects.

Now Earth travels once around the sun every Earth year. The distance of travel is about $2\pi R$, where R is roughly 150,000,000,000 meters. Therefore, the orbital velocity of the Earth is approximately (150,000,000,000 meters)(2π)/(31,000,000 seconds) = 30,400 meters per second = 30.4 km second.

The velocity vectors of Earth amid successive half-year intervals are antiparallel in orientation. Thus, the effective acceleration of Earth is approximately (4)(30,400 m/s)/(31,000,000 s) = 0.0039229 m/s^2 or 0.0003999 Earth Gs.

Note that the formula for non-varying centripetal force is as follows.

$F = \gamma m v^2/r$

Here, γ is the Lorentz factor of the moving body, m is the relativistc mass of the moving body, v is the velocity of the body, and r is the radius of revolution of the body.

In the following text, we use a different method of computing relativistic centripetal force which gives close but only approximate results compared to the above relativistic formulation. The approximations are chosen for convenience based on the context of the argument but otherwise provide a good estimate of centripetal force acting on the body in circular motion.

For non-relativistic centripetal force, the above formula simply reduces to,

$F = m v^2/r$

Here m is the invariant mass of the body.

As with the relativistic cases, the same approximate solutions are provided in the context of the presentation and the intuitive descriptions provided thereof. Note that the purpose of this book is to provide good understanding of the topics covered without being to mathematical. Afterall, the book is intended for a wider audience.

Thus, assuming such an acceleration, a cable made out of carbon nanotubes, graphene, carbon nitride fiber, diamond fiber, boron nitride nanotubes, carbon atom chains, and the like, having a volumetric density of 1,000 kilograms per cubic meter, which is one square centimeter in cross-section and 100 million kilometer long will have a tidal stress of (0.1 kg)(10^{11})(0.0039229 m/s^2) = 39,229,000 newtons, or 3,998,000 kilograms force.

In fact, the cable may be as long as roughly [(10,000,000)/(39,229,000)](100,000,000 km), or 25.49 billion kilometers, while undergoing an acceleration of 0.0000039229 m/s^2, or 0.0000003999 Earth Gs with the associated induced tidal stress without fatigue for scenarios where the cable has an average effective location at [1,000 EXP (1/2)] AU from the sun. This is due to the inverse squaring of gravitational attraction as a function of radius. So a network of travel tube located in the Kuiper Belt can support a huge human and material goods transport infrastructure.

Thus, carbonaceous supermaterials-based cables can be used for transportation throughout the planetary solar system from orbital radii ranging from that of the planet Venus to locations well beyond that of Pluto because the centripetal accelerations associated with the orbital motions of the gas giants are radially progressively and strongly reduced with respect to that of Earth.

A large radially anchored ring-world infrastructure with a radial center of mass located about 1 billion kilometers from the sun and balanced by a radially external counterweight mechanism would have a circumference of about 6.28 billion kilometer. Assuming that each kilometer of the ring-world structure had a mass of 10 million metric tons, the entire ring would have a mass of 6.28 x 10^{16} metric tons.

Further assume that each kilometer arcuate element of the ring could house and support 100,000 persons simultaneously, the ring world could support 628 trillion persons.

More than one ring world may, in theory, be concentrically disposed. Such ring-worlds may be radially connected by travel tubes in the form of electromagnetic mass drivers. Travel tubes may span the entire circumference of the ring worlds as well. The mass drivers may optionally be configured similarly to modern-day Maglev trains. In one scenario, the tubes would pull the train, or cars, forward by electromagnets that would serially pull on ferromagnetic elements affixed to the cars. In another scenario, the cars would include permanent magnets to assist in the propulsion. In still another example, the cars would include electromagnetics for propulsion. For cases where the cars contain electromagnetics, the tubes may contain permanent magnets and/or iron or steel elements for which the cars would react against.

Assuming a comfortable 1 G of acceleration, the cars may obtain a velocity of [9.81 m/s^2](100,000 s) or 981,000 m/s in 100,000 seconds, which is about one day. The distance traveled will be 49.05 billion meters, or 49,050,000 kilometers in about one day. Increasing the acceleration to 10 Gs and the terminal velocity will be 9,810,000 m/s. The distance traveled in one day will be 490,500,000 kilometers. Increase the acceleration to 100 Gs, and the terminal velocity will be 98,100,000 m/s, or 98,100 kilometers per second in a Newtonian approximation. This is one-third the velocity of light! The relativistic kinetic energy of the spacecraft will be about 0.0606 times that of its total invariant energy. Such a terminal velocity can be achieved for roughly one revolution around a ring world, having a radius of 1 billion kilometers, assuming roughly 100 Gs of tangential acceleration.

Assume ten complete revolutions around a solar system ring world of the above radius where the spacecraft gains about 0.0606 times its total invariant energy in relativistic kinetic energy for each revolution. The spacecraft Lorentz factor will be about 1.606, with a velocity of 0.7827 C.

Assume 100 complete revolutions around the above solar system ring world, where the spacecraft gains about 0.0606 times its total invariant energy in relativistic kinetic energy for each revolution. The spacecraft Lorentz factor will be about 1 + 6.06, or 7.06 with a velocity of 0.98994 C.

Assume 1,000 complete revolutions around the solar system ring world where the spacecraft gains about 0.0606 times its total invariant energy in relativistic kinetic energy for each revolution. The spacecraft Lorentz factor will be about 1 + 60.6, or 61.6 with a velocity of 0.999864 C.

Assume 10,000 complete revolutions around the solar-system ring world, where the spacecraft gains about 0.0606 times its total invariant energy in relativistic kinetic energy for each revolution. The spacecraft Lorentz factor will be about 1 + 606, or 607 with a velocity of 0.99999864 C.

Assume 100,000 complete revolutions around the solar system ring-world where the spacecraft gains about 0.0606 times its total invariant energy in relativistic kinetic energy for each revolution. The spacecraft Lorentz factor will be about 1 + 6,060, or 6,061 with a velocity of 0.99999864 C.

At a Lorentz factor of 100,000, the car occupants would experience a forward time travel of 100,000 years with respect to the stationary ring world frame. In 10 years, the occupants would experience a forward time travel of 1 million years. In 100 years, the occupants would experience a forward time travel of 10 million years. Assuming human life expectancy can be augmented to over 1,000 years, human persons could experience alive and awake forward time travel of 100 million years. Cryogenic sleep preservation and/or near-freezing hibernation states can enable much further forward time travel for the car occupants.

For reduced radial acceleration and centripetal G-forces, ring worlds having a radius of 10 billion kilometers, or perhaps even 100 billion kilometers, can be used. Accordingly, the energy gain per revolution could be much higher than that for the 1-billion-kilometer-radius ring world.

Alternatively, once the cars reach the desired terminal Lorentz factors, the cars could be released into interstellar space. At a Lorentz factor of 100,000, the car occupants would experience a forward time travel of 100,000 years with respect to the stationary ring world frame and travel 100,000 light-years through intergalactic space. In 10 years, the occupants would experience a forward time travel of

1,000,000 years and travel 1 million light-years through intergalactic space. In 100 years, the occupants would experience a forward time travel of 10 million years and travel 10 million light-years through intergalactic space. Assuming human life expectancy can be augmented to over 1,000 years, persons could experience alive and awake forward time travel of 100 million years and travel 100 million light-years through intergalactic space. Cryogenic sleep preservation and/or near-freezing hibernation states can enable much-farther-forward time travel for the car occupants.

The ratio of spacecraft kinetic energy gain and its total invariant energy can vary from one ring world to another, as well as within a given ring world. Larger ring worlds are more facilitative of high ratios as such than ring worlds of smaller radii. This is true for three main reasons. First, the acceleration circumferential path per revolution is greater for larger rings, thus enabling longer acceleration paths. Secondly, the radial acceleration for a given Lorentz factor is less for cars traversing larger ring-worlds. Thirdly, the tangential acceleration for a very large ring world can be significantly less than that for a much smaller ring world and yet enable the car traversing one revolution of the larger ring world to obtain a significantly higher Lorentz factor than that which is practical for one loop around the smaller ring world.

We now consider a locally interstellar subway system.

An interstellar subway system as such may reasonably extend out to a distance of about 20 light-years from o'l sol. The numbers of stars and brown dwarfs within 20 light-years of Earth is estimated to be about 135. Most of this mass within the local interstellar neighborhood exist as hydrogen and helium. Perhaps as much as 0.0001 of the mass exists as heavy metals. In astronomic literature, heavy metals are elements and isotopes heavier than helium atoms or nuclei.

Assume that the heavy metal composition is conducive to the formation of carbonaceous supermaterials, super alloys, rock, concretes, and the like in the amount of 0.00001 of the total baryonic mass content within the local interstellar neighborhood. Since the free interstellar gas is generally somewhere around two to ten times more plentiful than the stellar matter in a given local region of interstellar space, we can expect that perhaps the useful heavy metal composition of the local interstellar neighborhood would be roughly equal to 0.00002 to 0.0001 times the combined mass embodied in the interstellar objects within the local interstellar neighborhood. Since the average star is about 1/5 of the mass of the sun, the total available construction material ranges from (0.00002)(0.2)(135) to (0.0001)(0.2)(135) solar masses, or 0.00054 to 0.0027 solar masses. This would be about $[2 \times (10^{27})](0.00054)$ to $[2 \times (10^{27})](0.0027)$ metric tons or $[1.08 \times (10^{24})]$ to $[5.4 \times (10^{24})]$ metric tons.

We now assume that each person needs 1,000 metric tons for life support. The total number of simultaneously supportable persons will be $[1.08 \times (10^{21})]$ to $[5.4 \times (10^{21})]$. Since human metabolism radiates about 80 watts, the above human population will produce a mere $(80)[1.08 \times (10^{21})]$ to $(80)[5.4 \times (10^{21})]$ watts of radiant power, or about $[8.64 \times (10^{22})]$ to $[4.32 \times (10^{23})]$ watts of radiative power. This is equivalent to that of a red dwarf star, or 0.0002 to 0.001 times the solar output.

Since the total stellar output within the interstellar neighborhood or 20 light-years radial extension from the Sun is several orders of magnitude greater than the human metabolic power for the above conjectured population, stellar power alone can plausibly provide life support for such a huge human population.

Even the CMBR light that passes through the local interstellar neighborhood can provide enough power for such life-support activities. In fact, since the effective temperature of starlight alone with respect to the surface of Earth is of the same order of magnitude as that of the incident CMBR. Capturing either source of radiant energy would enable easy support of the above conjectured populations.

It is possible that a Dewar type of multihull system can capture human metabolic waste heat and other industrial thermal and electromagnetic emission. The entire hab and tube structure of the local interstellar system may accordingly be enclosed in conductive multihull shells to capture almost all waste heat and magnetic emissions from the subway tubes.

The travel tubes can include linear and curvilinear forms. For extreme spacecraft Lorentz factor–capable cars, mass driver tubes can take any of the former topologies.

Chapter 9

Cosmic Lifeboats

This chapter includes a brief digression on so-called lifeboats and pods with cryogenically asleep humans for which the pods are designed to pass through cosmic eras before the next phase changes, as well as the next and all future phase changes of our universe, multiverse, forest, biosphere, etc. The following paragraphs are clause-like conjectures on the nature of these lifeboats.

Lifeboats and pods with cryogenically asleep humans for which the pods are designed to pass through cosmic eras before the next phase changes as well as the next and all future phase changes of our universe, multiverse, forest, biosphere, etc., where associated designs include electronic charging of lifeboats for which positrons are stored in a location on the lifeboat and made to interact with background electrons for which the associated gamma ray energy is absorbed and used to power accelerators that generate matter and antimatter protons, matter and antimatter deuterons, matter and antimatter alpha particles, matter and antimatter atomic nuclei, matter and antimatter ions, matter and antimatter mesons, matter and antimatter baryons, etc., where the normal-matter stable particles are retained and the antimatter stable particles are retained in a charge-specific location aboard the lifeboats, and the rest of the particles are inter-pair-annihilated or left to decay.

But either way, the annihilation energy and decay energies are fed back into an accelerator to produce more matter and antimatter protons, matter and antimatter deuterons, matter and antimatter alpha particles, matter and antimatter atomic nuclei, matter and antimatter ions, matter and antimatter mesons, matter and antimatter baryons, etc., where the normal-matter stable particles are retained and the antimatter stable particles are retained in charge=specific location aboard the lifeboats, and the rest of the particles are inter-pair-annihilated or left to decay. But either way, the annihilation energy and decay energies are fed back into an accelerator to produce in a second stage more matter and antimatter protons, matter and antimatter deuterons, matter and antimatter alpha particles, matter and antimatter atomic nuclei, matter and antimatter ions, matter and antimatter mesons, matter and antimatter baryons, etc., where the normal-matter stable particles are retained and the antimatter stable particles are retained in a charge-specific location aboard the lifeboats, and the rest of the particles are inter-pair-annihilated or left to decay.

But either way, the annihilation energy and decay energies are fed back into an accelerator to produce in a third stage more matter and antimatter protons, matter and antimatter deuterons, matter and antimatter alpha particles, matter and antimatter atomic nuclei, matter and antimatter ions, matter and antimatter mesons, matter and antimatter baryons, etc., where the normal-matter stable particles are retained and the antimatter stable particles are retained in a charge-specific location aboard the lifeboats, and the rest of the particles are inter-pair-annihilated or left to decay. But either way, the annihilation energy and decay energies are fed back into an accelerator to produce in a fourth stage more matter and antimatter protons, matter and antimatter deuterons, matter and antimatter alpha particles, matter and antimatter atomic nuclei, matter and antimatter ions, matter and antimatter mesons, matter and antimatter baryons, etc., where the normal-matter stable particles are retained and the antimatter stable particles are retained in a charge-specific location aboard the lifeboats, and the rest of the particles are inter-pair-annihilated or left to decay.

But either way, the annihilation energy and decay energies are fed back into an accelerator to produce in a fifth stage more … matter and antimatter protons, matter and antimatter deuterons, matter and antimatter alpha particles, matter and antimatter atomic nuclei, matter and antimatter ions, matter and antimatter mesons, matter and antimatter baryons, etc., where the normal-matter stable particles are retained and the antimatter stable particles are retained in a charge-specific location aboard the lifeboats, and the rest of the particles are inter-pair-annihilated or left to decay, but either way, the annihilation energy and decay energies are fed back into an accelerator to produce in an (n-1)th stage more matter and antimatter protons, matter and antimatter deuterons, matter and antimatter alpha particles, matter and antimatter atomic nuclei, matter and antimatter ions, matter and antimatter mesons, matter and antimatter baryons, etc., where the normal-matter stable particles are retained and the antimatter stable particles are retained in a charge-specific location aboard the lifeboats, and the rest of the particles are inter-pair-annihilated or left to decay, but either way, the annihilation energy and decay energies are fed back into an accelerator to produce in an nth stage more … and so on forever. Onward ho!

Such a mechanism may be able to produce more normal-matter and antimatter or more antimatter than normal matter or equal amounts of both matter and antimatter of the same and different species of particles so as to gradually restock the cosmos with baryonic matter and cold dark matter as well as hot dark matter of Standard Model and non–Standard Model types.

The lifeboats might reconfigure a cosmos over cosmic time periods to take the form of a matter-antimatter–compatible universe, multiverse, forest, biosphere, and the like by using accelerators and similar filtering mechanisms to convert future phase change particles back into Standard Model and non–Standard Model present-era cold and hot matter and associated light-speed field quanta.

Propulsion methods based on accelerators which produce matter and antimatter electrons, muons, tauons, protons, alpha particles, higher-atomic-number nuclei and ions, stable and unstable matter and antimatter exotic mesons and baryons, and the like for which the antimatter versions are accelerated into space away from the spacecraft while the normal-matter versions build up to produce a net electric charge on the spacecraft for propulsion.

Propulsion methods based on accelerators that produce matter and antimatter muons, tauons, and charged mesons and charged unstable baryons for which the normal-matter-charged particles are filtered and accelerated in a first stage in cases where they live a long time in an accelerator and which are then used as feedstock to produce more matter and antimatter particles for which the antimatter versions are allowed to decay while the normal-matter versions are retained, filtered, and accelerated in a second stage in cases where they live a long time in an accelerator and which are then used as feedstock to produce more matter and antimatter particles for which the antimatter versions are allowed to decay while the normal-matter versions are retained, filtered, and accelerated in a third stage in cases where they live a long time in an accelerator and which are then used as feedstock to produce more matter and antimatter particles for which the antimatter versions are allowed to decay while the normal-matter versions are retained, filtered, and accelerated in a fourth stage in cases where they live a long time in an accelerator and which are then used as feedstock to produce more matter and antimatter particles for which the antimatter versions are allowed to decay while the normal-matter versions are retained, filtered, and accelerated in a fifth stage … in cases where they live a long time in an accelerator and which are then used as feedstock to produce more matter and antimatter particles for which the antimatter versions are allowed to decay while the normal-matter versions are retained, filtered, and accelerated in a (n-1)th stage in cases where they live a long time in an accelerator and which then used as feedstock to produce more matter and antimatter particles for which the antimatter versions are allowed to decay while the normal-matter versions are retained, filtered, and accelerated in an nth stage. Where *n* may be as great as an ensemble, an infinity scraper, an Aleph 0, a greater than one exponentated Aleph 0, an Aleph 1, a greater than one exponentated Aleph 1, an Aleph 2, a greater than one exponentiated Aleph 2, an Aleph 3, a greater than one exponentiated Aleph 3, and so on. Onward ho! Forevermore!

Lifeboats and pods with cryogenically asleep humans for which the pods are designed to pass through cosmic eras before the next phase changes, as well as the next and all future phase changes of our universe, multiverse, forest, biosphere, etc., where associated designs include the electronic charging of lifeboats for which positrons are stored in a location on the lifeboat and made to interact with background electrons for which the associated gamma ray energy is absorbed and used to power accelerators that generate matter and antimatter protons, matter and antimatter deuterons, matter and antimatter alpha particles, matter and antimatter atomic nuclei, matter and antimatter ions, matter and antimatter mesons, matter and antimatter baryons, etc., where the statistically randomly longer-living normal-matter stable particles are retained and the antimatter stable particles are retained in a charge-specific location aboard the lifeboats, and the rest of the particles are inter-pair-annihilated or left to decay.

But either way, the annihilation energy and decay energies are fed back into an accelerator to produce more matter and antimatter protons, matter and antimatter deuterons, matter and antimatter alpha particles, matter and antimatter atomic nuclei, matter and antimatter ions, matter and antimatter mesons, matter and antimatter baryons, etc., where the statistically randomly longer-living normal-matter stable particles are retained and the antimatter stable particles are retained in a charge-specific location aboard the lifeboats, and the rest of the particles are inter-pair-annihilated or left to decay, but either way, the annihilation energy and decay energies are fed back into an accelerator to produce in a second stage more matter and antimatter protons, matter and antimatter deuterons, matter and antimatter alpha particles, matter and antimatter atomic nuclei, matter and antimatter ions, matter and antimatter mesons, matter and antimatter baryons, etc., where the statistically randomly longer-living normal-matter stable particles are retained and the antimatter stable particles are retained in a charge-specific location aboard the lifeboats, and the rest of the particles are inter-pair-annihilated or left to decay

But either way, the annihilation energy and decay energies are fed back into an accelerator to produce in a third stage more matter and antimatter protons, matter and antimatter deuterons, matter and antimatter alpha particles, matter and antimatter atomic nuclei, matter and antimatter ions, matter and antimatter mesons, matter and antimatter baryons, etc., where the statistically randomly longer-living normal-matter stable particles are retained and the antimatter stable particles are retained in a charge-specific location aboard the lifeboats, and the rest of the particles are inter-pair-annihilated or left to decay, but either way, the annihilation energy and decay energies are fed back into an accelerator to produce in a fourth stage more matter and antimatter protons, matter and antimatter deuterons, matter and antimatter alpha particles, matter and antimatter atomic nuclei, matter and antimatter ions, matter and antimatter mesons, matter and antimatter baryons, etc., where the statistically randomly longer-living normal-matter stable particles are retained and the antimatter stable particles are retained in a charge-specific location aboard the lifeboats, and the rest of the particles are inter-pair-annihilated or left to decay, but either way, the annihilation energy and decay energies are fed back into an accelerator to produce in a fifth stage more … matter and antimatter protons, matter and antimatter deuterons, matter and antimatter alpha particles, matter and antimatter atomic nuclei, matter and antimatter ions, matter and antimatter mesons, matter and antimatter baryons, etc., where the statistically randomly longer-living normal-matter stable particles are retained and the antimatter stable particles are retained in a charge-specific location aboard the lifeboats, and the rest of the particles are inter-pair-annihilated or left to decay, but either way, the annihilation energy and decay energies are fed back into an accelerator to produce in an (n-1)th stage more matter and antimatter protons, matter and antimatter deuterons, matter and antimatter alpha particles, matter and antimatter atomic nuclei, matter and antimatter ions, matter and antimatter mesons, matter and antimatter baryons, etc., where the statistically randomly longer-living normal-matter stable particles are retained and the antimatter stable particles are retained in a charge-specific location aboard the lifeboats, and the rest of the particles are inter-pair-annihilated or left to decay, but either way, the annihilation energy and decay energies are fed back into an accelerator to produce in an nth stage more … on so on forever. Onward ho!

Such a mechanism and similar mechanisms may be able to produce more normal-matter and antimatter or more antimatter than normal-matter or equal amounts of both matter and antimatter of the same and different species of particles so as to gradually restock the cosmos with baryonic matter and cold dark matter as well as hot dark matter of Standard Model and non–Standard Model types.

The lifeboats might reconfigure a cosmos over cosmic time periods to take the form of a matter-antimatter compatible universe, multiverse, forest, biosphere, and the like by using accelerators and similar filtering mechanisms to convert future-phase change particles back into Standard Model and non–Standard Model present-era cold and hot matter and associated light-speed field quanta.

Propulsion methods based on accelerators that produce matter and antimatter electrons, muons, tauons, protons, alpha particles, higher-atomic-number nuclei, and ions, stable and unstable matter and antimatter exotic mesons and baryons, and the like for which the antimatter versions are accelerated into space away from the spacecraft while the normal-matter versions build up to produce a net electric charge on the spacecraft for propulsion where the statistically randomly longer-living normal-matter stable particles are retained.

Propulsion methods based on accelerators which produce matter and antimatter muons, tauons, and charged mesons and charged unstable baryons for which the normal-matter charged particles are filtered and accelerated in a first stage in cases where they live a long time in an accelerator and which are then used as feed-stock to produce more matter and antimatter particles for which the antimatter versions are allowed to decay while the normal-matter versions are retained, filtered, and accelerated in a second stage in cases where they live a long time in an accelerator where the statistically randomly longer-living normal-matter stable particles are partially or fully retained and which are then used as feedstock to produce more matter and antimatter particles for which the antimatter versions are allowed to decay while the normal-matter versions are retained, filtered, and accelerated in a third stage in cases where they live a long time in an accelerator where the statistically randomly longer-living normal-matter stable particles are partially or fully retained and which are then used as feed-stock to produce more matter and antimatter particles for which the antimatter versions are allowed to decay while the normal-matter versions are retained, filtered, and accelerated in a fourth stage in cases where they live a long time in an accelerator where the statistically randomly longer-living normal-matter stable particles are partially or fully retained and which are then used as feed-stock to produce more matter and antimatter particles for which the antimatter versions are allowed to decay while the normal-matter versions are retained, filtered, and accelerated in a fifth stage … in cases where they live a long time in an accelerator and which are then used as feed-stock to produce more matter and antimatter particles for which the antimatter versions are allowed to decay while the normal-matter versions are retained, filtered, and accelerated in a (n-1)th stage in cases where they live a long time in an accelerator where the statistically randomly longer-living normal-matter stable particles are partially or fully retained and which are then used as feed-stock to produce more matter and antimatter particles for which the antimatter versions are allowed to decay while the normal-matter versions are retained, filtered, and accelerated in an nth stage. Where *n* may be as great as an ensemble, an infinity scraper, an Aleph 0, a greater than one exponentiated Aleph 0, an Aleph 1, a greater than one exponentated Aleph 1, an Aleph 2, a greater than one exponentated Aleph 2, an Aleph 3, a greater than one exponentated Aleph 3, and so on Onward ho! Forevermore!

Lifeboats and pods with cryogenically asleep humans for which the pods are designed to pass-through cosmic eras before the next phase changes as well as the next and all future phase changes of our universe, multi-verse, forest, biosphere, etc. where associated designs include the electronic charging of lifeboats for which positrons are stored in a location on the lifeboat and made to interact with background electrons for which the associated gamma-ray energy is absorbed and used to power accelerators that generate matter and antimatter protons, matter and antimatter deuterons, matter and antimatter alpha particles, matter and antimatter atomic nuclei, matter and antimatter ions, matter and antimatter mesons, matter and antimatter baryons, etc., where the statistically randomly longer-living normal-matter stable particles are retained and thus used as breeding stock to produce more longer-living normal-matter particles.

And the process is repeated over and over, and the antimatter stable particles are retained in a charge-specific location aboard the lifeboats, and the rest of the particles are inter-pair-annihilated or left to decay, but either way, the annihilation energy and decay energies are fed back into an accelerator to produce more matter and antimatter protons, matter and antimatter deuterons, matter and antimatter alpha particles, matter and antimatter atomic nuclei, matter and antimatter ions, matter and antimatter mesons, matter and antimatter baryons, etc., where the statistically randomly longer-living normal-matter stable particles are retained and thus used as breeding stock to produce more longer-living normal-matter particles and the process is repeated over and over, in a charge-specific location aboard the lifeboats, and the rest of the particles are inter-pair-annihilated or left to decay, but either way, the annihilation energy and decay energies are fed back into an accelerator to produce in a second stage more matter and antimatter protons, matter and antimatter deuterons, matter and antimatter alpha particles, matter and antimatter atomic nuclei, matter and antimatter ions, matter and antimatter mesons, matter and antimatter baryons, etc., where the statistically randomly longer-living normal-matter stable particles are retained and thus used as breeding stock to produce more longer-living normal-matter particles.

And the process is repeated over and over, and the antimatter stable particles are retained in a charge-specific location aboard the lifeboats, and the rest of the particles are inter-pair-annihilated or left to decay, but either way, the annihilation energy and decay energies are fed back into an accelerator to produce in a third stage more matter and antimatter protons, matter and antimatter deuterons, matter and antimatter alpha particles, matter and antimatter atomic nuclei, matter and antimatter ions, matter and antimatter mesons, matter and antimatter baryons, etc., where the statistically randomly longer-living normal-matter stable particles are retained and thus used as breeding stock to produce more longer-living normal-matter particles.

And the process is repeated over and over, and the antimatter stable particles are retained in a charge-specific location aboard the lifeboats, and the rest of the particles are inter-pair-annihilated or left to decay, but either way, the annihilation energy and decay energies are fed back into an accelerator to produce in a fourth stage more matter and antimatter protons, matter and antimatter deuterons, matter and antimatter alpha particles, matter and antimatter atomic nuclei, matter and antimatter ions, matter and antimatter mesons, matter and antimatter baryons, etc., where the statistically randomly longer-living normal-matter stable particles are retained and thus used as breeding stock to produce more longer-living normal-matter particles, and the process is repeated over and over, and thus used as breeding stock to produce more longer-living normal-matter particles.

And the process is repeated over and over in a charge-specific location aboard the lifeboats, and the rest of the particles are inter-pair-annihilated, or left to decay, but either way, the annihilation energy and decay energies are fed back into an accelerator to produce in a fifth stage more … matter and antimatter protons, matter and antimatter deuterons, matter and antimatter alpha particles, matter and antimatter atomic nuclei, matter and antimatter ions, matter and antimatter mesons, matter and antimatter baryons, etc., where the statistically randomly longer-living normal-matter stable particles are retained and thus used as breeding stock to produce more longer-living normal-matter particles. And the process is repeated over and over, and the antimatter stable particles are retained in a charge-specific location aboard the lifeboats, and the rest of the particles are inter-pair-annihilated or left to decay, but either way, the annihilation energy and decay energies are fed back into an accelerator to produce in an (n-1)th stage more matter and antimatter protons, matter and antimatter deuterons, matter and antimatter alpha particles, matter and antimatter atomic nuclei, matter and antimatter ions, matter and antimatter mesons, matter and antimatter baryons, etc., where the statistically randomly longer-living normal-matter stable particles are retained and thus used as breeding stock to produce more longer-living normal-matter particles. And the process is repeated over and over, and the antimatter stable particles are retained in a charge-specific location aboard the lifeboats, and the rest of the particles are inter-pair-annihilated or left to decay, but either way, the annihilation energy and decay energies are fed back into an accelerator to produce in an nth stage more … and so on forever. Onward ho!

Such a mechanism and similar mechanisms may be able to produce more normal-matter and antimatter or more antimatter than normal-matter or equal amounts of both matter and antimatter of the same and different species of particles where the statistically randomly longer-living normal-matter stable particles are retained and thus used as breeding stock to produce more longer-living normal-matter particles. And the process is repeated over and over, so as to gradually restock the cosmos with baryonic matter and cold dark matter as well as hot dark matter of Standard Model and non–Standard Model types.

The lifeboats might reconfigure a cosmos over cosmic time periods to take the form of a matter-antimatter–compatible universe, multiverse, forest, biosphere and the like by using accelerators and similar filtering mechanisms to convert future-phase change particles back into Standard Model and non–Standard Model present-era cold and hot matter and associated light-speed field quanta.

Propulsion methods based on accelerators which produce matter and antimatter electrons, muons, tauons, protons, alpha particles, higher-atomic-number nuclei, and ions, stable and unstable matter and antimatter exotic mesons and baryons, and the like for which the antimatter versions are accelerated into space away from the spacecraft while the normal-matter versions build up to produce a net electric charge on the spacecraft for propulsion where the statistically randomly longer-living normal-matter stable particles are retained and thus used as breeding stock to produce more longer-living normal-matter particles, and the process is repeated over and over.

Propulsion methods based on accelerators which produce matter and antimatter muons, tauons, and charged mesons and charged unstable baryons for which the normal-matter charged particles are filtered and accelerated in a first stage in cases where they live a long time in an accelerator and which are then used as feedstock to produce more matter and antimatter particles for which the antimatter versions are allowed to decay where the statistically randomly longer-living normal-matter stable particles are retained and thus used as breeding stock to produce more longer-living normal-matter particles. And the process is repeated over and over, and is thus used as breeding stock to produce more longer-living normal-matter particles. And the process is repeated over and over, filtered, and accelerated in a second stage in cases where they live a long time in an accelerator where the statistically randomly longer-living normal-matter stable particles are partially or fully retained and which are then used as feed-stock to produce more matter and antimatter particles for which the antimatter versions are allowed to decay where the statistically randomly longer-living normal-matter stable particles are retained and thus used as breeding stock to produce more longer-living normal-matter particles.

And the process is repeated over and over and are then used as feedstock to produce more matter and antimatter particles for which the antimatter versions are allowed to decay where the statistically randomly longer-living normal-matter stable particles are retained and thus used as breeding stock to produce more longer-living normal-matter particles. And the process is repeated over and over and are then used as feedstock to produce more matter and antimatter particles for which the antimatter versions are allowed to decay while the normal-matter versions are retained, filtered, and accelerated in a fifth stage … In cases where they live a long time in an accelerator and are then used as feedstock to produce more matter and antimatter particles for which the antimatter versions are allowed to decay where the statistically randomly longer-living normal-matter stable particles are retained and thus used as breeding stock to produce more longer-living normal-matter particles.

And the process is repeated over and over in a (n-1)th stage in cases where they live a long time in an accelerator where the statistically randomly longer-living normal-matter stable particles are partially or fully retained and are then used as feedstock to produce more matter and antimatter particles for which the antimatter versions are allowed to decay while the normal-matter versions are retained, filtered, and accelerated in an nth stage. Where *n* may be as great as an ensemble, an infinity scraper, an Aleph 0, a greater than one exponentated Aleph 0, an Aleph 1, a greater than one exponentiated Aleph 1, an Aleph 2, a greater than one exponentated Aleph 2, an Aleph 3, a greater than one exponentated Aleph 3, and so on. Onward ho! Forevermore!

Lifeboats and pods with cryogenically asleep humans for which the pods are designed to pass through cosmic eras before the next phase changes as well as the next and all future phase changes of our universe, multiverse, forest, biosphere, etc., where associated designs include the electronic charging of lifeboats for which positrons are stored in a location on the lifeboat and made to interact with background electrons for which the associated gamma-ray energy is absorbed and used to power accelerators that generate matter and antimatter protons, matter and antimatter deuterons, matter and antimatter alpha particles, matter and antimatter atomic nuclei, matter and antimatter ions, matter and antimatter mesons, matter and antimatter baryons, etc., where the statistically randomly longer-living antimatter stable particles are retained and thus used as breeding stock to produce more longer-living antimatter particles.

And the process is repeated over and over, and the antimatter stable particles are retained in a charge-specific location aboard the lifeboats, and the rest of the particles are inter-pair-annihilated or left to decay, but either way, the annihilation energy and decay energies are fed back into an accelerator to produce more matter and antimatter protons, matter and antimatter deuterons, matter and antimatter alpha particles, matter and antimatter atomic nuclei, matter and antimatter ions, matter and antimatter mesons, matter and antimatter baryons, etc., where the statistically randomly longer-living antimatter stable particles are retained and thus used as breeding stock to produce more longer-living antimatter particles.

And the process is repeated over and over, in a charge-specific location aboard the lifeboats, and the rest of the particles are inter-pair-annihilated or left to decay, but either way, the annihilation energy and decay energies are fed back into an accelerator to produce in a second stage more matter and antimatter protons, matter and antimatter deuterons, matter and antimatter alpha particles, matter and antimatter atomic nuclei, matter and antimatter ions, matter and antimatter mesons, matter and antimatter baryons, etc., where the statistically randomly longer-living antimatter stable particles are retained and thus used as breeding stock to produce more longer-living antimatter particles.

And the process is repeated over and over, and the antimatter stable particles are retained in a charge-specific location aboard the lifeboats, and the rest of the particles are inter-pair-annihilated or left to decay, but either way, the annihilation energy and decay energies are fed back into an accelerator to produce in a third stage more matter and antimatter protons, matter and antimatter deuterons, matter and antimatter alpha particles, matter and antimatter atomic nuclei, matter and antimatter ions, matter and antimatter mesons, matter and antimatter baryons, etc., where the statistically randomly longer-living antimatter stable particles are retained and thus used as breeding stock to produce more longer-living antimatter particles.

And the process is repeated over and over, and the antimatter stable particles are retained in a charge-specific location aboard the lifeboats, and the rest of the particles are inter-pair-annihilated or left to decay, but either way, the annihilation energy and decay energies are fed back into an accelerator to produce in a fourth stage more matter and antimatter protons, matter and antimatter deuterons, matter and antimatter alpha particles, matter and antimatter atomic nuclei, matter and antimatter ions, matter and antimatter mesons, matter and antimatter baryons, etc., where the statistically randomly longer-living antimatter stable particles are retained and thus used as breeding stock to produce more longer-living antimatter particles. And the process is repeated over and over, and thus used as breeding stock to produce more longer-living antimatter particles.

And the process is repeated over and over in a charge-specific location aboard the lifeboats, and the rest of the particles are inter-pair-annihilated or left to decay, but either way, the annihilation energy and decay energies are fed back into an accelerator to produce in a fifth stage more … matter and antimatter protons, matter and antimatter deuterons, matter and antimatter alpha particles, matter and antimatter atomic nuclei, matter and antimatter ions, matter and antimatter mesons, matter and antimatter baryons, etc., where the statistically randomly longer-living antimatter stable particles are retained and thus used as breeding stock to produce more longer-living antimatter particles.

And the process is repeated over and over, and the antimatter stable particles are retained in a charge-specific location aboard the lifeboats, and the rest of the particles are inter-pair-annihilated or left to decay, but either way, the annihilation energy and decay energies are fed back into an accelerator to produce in an (n-1)th stage more matter and antimatter protons, matter and antimatter deuterons, matter and antimatter alpha particles, matter and antimatter atomic nuclei, matter and antimatter ions, matter and antimatter mesons, matter and antimatter baryons, etc., where the statistically randomly longer-living antimatter stable particles are retained and thus used as breeding stock to produce more longer-living antimatter particles. And the process is repeated over and over, and the antimatter stable particles are retained in a charge-specific location aboard the lifeboats, and the rest of the particles are inter-pair-annihilated or left to decay, but either way, the annihilation energy and decay energies are fed back into an accelerator to produce in an nth stage more … and so on forever. Onward ho!

Such a mechanism and similar mechanisms may be able to produce more normal matter and antimatter or more antimatter than normal matter or equal amounts of both matter and antimatter of the same and different species of particles where the statistically randomly longer-living normal-matter stable particles are retained and thus used as breeding stock to produce more longer-living normal-matter particles. And the process is repeated over and over, so as to gradually restock the cosmos with baryonic matter and cold dark matter as well as hot dark matter of Standard Model and non–Standard Model types.

The lifeboats might reconfigure a cosmos over cosmic time periods to take the form of a matter-antimatter compatible universe, multiverse, forest, biosphere and the like by using accelerators and similar filtering mechanisms to convert future phase change particles back into Standard Model and non–Standard Model present-era cold and hot matter and associated light-speed field quanta.

Propulsion methods based on accelerators which produce matter and antimatter electrons, muons, tauons, protons, alpha particles, higher-atomic-number nuclei, and ions, stable and unstable matter and antimatter exotic mesons and baryons, and the like for which the antimatter versions are accelerated into space away from the spacecraft while the normal-matter versions build up to produce a net electric charge on the spacecraft for propulsion where the statistically randomly longer-living antimatter stable particles are retained and thus used as breeding stock to produce more longer-living normal-matter particles, and the process is repeated over and over.

Propulsion methods based on accelerators that produce matter and antimatter muons, tauons, and charged mesons and charged unstable baryons for which the antimatter charged particles are filtered and accelerated in a first stage in cases where they live a long time in an accelerator and which are then used as feedstock to produce more matter and antimatter particles for which the antimatter versions are allowed to decay where the statistically randomly longer-living antimatter stable particles are retained and thus used as breeding stock to produce more longer-living antimatter particles. And the process is repeated over and over, and is thus used as breeding stock to produce more longer-living antimatter particles. And the process is repeated over and over, the antimatter particles filtered and accelerated in a second stage in cases where they live a long time in an accelerator where the statistically randomly longer-living antimatter stable particles are partially or fully retained and are then used as feedstock to produce more matter and antimatter particles for which the normal-matter versions are allowed to decay where the statistically randomly longer-living antimatter stable particles are retained and thus used as breeding stock to produce more longer-living antimatter particles.

And the process is repeated over and over, where the antimatter particles are then used as feed-stock to produce more matter and antimatter particles for which the normal versions are allowed to decay where the statistically randomly longer-living antimatter stable particles are retained and thus used as breeding stock to produce more longer-living antimatter particles. And the process is repeated over and over, where the antimatter particles are then used as feedstock to produce more matter and antimatter particles for which the normal-matter versions are allowed to decay while the antimatter versions are retained, filtered, and accelerated in a fifth stage … in cases where they live a long time in an accelerator and are then used as feedstock to produce more matter and antimatter particles for which the normal-matter versions are allowed to decay, where the statistically randomly longer-living antimatter stable particles are retained and thus used as breeding stock to produce more longer-living antimatter particles.

And the process is repeated over and over in a (n-1)th stage in cases where they live a long time in an accelerator where the statistically randomly longer-living antimatter stable particles are partially or fully retained and are then used as feedstock to produce more matter and antimatter particles for which the normal-matter versions are allowed to decay while the antimatter versions are retained, filtered, and accelerated in an nth stage. Where *n* may be as great as an ensemble, an infinity scraper, an Aleph 0, a greater than one exponentated Aleph 0, an Aleph 1, a greater than one exponentated Aleph 1, an Aleph 2, a greater than one exponentated Aleph 2, an Aleph 3, a greater than one exponentated Aleph 3, and so on. Onward ho! Forevermore!

Chapter 10

Relatively Near-Term Antimatter Rocket Prospects

This chapter includes descriptions of pure matter-antimatter rocket or reactor-powered manned interstellar spacecraft scenarios. Antimatter is the most mass-specific energy-dense known fuel. When it's mixed with equal portions of its opposite or normal-matter fuel of the same species, pure energy is the result. Although some reactions have significant neutrino losses, it is conceivable that the neutrinos produced as part of the reaction products of some forms of matter-antimatter reactions could be harnessed for enhanced propulsion energy. Other forms of matter-antimatter fuel are ideal and result in the production of pure photonic exhaust, which can be readily directed in, or converted to, a collimated light-speed exhaust plume.

In this digression, we cover just a little bit of math. For those not desiring the brief mathematical equations, I ask you to see the computed values at the end of these equations and to consider them as factual to the number of significant digits provided.

For matter-antimatter reactor-powered, manned, interstellar spacecraft, or perhaps manned matter-antimatter rocket spacecraft, we will first assume a mass ratio of 10. Using the relativistic rocket equation:

$$\Delta v = C\ tanh\ [(I_{sp}/C)\ ln\ (M_0/M_1)] = 0.98\ C$$

For $M_0/M_1 = 100$:

$$\Delta v = C\ tanh\ [(I_{sp}/C)\ ln\ (100)] = 0.9998\ C$$

The Lorentz factor of 0.98 C is:

$$1/\{1 - [(v/C)^2]\}^{1/2} = 1/\{1 - [(0.98\ C/C)^2]\}^{1/2} = 5.0252$$

For v = 0.9998 C, the Lorentz factor is:

$$1/\{1 - [(0.9998\ C/C)^2]\}^{1/2} = 50.0025$$

The above examples assume that both the matter and antimatter portions of the fuel are carried along from the start of the mission and that the efficiency of the system is very close to 100 percent. Such an assumption is perhaps a tall order with current matter-antimatter rocket concepts.

We can go to further extremes on the relativistic rocket themes by speculating that antimatter rockets might derive their normal-matter reactants from the interstellar medium where the propulsion system efficiency would approach 100 percent, exactly, and where the I_{sp} therefore is greater than C.

Although the exhaust velocity cannot be greater than C, for the case where normal matter can be extracted from space in such a manner that the drag energy can be recycled, the momentum delivered to the spacecraft per unit of utilized onboard fuel is twice that obtainable for cases where both components are carried on board from the start. Specific impulse can be viewed as the quantity of momentum delivered to the spacecraft per unit of onboard fuel.

About 10^{15} watts could be collected in the case where 100 million membranous solar concentrators are deployed in a 1 AU solar orbit and where each solar concentrator has a capture area of 10,000 square meters. Each of these concentrators could have a high-power density PV cell that may be as much as 40 percent or more efficient. This would result in 400 TW of electrical power being generated. In one year, $[4 \times (10^{14})][3 \times (10^{7})]$ joules of electrical power may be produced or the equivalent of about 120 metric tons of matter converted into energy. Twelve metric tons of antimatter per year could be produced by the above systems if the antimatter can be produced with 20 percent efficiency. At nearly 100 percent efficiency, 60 metric tons per year could be produced. The antimatter generators would always produce an equal amount of normal matter except for reactions that violate CPT invariance since antimatter is always produced along with normal matter. However, the details of CPT invariance are another story and so are not included here.

Twelve thousand metric tons of antimatter could be generated per year at 20 percent efficiency from 1,000 such stations. Sixty thousand metric tons per year could be produced at nearly 100 percent efficiency. The antimatter generation could be ramped up several more orders of magnitude in order to produce millions, if not hundreds, of millions of tons of antimatter per year, provided we can develop a workable infrastructure. In one hundred years, this would amount to tens of billions of tons.

One caveat is the production of very low-cost and durable reflectors with current technology and cheap abundant materials or technology and materials to be developed that are suitably light-weight and robust in the environment of outer space.

My brother John and I have invented patented apparatuses that include, but are not limited to, very low-cost, high-mass specific-power-output inflatable reflectors made of durable, high-modulus, reflective, membranous materials. We managed to produce reflectors that have a mass specific power output on Earth's surface of up to 10 kilowatts/kilogram using 0.5 mil metalized Mylar or 0.5 mil metalized nylon. The method of manufacture involves efficient flat sheet manufacturing patterns using mainly 4, 6, or 8 sheets of thermally bonded, adhesively bonded, or otherwise-bonded materials. For our first prototypes, we used a clothes iron to thermally bond metalized polymer film-based toy balloon cutouts of various inner and outer radii. We were more than able to cook hot dogs to a char even in intermittent sunlight using the devices in a total of about fifteen minutes, or with about seven minutes of sunlight.

The mass-specific power yield of our reflectors will increase as thin film materials of greater strength are developed.

Some potential exists for using exotic, super-high-strength materials, for making the devices such as carbon nanotube membranous sheets of anywhere from a few nanometers in thickness to tens of nanometers in thickness, thus increasing the mass-specific power yield of the reflectors by four orders of magnitude. Other potential super-high-strength materials of construction include the following:

1. graphene oxide paper
2. boron-nitride nanotubes
3. graphene
4. carbon atom chains
5. diamond fibers
6. Beta carbon nitride fibers

Such apparatuses can also be useful as solar sails and beam sails. The thinner the sail, the lower the mass per unit area of sail, and, therefore, thinner sails can be used to capture more energy because they can be made larger than thicker sails of the same mass. There are some lower limits to sail thickness because sails that are too thin or that are not made of the proper materials would allow too much sunlight to pass through and thus result in a huge loss in efficiency.

All that would be required from the reflector material standpoint to collect 10^{15} watts with our current technology is 100 billion kilograms of material, or 100 million metric tons. Ten billion metric tons would suffice for 1,000 such stations. Perhaps the building and deployment of such reflectors would provide a carbon sink due to the carbon requirement for the production of any carbonaceous high-strength polymeric materials used to fabricate the concentrators. This could result in reduced atmospheric carbon, and thus a mitigated problem of global warming.

For reflectors made of carbon nanotube materials, or perhaps the even stronger carbon graphene, the combined mass required for the reflector portions of the 1,000 conjectured stations is only 10 million metric tons. A mere 10 million metric tons could be used to collect 10^{18} watts of solar energy.

Note that some mechanism for allowing the collection stations to remain in steady positions around the sun would be required. Perhaps some sort of angled adjustment procedure, ion or electron rocket thrust, electrodynamic-hydrodynamic-plasma-drive thrust, or other means can be used to keep the antimatter-generation stations in stable orbit about the sun. A very interesting mechanism would entail the deployment of a negative index of refraction type of material affixed to the solar radiation-concentrating stations. A negative refraction index material has a strange property by which such materials are pulled forward by impinging light instead of being pushed by the light. More will be said about negative refraction index materials again in later sections of this book.

Note that we achieved full stable deployment of our devices, which were usually only about one meter in diameter with a relative pressure of about 0.1 PSI or less. The larger the device, the lower the internal pressure can be to deploy the devices.

Another caveat is the ability to launch the collection stations. I think the problem is tractable this very century if we can get the hardware in solar orbit at 1 AU. At 0.1 AU, the required mass of the reflective materials drops by a hundredfold. However, low-cost and efficient access to solar orbit is needed in order to launch and deploy the systems at 1 AU.

What could be done with 60,000 metric tons of matter-antimatter fuel that would somehow be utilized with almost 100 percent efficiency? Assume a spacecraft with a final payload mass of 6,000 metric tons.

Using the relativistic rocket equation:

$$\Delta v = C\ tanh\ [(I_{sp}/C)\ ln\ (M_0/M_1)] = 0.98\ C$$

Where the mass ratio is equal to 10. This corresponds to a relativistic Lorentz factor of about 5.

For a mass ratio of 100, e.g., a fully fueled mass to dry weight of the vehicle of (60,000 metric tons)/(600 metric tons), Δv is equal to 0.9998 C. This corresponds to a relativistic Lorentz factor of 50.

Consider the case where only the antimatter fuel component is taken along from the start, perhaps in the form of anti-hydrogen ice. Consequently, the effective I_{sp} is greater than C for systems that operate at near-100-percent efficiency where the matter fuel is collected in route.

Assume that the human life expectancy can be augmented to 1,100 years in duration. A Lorentz factor of 5 would permit roughly 5,000 light-year trips for the original living crew: a Lorentz factor of 50, 50,000 light-year trips; a Lorentz factor of 500, 500,000 light-year trips; and a Lorentz factor of 5,000, 5 million light-year trips. A life expectancy of 1,100 years would permit the travelers to live out the remaining 100 years of their lives in relaxation and style on the beaches, lakesides, and mountain resorts on any of an untold number of beautiful worlds yet to be discovered.

I feel that provided an expansive and bold funding initiative could be established, we could launch such missions within the next two centuries in droves, and for missions that are limited to terminal Lorentz factors of, say, between 5 and 50, this very century, perhaps within the lifetimes of some of our children.

We need not use rocket thrust with the need to carry extra fuel in order to arrive safely at the points of destination. As indicated before, the spacecraft could be slowed by any of the previously listed electrodynamic braking mechanisms.

Magnetic braking could be accomplished by deploying a large superconducting coil that would build up extremely high current as it passed through the interstellar or intergalactic magnetic fields, thus producing a magnetic field to react against the space-based fields in a drag-inducing manner. Electrodynamic-hydrodynamic-plasma braking could be accomplished by a reverse interstellar ramjet type of mechanism. Magnetic-sail-based braking could be accomplished through the deployment of a magnetic bottle consisting of plasma deployed around the ship and held fixed by electrodynamic fields.

As with most relativistic rocket concepts, the above electrodynamic deceleration mechanisms can be augmented if needed by reverse rocket thrust of the same form(s) as that used to accelerate the spacecraft.

I think of the meager infrastructure that the New World settlers had here in what would become the United States, and now we have superhighways, one hundred story buildings, hundreds of airports, a roughly three hundred advanced ship navy, dozens of large cities, and the list goes on and on.

I have a gut feeling that we can produce vast quantities of antimatter from the sun, and we still do not know what the properties of bulk quantities of antimatter are due to CPT violation that occurs in certain particle pair creations. We have not yet determined whether antimatter possesses antigravity as Frank Close speculates in his book entitled *Antimatter* (Close, 2009) .

One way or another, I feel that antimatter sequestration from natural sources, such as within the magnetosphere of the gas giant planets within our solar system, or its artificial production in bulk quantities, will prove extremely useful for our manned interstellar missions within the next two centuries, and even bolder missions beyond.

Chapter 11

Strange Matter Reactors and Bombs and Other Things Nice

This chapter includes a description of highly conjectural forms of QCD/QED reactors and bombs. Such reactors and bombs would be used to power highly relativistic starships by converting a large-fuel mass at least partially into energy and/or exothermic conversion of interstellar ordinary baryonic matter into strange, charmed, and/or bottom quark–based matter. Herein, attractive use of Times New Roman font enlarged to 14 point size made for clarity.

Now that the nuclear and the eventual exotic QCD energy applications genie is permanently out of the bottle and exotic QCD energy applications loom on the horizon,, how might such applications be used to reach the stars?

Below is a brief summary of thoughts on how exotic applications of nuclear and subnuclear reactions can open up the cosmos for colonization and exploration by humanity. The technologies considered have likely already been pondered and studied to some degree, but if not, they definitely will be in the future.

Imagine a technology that could convert the elements in the periodic table atomic to other exotic forms of baryonic matter releasing prodigious quantities of energy in a mass-specific reaction energy that is higher than that released by nuclear fusion. It has been proposed that certain species of some stable strangelets that might be produced in particle accelerators may be able to convert periodic table atoms into strange matter. An unfortunate consequence of this may be a runaway reaction that could cause the planet to be exothermically converted into strange matter or matter comprised mostly of strange quarks with some additional species of quarks mixed in.

Stable catalytic strangelets may be ideal for powering future interstellar ramjet starships (ISRs). Imagine a large mass of stable strange matter for which interstellar and intergalactic matter would be funneled into the intake of a reaction chamber, where the intake mass would be converted to quickly decaying strangelets, which would convert even more of the intake mass into energy through decay processes. The drag energy in the form of heat could be recycled to power photon, ion, electron, positron, muon, antimuon, tauon, antitauon, proton, antiproton, exotic charged meson, and/or exotic charged baryon rockets. The stable or metastable strangelets produced could be used as an accelerated thrust stream. A starship using this form of propulsion might accelerate and achieve high Lorentz factors and/or near light speeds in any hyperspatial dimension where periodic table atomic matter or other convertible baryonic matter is present.

Let us consider a scenario where a species of strangelet would not decay on average *n* percent of the time before making contact with an ordinary baryon or atomic nucleus and where the probability of the conversion of the baryon or atomic nucleus is 100 percent. The average probability of the initial strangelet converting an ordinary nucleon or atomic nucleus to another catalytic strangelet is n/100. After *m* time steps where, for which each time step, the probability of the average catalytic strangelet converting a baryon or nucleus it first contacts is n/100, the number of catalyzed baryons or nuclei is equal to in first order:

$$N = G_{strange}\{n/100\}_1 + G_{strange}\{n/100\}_1 \{n/100\}_2 + G_{strange}\{n/100\}_2 \{n/100\}_2\{n/100\}_3 + \ldots + G_{strange}\{n/100\}_1 \{n/100\}_2 \ldots \{ n/100\}_m$$

Where *m* is the number of time steps for the overall reaction extinction and $G_{strange}$ is the initial population of strangelets.

Another way of presenting reaction wave propagation is to consider the average lifetime of the strangelets and the average distance traveled by the strangelets relative to the mean free path for strangelet and nucleon or atomic nuclei contact. For cases where the average strangelet would not decay with a probability of n/100 before completing the travel of its mean free path, the number of catalyzed baryons or nuclei is also equal to in first order:

$$N = G_{strange}\{n/100\}_1 + G_{strange}\{n/100\}_2 \{n/100\}_2 + G_{strange}\{n/100\}_2 \{n/100\}_2\{n/100\}_3 + \ldots + G_{strange}\{n/100\}_1 \{n/100\}_2 \ldots \{ n/100\}_m$$

Where *m* is the number of time steps for the overall reaction extinction.

Considering the length of time, t_m, traveled per mean free path where the average nondecay probability is n/100 for each mean path traveled, the velocity of the strangelet is L/t_m, where the velocity of the strangelet is negligibly relativistic.

For cases where the strangelets produced are relativistic, the average distance traveled by such strangelets before it decays is given by:

$$D = [L/t_m](\gamma)\ (n)/(100) = [L/t_m]\{1/\{1 - [(v/C)^2]\}^{1/2}\}(n)/(100)$$

Now, applying the exponetial decay formula:

$$N(t) = N_0\, e^{-t,strange/Tstrange} = N_0\,(1/2)e^{-t,strange/t,strange1/2}$$

Where T is the mean lifetime of a particle (such as an atom or a subatomic particle), $t_{1/2}$ is the half-life of the particle, and N_0 is the starting population. $N(t)$ is the quantity that still remains and has not yet decayed after a time t.

Therefore the half-life is expressed as follows:

$$t_{1/2} = \ln(2)/\lambda = T \ln(2)$$

Here, λ is a positive number called the decay constant of the decaying quantity.

For a strangelet having a probability of non-decay along a mean path length of P = n/100, the number of strangelets out of a starting population of $N_0 = G_{strange}$ that contact a baryon or atomic nuclei is $N_0\{n/100\}$.

Where $\{n/100\} = e^{-t,strange/Tstrange} = (1/2)\, e^{-t,strange/t,strange1/2}$

Therefore, the number of catalyzed baryons becomes:

$$N = G_{strange}\{e^{-t,strange/Tstrange}\}_1 + G_{strange}\{e^{-t,strange/Tstrange}\}\{e^{-t,strange/Tstrange}\}_2 + G_{strange}\, e^{-t,strange/Tstrange}\}\{e^{-t,strange/Tstrange}\}\{e^{-t,strange/Tstrange}\}_3 + \ldots + G_{strange}\{e^{-t,strange/Tstrange}\}\{e^{-t,strange/Tstrange}\}\ldots\{e^{-t,strange/Tstrange}\}_m$$

$$= G_{strange}\{(1/2)\, e^{-t,strange/t,strange1/2}\}_1 + G_{strange}\{(1/2)\, e^{-t,strange/t,strange1/2}\}\{(1/2)\, e^{-t,strange/t,strange1/2}\}_2 + G_{strange}\{(1/2)\, e^{-t,strange/t,strange1/2}\}\{(1/2)\, e^{-t,strange/t,strange1/2}\}\{(1/2)\, e^{-t,strange/t,strange1/2}\}_3 + \ldots + G_{strange}\{(1/2)\, e^{-t,strange/t,strange1/2}\}\{(1/2)\, e^{-t,strange/t,strange1/2}\}\ldots\{(1/2)\, e^{-t,strange/t,strange1/2}\}_m,$$

Where t is the travel time of the strangelets for the mean free path for strangelet to ordinary baryon contact.

Now, if $t_{1/2}$ and T are background stationary frames and the strangelets are relativistic, the number of catalyzed baryons becomes in first order:

$$N = G_{strange}\{e^{-t,strange/(Tstrange\gamma}\}_1 + G_{strange}\{e^{-t,strange/(Tstrange\gamma)}\}\{e^{-t,strange/(Tstrange\gamma)}\}_2 + G_{strange}\{e^{-t,strange/(Tstrange\gamma)}\}\{e^{-t,strange/(Tstrange\gamma)}\}\{e^{-t,strange/(Tstrange\gamma)}\}_3 + \ldots + G_{strange}\{e^{-t,strange/(Tstrange\gamma)}\}\{e^{-t,strange/(Tstrange\gamma)}\}\ldots\{e^{-t,strange/(Tstrange\gamma)}\}_m$$

$$= G_{strange}\{(1/2)\quad e^{-t,strange/(\gamma t,strange1/2)}\}_1 + G_{strange}\{(1/2)\quad e^{-t,strange/(\gamma t,strange1/2)}\}\{(1/2)\quad e^{-t,strange/(\gamma t,strange1/2)}\}_2 + G_{strange}\{(1/2)\quad e^{-}$$

$^{t,strange/(\gamma t,strange1/2)}\}\{(1/2)\ e^{-t,strange/(\gamma t,strange1/2)}\}\{(1/2)\ e^{-t,strange/(\gamma t,strange1/2)}\}_3 + \ldots + G_{strange}\{(1/2)\ e^{-t,strange/(\gamma t,strange1/2)}\}\{(1/2)\ e^{-t,strange/(\gamma t,strange1/2)}\} \ldots \{(1/2)\ e^{-t,strange/(\gamma t,strange1/2)}\}_m$

For cases where the average strangelet would not decay with a probability of n/100 after traveling the mean free path, and the probability of baryon or nuclei conversion upon contact is k/100, but where the strangelets are always destroyed when contacting the nuclei or converted into a non-strange-transmutative form with a probability of unity, the number of catalyzed baryons or nuclei is equal to in first order:

$$N = G_{strange}\{n/100\}\{k/100\}_1 + G_{strange}\{n/100\}\{k/100\}_1\{n/100\}\{k/100\}_2 + G_{strange}\{n/100\}\{k/100\}_1\ \{n/100\}\{k/100\}_2\{n/100\}\{k/100\}_3 + \ldots + G_{strange}\{n/100\}\{k/100\}_1\{n/100\}\{k/100\}_2 \ldots \{n/100\}\{k/100\}_m,$$

Here, *m* is the number of time steps for the overall reaction extinction.

Considering the length of time, t_m, traveled per mean free path where the average non-decay probability is n/100 for each mean path traveled, the velocity of the strangelet is L/t_m where the velocity of the strangelet is negligibly relativistic.

For cases where the strangelets produced are relativistic, the distance traveled by a given strangelet before it decays is given by:

$$[L/t_m](\gamma) = [L/t_m]\{1/\{1 - [(v/C)^2]\}^{1/2}\}$$

Once again, $N(t) = N_0\ e^{-t,strange/Tstrange} = N_0\ (1/2)e^{-t,strange/t,strange1/2}$, where T is the mean lifetime of a particle (such as an atom or subatomic particle), $t_{1/2}$ is the half-life of the particle, and N_0 is the starting population. $N(t)$ is the quantity that still remains and has not yet decayed after a time t, and half-life $t_{1/2} = \ln(2)/\lambda = T \ln(2)$, where λ is a positive number called the decay constant of the decaying quantity.

In the case of a strangelet species having a probability of non-decay along a mean path length of P = n/100, the number of strangelets out of a starting population of $N_0 = G_{strange}$ that contact a baryon or atomic nuclei is $N_0\{n/100\}$ where $\{n/100\} = e^{-t,strange/Tstrange} = (1/2)\ e^{-t,strange/t,strange1/2}$. Therefore, the number of catalyzed baryons becomes in first order:

$$N = G_{strange}\{e^{-t,strange/Tstrange}\}\{k/100\}_1 + G_{strange}\{e^{-t,strange/Tstrange}\}\{k/100\}_1\{e^{-t,strange/Tstrange}\}\{k/100\}_2 + G_{strange}\ \{e^{-t,strange/Tstrange}\}\{k/100\}_1\ \{e^{-t,strange/Tstrange}\}\{k/100\}_2\ \{e^{-t,strange/Tstrange}\}\{k/100\}_3 + \ldots + G_{strange}\{\ e^{-}$$

$$^{t,strange/Tstrange}\}\{k/100\}_1\{\ e^{-t,strange/Tstrange}\}\{k/100\}_2\ \ldots\{\ e^{-t,strange/Tstrange}\}\{k/100\}_m$$

$$= G_{strange}\{(1/2)\ e^{-t,strange/t,strange1/2}\}\{k/100\}_1 + G_{strange}\{(1/2)\ e^{-t,strange/t,strange1/2}\}\{k/100\}_1\{(1/2)\ e^{-t,strange/t,strange1/2}\}\{k/100\}_2 + G_{strange}\{(1/2)\ e^{-t,strange/t,strange1/2}\}\{k/100\}_1\{(1/2)\ e^{-t,strange/t,strange1/2}\}\{k/100\}_2\ \{(1/2)\ e^{-t,strange/t,strange1/2}\}\{k/100\}_3 + \ldots + G_{strange}\{(1/2)\ e^{-t,strange/t,strange1/2}\}\{k/100\}_1\{(1/2)\ e^{-t,strange/t,strange1/2}\}\{k/100\}_2\ldots\{(1/2)\ e^{-t,strange/t,strange1/2}\}\{k/100\}_m$$

If $t_{1/2}$ and T are background stationary frames and the strangelets are relativistic, the number of catalyzed baryons becomes in first order:

$$N = G_{strange}\{e^{-t,strange/(Tstrange\gamma}\}\{k/100\}_1 + G_{strange}\{e^{-t,strange/(Tstrange\gamma)}\}\{k/100\}_1\{\ e^{-t,strange/(Tstrange\gamma)}\}\{k/100\}_2 + G_{strange}\{\ e^{-t,strange/(Tstrange\gamma)}\}\{k/100\}_1\{\ e^{-t,strange/(Tstrange\gamma)}\}\{k/100\}_2\{\ e^{-t,strange/(Tstrange\gamma)}\}\{k/100\}_3 + \ldots + G_{strange}\{\ e^{-t,strange/(Tstrange\gamma)}\}\{k/100\}_1\{\ e^{-t,strange/(Tstrange\gamma)}\}\{k/100\}_2\ \ldots\{\ e^{-t,strange/(Tstrange\gamma)}\}\{k/100\}_m$$

$$= G_{strange}\{(1/2)\ e^{-t,strange/(\gamma t,strange1/2)}\}\ \{k/100\}_1 + G_{strange}\{(1/2)\ e^{-t,strange/(\gamma t,strange1/2)}\}\{k/100\}_1\{(1/2)\ e^{-t,strange/(\gamma t,strange1/2)}\}\ \{k/100\}_2 + G_{strange}\{(1/2)\ e^{-t,strange/(\gamma t,strange1/2)}\}\{k/100\}_1\{(1/2)\ e^{-t,strange/(\gamma t,strange1/2)}\}\{k/100\}_2\{(1/2)\ e^{-t,strange/(\gamma t,strange1/2)}\}\{k/100\}_3 + \ldots + G_{strange}\{(1/2)\ e^{-t,strange/(\gamma t,strange1/2)}\}\{k/100\}_1\{(1/2)\ e^{-t,strange/(\gamma t,strange1/2)}\}\{k/100\}_2\ldots\{(1/2)\ e^{-t,strange/(\gamma t,strange1/2)}\}\ \{k/100\}_m$$

Let us consider a scenario where a species, i, of strangelet would not decay on average *n* percent of the time before making contact with an ordinary baryon or atomic nucleus, and where the probability of the conversion of the baryon or atomic nucleus is 100 percent. The average probability of the initial strangelet converting an ordinary nucleon or atomic nucleus to another catalytic strangelet is $n_{,i}/100$. After *m* time steps where, for which each time step, the probability of the average catalytic strangelet converting a baryon or nucleus in its first contacts is $n_{,i}/100$, the number of catalyzed baryons or nuclei is equal to in first order:

$$N_{,i} = G_{strange,i}\{n_{,i}/100\}_1 + G_{strange,i}\{n_i/100\}_1\ \{n_{,i}/100\}_2 + G_{strange,\ i}\{n_{,i}/100\}_2\ \{n_{,i}/100\}_2\{n_{,i}/100\}_3 + \ldots + G_{strange,i}\{n_{,i}/100\}_1\ \{n_{,i}/100\}_2\ldots\{\ n_{,i}/100\}_{m,i}$$

Where $m_{,i}$ is the number of time steps for the overall reaction extinction and $G_{strange,i}$ is the initial population of strangelets.

So, for M species of strangelets, where each species has its unique mean free-path, initial population, and number of time steps for overall reaction extinctions, the number of catalyzed baryons or nuclei is equal to in first order:

$$\sum(i = 1; i = M) N_i = \sum(i = 1; i = M) \{G_{strange,i}\{n_{,i}/100\}_1 + G_{strange,i}\{n_{,i}/100\}_1 \{n_{,i}/100\}_2 + G_{strange, i}\{n_{,i}/100\}_2 \{n_{,i}/100\}_2\{n_{,i}/100\}_3 + \ldots + G_{strange,i}\{n_{,i}/100\}_1 \{n_{,i}/100\}_2 \ldots\{ n_{,i}/100\}_{m,i}\}$$

Where the average yield energy per nucleon or nucleus conversion caused by transformation by particles if the ith species is E_i;

The total yield energy for a given QCD bomb is in first order:

$$\sum(i = 1; i = M) [(N_i)(E_i)] = \sum(i = 1; i = M) (E_i) \{\{G_{strange,i}\{n_{,i}/100\}_1 + G_{strange,i}\{n_{,i}/100\}_1 \{n_{,i}/100\}_2 + G_{strange, i}\{n_{,i}/100\}_2 \{n_{,i}/100\}_2\{n_{,i}/100\}_3 + \ldots + G_{strange,i}\{n_{,i}/100\}_1 \{n_{,i}/100\}_2 \ldots\{ n_{,i}/100\}_{m,i}\}\}$$

The time averaged explosive power of a given QCD bomb in the ship frame is in first order:

$$<d\{\sum(i = 1; i = M) [(N_i)(E_i)]\} /dt_{ship} > = <d\{\sum(i = 1; i = M) (E_i)\{\{G_{strange,i}\{n_{,i}/100\}_1 + G_{strange,i}\{n_{,i}/100\}_1 \{n_{,i}/100\}_2 + G_{strange, i}\{n_i/100\}_2 \{n_{,i}/100\}_2\{n_{,i}/100\}_3 + \ldots + G_{strange,i}\{n_{,i}/100\}_1 \{n_{,i}/100\}_2 \ldots\{ n_{,i}/100\}_{m,i}\}\}/dt_{ship} >$$

Another way of presenting reaction wave propagation is to consider the average lifetime of the strangelets and the average distance traveled by the strangelets relative to the mean free path for strangelet and nucleon or atomic nuclei contact. For cases where the strangelet of the ith species of strangelets would not decay with a probability of $n_{,i}/100$ before completing the travel of its mean free path, the number of catalyzed baryons or nuclei is also equal to in first order:

$$N_i = G_{strange,i}\{n_{,i}/100\}_1 + G_{strange,i}\{n_{,i}/100\}_1 \{n_{,i}/100\}_2 + G_{strange, i}\{n_{,i}/100\}_2 \{n_{,i}/100\}_2\{n_{,i}/100\}_3 + \ldots + G_{strange,i}\{n_{,i}/100\}_1 \{n_{,i}/100\}_2 \ldots\{ n_{,i}/100\}_{m,i}$$

Where $m_{,i}$ is the number of time steps for the overall reaction extinction for the ith species of strangelet.

So, for M species of strangelets, where each species has its unique mean free-path, initial population, and number of time steps for overall reaction extinctions, the number of catalyzed baryons or nuclei is equal to in first order:

$$\sum(i = 1; i = M) N_i = \sum(i = 1; i = M) \{G_{strange,i}\{n_{,i}/100\}_1 + G_{strange,i}\{n_{,i}/100\}_1 \{n_{,i}/100\}_2 + G_{strange,\ i}\{n_{,i}/100\}_2 \{n_{,i}/100\}_2\{n_{,i}/100\}_3 + \ldots + G_{strange,i}\{n_{,i}/100\}_1 \{n_{,i}/100\}_2 \ldots\{ n_{,i}/100\}_{m,i}\}$$

Where the average yield energy per nucleon or nucleus conversion caused by transformation by particles if the ith species is E_i:

The total yield energy for a given QCD bomb is in first order:

$$\sum(i = 1; i = M) [(N_i)(E_i)] = \sum(i = 1; i = M) (E_i)\{\{G_{strange,i}\{n_{,i}/100\}_1 + G_{strange,i}\{n_{,i}/100\}_1 \{n_{,i}/100\}_2 + G_{strange,\ i}\{n_{,i}/100\}_2 \{n_{,i}/100\}_2\{n_{,i}/100\}_3 + \ldots + G_{strange,i}\{n_{,i}/100\}_1 \{n_{,i}/100\}_2 \ldots\{ n_{,i}/100\}_{m,i}\}\}$$

The time averaged explosive power of a given QCD bomb in the ship frame is in first order:

$$<d\{\sum(i = 1; i = M) [(N_i)(E_i)]\} /dt_{ship} > = <d\{\sum(i = 1; i = M) (E_i)\{\{G_{strange,i}\{n_{,i}/100\}_1 + G_{strange,i}\{n_{,i}/100\}_1 \{n_{,i}/100\}_2 + G_{strange,\ i}\{n_{,i}/100\}_2 \{n_{,i}/100\}_2\{n_{,i}/100\}_3 + \ldots + G_{strange,i}\{n_{,i}/100\}_1 \{n_{,i}/100\}_2 \ldots\{ n_{,i}/100\}_{m,i}\}\}\}/dt_{ship} >$$

Considering the length of time, $t_{m,i}$, traveled per mean free path where the average nondecay probability is $n_{,i}/100$ for each mean path traveled, the velocity of the strangelet is $L_i/t_{m.i}$ where the velocity of the strangelet is negligibly relativistic for the ith species of strangelets.

For cases where the ith species of strangelets produced are relativistic, the average distance traveled by such strangelets before it decays is given by:

$$D_i = [L_i/t_{m,i}](\gamma_{,i}) (n_{,i})/(100) = [L_i/t_{m,i}]\{1/\{1 - [(v_{,i}/C)^2]\}^{1/2}\}(n_{,i})/(100)$$

Now, applying the exponetial decay formula:

$$N_i(t_i) = N_{0i}\, e^{-t,strange.i/Tstrange,i} = N_{0i}\, (1/2)e^{-t,strange,i/t,strange,i1/2}$$

Where T_i is the mean lifetime of a particle (such as an atom or subatomic particle), $t_{i1/2}$ is the half-life of the particle, and N_{0i} is the starting population. $N_i(t_i)$ is the quantity that still remains and has not yet decayed after a time t_i.

Therefore the half-life is expressed as follows:

$$t_{i1/2} = \ln(2)/\lambda_i = T_i \ln(2)$$

Here, λ_i is a positive number called the decay constant of the decaying quantity.

For a strangelet of the ith strangelet species having a probability of nondecay along a mean path length of $P_i = n_{,i}/100$, the number of strangelets out of a starting population of $N_{0i} = G_{strange,i}$ that contact a baryon or atomic nuclei is

$$N_{0i}\{n_{,i}/100\} \text{ where } \{n_{,i}/100\} = e^{-t,strange,i/Tstrange,i} = (1/2)\, e^{-t,strange,i/t,strange,i1/2}$$

Therefore, the number of catalyzed baryons becomes in first order:

$$N_i = G_{strange,i}\{e^{-t,strange,i/Tstrange,i}\}_1 + G_{strange,i}\{e^{-t,strange,i/Tstrange,i}\}\{e^{-t,strange,i/Tstrange,i}\}_2 + G_{strange,i}\, e^{-t,strange,i/Tstrange,i}\}\{e^{-t,strange,i/Tstrange,i}\}\{e^{-t,strange,i/Tstrange,i}\}_3 + \ldots + G_{strange,i}\{e^{-t,strange,i/Tstrange,i}\}\{e^{-t,strange,i/Tstrange,i}\}\ldots\{e^{-t,strange,i/Tstrange,i}\}_{m,i}$$

$$= G_{strange,i}\{(1/2)\, e^{-t,strange,i/t,strange,i1/2}\}_1 + G_{strange,i}\{(1/2)\, e^{-t,strange,i/t,strange,i1/2}\}\{(1/2)\, e^{-t,strange,i/t,strange,i1/2}\}_2 + G_{strange,i}\{(1/2)\, e^{-t,strange,i/t,strange,i1/2}\}\{(1/2)\, e^{-t,strange,i/t,strange,i1/2}\}\{(1/2)\, e^{-t,strange,i/t,strange,i1/2}\}_3 + \ldots + G_{strange,i}\{(1/2)\, e^{-t,strange,i/t,strange,i1/2}\}\{(1/2)\, e^{-t,strange,i/t,strange,i1/2}\}\ldots\{(1/2)\, e^{-t,strange,i/t,strange,i1/2}\}_{m,i},$$

Where t_i is the travel time of the ith species of strangelets for the mean free path for strangelet to ordinary baryon contact.

Where the average yield energy per nucleon or nucleus conversion caused by transformation by particles if the ith species is E_i:

The total yield energy for a given QCD bomb is in first order:

$$\sum(i = 1;\ i = M)\ [(N_i)(E_i)] = \sum(i = 1;\ i = M)\ (E_i)\ \{G_{strange,i}\{e^{-t,strange,i/Tstrange,i}\}_1 + G_{strange,i}\{e^{-t,strange,i/Tstrange,i}\}\{e^{-t,strange,i/Tstrange,i}\}_2 + G_{strange,i}\, e^{-t,strange,i/Tstrange,i}\}\{e^{-t,strange,i/Tstrange,i}\}\{e^{-t,strange,i/Tstrange,i}\}_3 + \ldots + G_{strange,i}\{e^{-t,strange,i/Tstrange,i}\}\{e^{-t,strange,i/Tstrange,i}\}\ldots\{e^{-t,strange,i/Tstrange,i}\}_{m,i}\}$$

$$= \sum(i = 1;\ i = M)\ (E_i)\{G_{strange,i}\{(1/2)\, e^{-t,strange,i/t,strange,i1/2}\}_1 + G_{strange,i}\{(1/2)\, e^{-t,strange,i/t,strange,i1/2}\}\{(1/2)\, e^{-t,strange,i/t,strange,i1/2}\}_2 + G_{strange,i}\{(1/2)\, e^{-t,strange,i/t,strange,i1/2}\}\{(1/2)\, e^{-t,strange,i/t,strange,i1/2}\}\{(1/2)\, e^{-t,strange,i/t,strange,i1/2}\}_3 + \ldots + G_{strange,i}\{(1/2)\, e^{-t,strange,i/t,strange,i1/2}\}\{(1/2)\, e^{-t,strange,i/t,strange,i1/2}\}\ldots\{(1/2)\, e^{-t,strange,i/t,strange,i1/2}\}_{m,i}\},$$

The time averaged explosive power of a given QCD bomb in the ship frame is in first order:

$$<d\{\sum(i = 1; i = M) [(N_i)(E_i)]\}/dt_{ship}> = <d\{\sum(i = 1; i = M) (E_i) \{G_{strange,i}\{e^{-t,strange,i/Tstrange,i}\}_1 + G_{strange,i}\{e^{-t,strange,i/Tstrange,i}\} \{e^{-t,strange,i/Tstrange,i}\}_2 + G_{strange,i}\, e^{-t,strange,i/Tstrange,i}\} \{e^{-t,strange,i/Tstrange,i}\} \{ e^{-t,strange,i/Tstrange,i}\}_3 + \ldots + G_{strange,i}\{ e^{-t,strange,i/Tstrange,i}\} \{ e^{-t,strange,i/Tstrange,i}\} \ldots\{ e^{-t,strange,i/Tstrange,i}\}_{m,i}\}\}/dt_{ship} >$$

$$= <d\{\sum(i = 1; i = M) (E_i)\{G_{strange,i}\{(1/2) e^{-t,strange,i/t,strange,i1/2}\}_1 + G_{strange,i}\{(1/2) e^{-t,strange,i/t,strange,i1/2}\} \{(1/2) e^{-t,strange,i/t,strange,i1/2}\}_2 + G_{strange,i}\{(1/2) e^{-t,strange,i/t,strange,i1/2}\} \{(1/2) e^{-t,strange,i/t,strange,i1/2}\} \{(1/2) e^{-t,strange,i/t,strange,i1/2}\}_3 + \ldots + G_{strange,i}\{(1/2) e^{-t,strange,i/t,strange,i1/2}\} \{(1/2) e^{-t,strange,i/t,strange,i1/2}\} \ldots\{(1/2) e^{-t,strange,i/t,strange,i1/2}\}_{m,i}\}\}/dt_{ship}>$$

Now, if $t_{i1/2}$ and T_i are background stationary frames and the ith species of strangelets are relativistic, the number of catalyzed baryons becomes in first order:

$$N = G_{strange,i}\{e^{-t,strange,i/(Tstrange,i\gamma}\}_1 + G_{strange,i}\{e^{-t,strange,i/(Tstrange,i\gamma)}\}\{ e^{-t,strange,i/(Tstrange,i\gamma)}\}_2 + G_{strange,i}\{ e^{-t,strange,i/(Tstrange,i\gamma)}\}\{ e^{-t,strange,i/(Tstrange,i\gamma)}\}\{ e^{-t,strange,i/(Tstrange,i\gamma)}\}_3 + \ldots + G_{strange,i}\{ e^{-t,strange,i/(Tstrange,i\gamma)}\}\{ e^{-t,strange,i/(Tstrange,i\gamma)}\} \ldots\{ e^{-t,strange,i/(Tstrange,i\gamma)}\}_{m,i}$$

$$= G_{strange,i}\{(1/2) e^{-t,strange,i/(\gamma t,strange,i1/2)}\}_1 + G_{strange,i}\{(1/2) e^{-t,strange,i/(\gamma t,strange,i1/2)}\}\{(1/2) e^{-t,strange,i/(\gamma t,strange,i1/2)}\}_2 + G_{strange,i}\{(1/2) e^{-t,strange,i/(\gamma t,strange,i1/2)}\}\{(1/2) e^{-t,strange,i/(\gamma t,strange,i1/2)}\}\{(1/2) e^{-t,strange,i/(\gamma t,strange,i1/2)}\}_3 + \ldots + G_{strange,i}\{(1/2) e^{-t,strange,i/(\gamma t,strange,i1/2)}\}\{(1/2) e^{-t,strange,i/(\gamma t,strange,i1/2)}\} \ldots\{(1/2) e^{-t,strange,i/(\gamma t,strange,i1/2)}\}_{m,i}$$

Where the average yield energy per nucleon or nucleus conversion caused by transformation by particles if the ith species is E_i:

The total yield energy for a given QCD bomb is in first order:

$$\sum(i = 1; i = M) [(N_i)(E_i)] = \sum(i = 1; i = M) (E_i) \{G_{strange,i}\{e^{-t,strange,i/(Tstrange,i\gamma}\}_1 + G_{strange,i}\{e^{-t,strange,i/(Tstrange,i\gamma)}\}\{ e^{-t,strange,i/(Tstrange,i\gamma)}\}_2 + G_{strange,i}\{ e^{-t,strange,i/(Tstrange,i\gamma)}\}\{ e^{-t,strange,i/(Tstrange,i\gamma)}\}\{ e^{-t,strange,i/(Tstrange,i\gamma)}\}_3 + \ldots + G_{strange,i}\{ e^{-t,strange,i/(Tstrange,i\gamma)}\}\{ e^{-t,strange,i/(Tstrange,i\gamma)}\} \ldots\{ e^{-t,strange,i/(Tstrange,i\gamma)}\}_{m,i}\}$$

$= \sum(i = 1; i = M) (E_{,i}) \{G_{strange,i}\{(1/2) e^{-t,strange,i/(\gamma t,strange,i1/2)}\}_1 + G_{strange,i}\{(1/2) e^{-t,strange,i/(\gamma t,strange,i1/2)}\}\{(1/2) e^{-t,strange,i/(\gamma t,strange,i1/2)}\}_2 + G_{strange,i}\{(1/2) e^{-t,strange,i/(\gamma t,strange,i1/2)}\}\{(1/2) e^{-t,strange,i/(\gamma t,strange,i1/2)}\}\{(1/2) e^{-t,strange,i/(\gamma t,strange,i1/2)}\}_3 + \ldots + G_{strange,i}\{(1/2) e^{-t,strange,i/(\gamma t,strange,i1/2)}\}\{(1/2) e^{-t,strange,i/(\gamma t,strange,i1/2)}\} \ldots \{(1/2) e^{-t,strange,i/(\gamma t,strange,i1/2)}\}_{m,i}\}$

The time-averaged explosive power of a given QCD bomb in the ship frame is in first order:

$<d\{\sum(i = 1; i = M) [(N_i)(E_i)]\}/dt> = <d\{\sum(i = 1; i = M) (E_i) \{G_{strange,i}\{e^{-t,strange,i/(Tstrange,i\gamma}\}_1 + G_{strange,i}\{e^{-t,strange,i/(Tstrange,i\gamma)}\}\{e^{-t,strange,i/(Tstrange,i\gamma)}\}_2 + G_{strange,i}\{e^{-t,strange,i/(Tstrange,i\gamma)}\}\{e^{-t,strange,i/(Tstrange,i\gamma)}\}\{e^{-t,strange,i/(Tstrange,i\gamma)}\}_3 + \ldots + G_{strange,i}\{e^{-t,strange,i/(Tstrange,i\gamma)}\}\{e^{-t,strange,i/(Tstrange,i\gamma)}\} \ldots \{e^{-t,strange,i/(Tstrange,i\gamma)}\}_{m,i}\}\}/dt_{ship}>$

$= <d\{\sum(i = 1; i = M) (E_i) \{G_{strange,i}\{(1/2) e^{-t,strange,i/(\gamma t,strange,i1/2)}\}_1 + G_{strange,i}\{(1/2) e^{-t,strange,i/(\gamma t,strange,i1/2)}\}\{(1/2) e^{-t,strange,i/(\gamma t,strange,i1/2)}\}_2 + G_{strange,i}\{(1/2) e^{-t,strange,i/(\gamma t,strange,i1/2)}\}\{(1/2) e^{-t,strange,i/(\gamma t,strange,i1/2)}\}\{(1/2) e^{-t,strange,i/(\gamma t,strange,i1/2)}\}_3 + \ldots + G_{strange,i}\{(1/2) e^{-t,strange,i/(\gamma t,strange,i1/2)}\}\{(1/2) e^{-t,strange,i/(\gamma t,strange,i1/2)}\} \ldots \{(1/2) e^{-t,strange,i/(\gamma t,strange,i1/2)}\}_{m,i}\}\}/dt_{ship}>$

For cases where the average strangelet of the ith species would not decay with a probability of $n_{,i}/100$ after traveling the mean free path, and the probability of baryon or nuclei conversion upon contact is $k_{,i}/100$, but where the strangelets are always destroyed when contacting the nuclei or converted into a non-strange-transmutative form with a probability of unity, the number of catalyzed baryons or nuclei is equal to in first order:

$N_i = G_{strange,i}\{n_{,i}/100\}\{k_i/100\}_1 + G_{strange,i}\{n_{,i}/100\}\{k_{,i}/100\}_1\{n_{,i}/100\}\{k_{,i}/100\}_2 + G_{strange,i}\{n_{,i}/100\}\{k_{,i}/100\}_1 \{n_{,i}/100\}\{k_{,i}/100\}_2\{n_{,i}/100\}\{k_{,i}/100\}_3 + \ldots + G_{strange,i}\{n_{,i}/100\}\{k_{,i}/100\}_1\{n_{,i}/100\}\{k_{,i}/100\}_2 \ldots \{n_{,i}/100\}\{k_{,i}/100\}_{m,i},$

Here, *m* is the number of time steps for the overall reaction extinction.

Where the average yield energy per nucleon or nucleus conversion caused by transformation by particles if the ith species is E_i:

The total yield energy for a given QCD bomb is in first order:

$\sum(i = 1; i = M) [(N_i)(E_i)] = \sum(i = 1; i = M) (E_i) \{G_{strange,i}\{n_{,i}/100\}\{k_{,i}/100\}_1 + G_{strange,i}\{n_{,i}/100\}\{k_{,i}/100\}_1\{n_{,i}/100\}\{k_{,i}/100\}_2 + G_{strange,i}\{n_{,i}/100\}\{k_{,i}/100\}_1 \{n_{,i}/100\}\{k_{,i}/100\}_2\{n_{,i}/100\}\{k_{,i}/100\}_3 + \ldots + G_{strange,i}\{n_{,i}/100\}\{k_{,i}/100\}_1\{n_{,i}/100\}\{k_{,i}/100\}_2 \ldots\{n_{,i}/100\}\{k_{,i}/100\}_{m,i}\}$,

The time-averaged explosive power of a given QCD bomb in the ship frame is in first order:

$<d\{\sum(i = 1; i = M) [(N_i)(E_i)]\} /dt_{ship} > = <d\{\sum(i = 1; i = M) (E_i) \{G_{strange,i}\{n_{,i}/100\}\{k_{,i}/100\}_1 + G_{strange,i}\{n_{,i}/100\}\{k_{,i}/100\}_1\{n_{,i}/100\}\{k_{,i}/100\}_2 + G_{strange,i}\{n_{,i}/100\}\{k_{,i}/100\}_1 \{n_{,i}/100\}\{k_{,i}/100\}_2\{n_{,i}/100\}\{k_{,i}/100\}_3 + \ldots + G_{strange,i}\{n_{,i}/100\}\{k_{,i}/100\}_1\{n_{,i}/100\}\{k_{,i}/100\}_2 \ldots\{n_{,i}/100\}\{k_{,i}/100\}_m\}\}/dt_{ship} >$

Considering the length of time, $t_{m,i}$, traveled per mean free path where the average nondecay probability is n/100 for each mean path traveled, the velocity of the ith species of strangelet is $L_i/t_{m,i}$ where the velocity of the strangelet is negligibly relativistic.

For cases where the ith species of strangelets produced are relativistic, the distance traveled by a given strangelet before it decays is given by:

$[L_i/t_{mi}](\gamma) = [L_i/t_{m,ii}]\{1/\{1 - [(v_{,i}/C)^2]\}^{1/2}\}$

Once again, $N_i (t_i) = N_{0i}\, e^{-t,strange,i/Tstrange,i} = N_{0i}\, (1/2)e^{-t,strange,i/t,strange,i1/2}$, where T_i is the mean lifetime of a particle such as an atom or subatomic particle, $t_{1/2}$ is the half-life of the particle, and N_{0i} is the starting population. $N_i (t_i)$ is the quantity that still remains and has not yet decayed after a time t; and half-life $t_{1/2} = \ln(2)/\lambda_i = T_i \ln (2)$, where λ_i is a positive number called the decay constant of the decaying quantity.

In the case of a strangelet species having a probability of nondecay along a mean path length of $P_i = n_i/100$, the number of strangelets out of a starting population of $N_{0i} = G_{strange,i}$ that contact a baryon or atomic nuclei is

$N_{0i}\{n_i/100\}$ where $\{n_{,i}/100\} = e^{-t,strange,i/Tstrange,i} = (1/2)\, e^{-t,strange,i/t,strange,i1/2}$

Therefore, the number of catalyzed baryons becomes in first order:

$$N_i = G_{strangei}\{e^{-t,strange,i/Tstrange,i}\}\{k_{,i}/100\}_1 + G_{strange,i}\{e^{-t,strange,i/Tstrange,i}\}\{k_{,i}/100\}_1\{e^{-t,strange,i/Tstrange,i}\}\{k_{,i}/100\}_2 + G_{strange,i}\{e^{-t,strange,i/Tstrange,i}\}\{k_{,i}/100\}_1\{e^{-t,strange,i/Tstrange,i}\}\{k_{,i}/100\}_2\{e^{-t,strange,i/Tstrange,i}\}\{k_{,i}/100\}_3 + \ldots + G_{strange,i}\{e^{-t,strange,i/Tstrange,i}\}\{k_{,i}/100\}_1\{e^{-t,strange,i/Tstrange,i}\}\{k_{,i}/100\}_2 \ldots \{e^{-t,strange,i/Tstrange,i}\}\{k_{,i}/100\}_{m,i}$$

$$= G_{strange,i}\{(1/2)\, e^{-t,strange,i/t,strange,i1/2}\}\{k_{,i}/100\}_1 + G_{strange,i}\{(1/2)\, e^{-t,strange,i/t,strange,i1/2}\}\{k_{,i}/100\}_1\{(1/2)\, e^{-t,strange,i/t,strange,i1/2}\}\{k_{,i}/100\}_2 + G_{strange,i}\{(1/2)\, e^{-t,strange,i/t,strange,i1/2}\}\{k_{,i}/100\}_1\{(1/2)\, e^{-t,strange,i/t,strange,i1/2}\}\{k_{,i}/100\}_2\{(1/2)\, e^{-t,strange,i/t,strange,i1/2}\}\{k_{,i}/100\}_3 + \ldots + G_{strange,i}\{(1/2)\, e^{-t,strange,i/t,strange,i1/2}\}\{k_{,i}/100\}_1\{(1/2)\, e^{-t,strange,i/t,strange,i1/2}\}\{k_{,i}/100\}_2 \ldots \{(1/2)\, e^{-t,strange,i/t,strange,i1/2}\}\{k_{,i}/100\}_{m,i}$$

Where the average yield energy per nucleon or nucleus conversion caused by transformation by particles if the ith species is E_i:

The total yield energy for a given QCD bomb is in first order:

$$\sum(i = 1;\ i = M)\ [(N_i)(E_i)] = \sum(i = 1;\ i = M)\ (E_i)\ \{G_{strange,i}\{e^{-t,strange,i/Tstrange,i}\}\{k_{,i}/100\}_1 + G_{strange,i}\{e^{-t,strange,i/Tstrange,i}\}\{k_{,i}/100\}_1\{e^{-t,strange,i/Tstrange,i}\}\{k_{,i}/100\}_2 + G_{strange,i}\{e^{-t,strange,i/Tstrange,i}\}\{k_{,i}/100\}_1\{e^{-t,strange,i/Tstrange,i}\}\{k_{,i}/100\}_2\{e^{-t,strange,i/Tstrange,i}\}\{k_{,i}/100\}_3 + \ldots + G_{strange,i}\{e^{-t,strange,i/Tstrange,i}\}\{k_{,i}/100\}_1\{e^{-t,strange,i/Tstrange,i}\}\{k_{,i}/100\}_2 \ldots \{e^{-t,strange,i/Tstrange,i}\}\{k_{,i}/100\}_{m,i}\}$$

$$= \sum(i = 1;\ i = M)\ (E_i)\ \{G_{strange,i}\{(1/2)\, e^{-t,strange,i/t,strange,i1/2}\}\{k_{,i}/100\}_1 + G_{strange,i}\{(1/2)\, e^{-t,strange,i/t,strange,i1/2}\}\{k_{,i}/100\}_1\{(1/2)\, e^{-t,strange,i/t,strange,i1/2}\}\{k_{,i}/100\}_2 + G_{strange,i}\{(1/2)\, e^{-t,strange,i/t,strange,i1/2}\}\{k_{,i}/100\}_1\{(1/2)\, e^{-t,strange,i/t,strange,i1/2}\}\{k_{,i}/100\}_2\{(1/2)\, e^{-t,strange,i/t,strange,i1/2}\}\{k_{,i}/100\}_3 + \ldots + G_{strange,i}\{(1/2)\, e^{-t,strange,i/t,strange,i1/2}\}\{k_{,i}/100\}_1\{(1/2)\, e^{-t,strange,i/t,strange,i1/2}\}\{k_{,i}/100\}_2 \ldots \{(1/2)\, e^{-t,strange,i/t,strange,i1/2}\}\{k_{,i}/100\}_{m,i}\}$$

The time-averaged explosive power of a given QCD bomb in the ship frame is in first order:

$$\langle d\{\sum(i = 1; i = M)\ [(N_i)(E_i)]\}\ /dt_{ship}\rangle = \langle d\{\sum(i = 1; i = M)\ (E_i)\ \{G_{strange,i}\{e^{-t,strange,i/Tstrange,i}\}\{k_{,i}/100\}_1 + G_{strange,i}\{e^{-t,strange,i/Tstrange,i}\}\{k_{,i}/100\}_1\{e^{-t,strange,i/Tstrange,i}\}\{k_{,i}/100\}_2 + G_{strange,i}\{e^{-t,strange,i/Tstrange,i}\}\{k_{,i}/100\}_1\{e^{-t,strange,i/Tstrange,i}\}\{k_{,i}/100\}_2\{e^{-t,strange,i/Tstrange,i}\}\{k_{,i}/100\}_3 + \ldots + G_{strange,i}\{e^{-t,strange,i/Tstrange,i}\}\{k_{,i}/100\}_1\{e^{-t,strange,i/Tstrange,i}\}\{k_{,i}/100\}_2 \ldots \{e^{-t,strange,i/Tstrange,i}\}\{k_{,i}/100\}_{m,i}\}\}/dt_{ship}\rangle$$

$$= <d\{\sum(i = 1;\ i = M)\ (E_i)\ \{G_{strange,i}\{(1/2)\ e^{-t,strange,i/t,strange,i1/2}\}\{k_{,i}/100\}_1 + G_{strange,i}\{(1/2)\ e^{-t,strange,i/t,strange,i1/2}\}\{k_{,i}/100\}_1\{(1/2)\ e^{-t,strange,i/t,strange,i1/2}\}\{k_{,i}/100\}_2 + G_{strange,i}\{(1/2)\ e^{-t,strange,i/t,strange,i1/2}\}\{k_{,i}/100\}_1\{(1/2)\ e^{-t,strange,i/t,strange,i1/2}\}\{k_{,i}/100\}_2\ \{(1/2)\ e^{-t,strange,i/t,strange,i1/2}\}\{k_{,i}/100\}_3 + \ldots + G_{strange,i}\{(1/2)\ e^{-t,strange,i/t,strange,i1/2}\}\{k_{,i}/100\}_1\{(1/2)\ e^{-t,strange,i/t,strange,i1/2}\}\{k_{,i}/100\}_2 \ldots\{(1/2)\ e^{-t,strange,i/t,strange,i1/2}\}\{k_{,i}/100\}_{m,i}\}\}/dt_{ship} >$$

If $t_{i1/2}$ and T_i are background stationary frames and the strangelets are relativistic, the number of catalyzed baryons becomes in first order:

$$N_i = G_{strange,i}\{e^{-t,strange,i/(Tstrange,i\gamma}\}\{k_{,i}/100\}_1 + G_{strange,i}\{e^{-t,strange,i/(Tstrange,i\gamma)}\}\{k_{,i}/100\}_1\{e^{-t,strange,i/(Tstrange,i\gamma)}\}\{k_{,i}/100\}_2 + G_{strange,i}\{e^{-t,strange,i/(Tstrange,i\gamma)}\}\{k/100\}_1\{e^{-t,strange,i/(Tstrange,i\gamma)}\}\{k_{,i}/100\}_2\{e^{-t,strange,i/(Tstrange,i\gamma)}\}\{k_{,i}/100\}_3 + \ldots + G_{strange,i}\{e^{-t,strange,i/(Tstrange,i\gamma)}\}\{k_{,i}/100\}_1\{e^{-t,strange,i/(Tstrange,i\gamma)}\}\{k_{,i}/100\}_2 \ldots\{e^{-t,strange,i/(Tstrange,i\gamma)}\}\{k_{,i}/100\}_{m,i}$$

$$= G_{strange,i}\{(1/2)\ e^{-t,strange,i/(\gamma t,strange,i1/2)}\}\ \{k_{,i}/100\}_1 + G_{strange,i}\{(1/2)\ e^{-t,strange,i/(\gamma t,strange,i1/2)}\}\{k_{,i}/100\}_1\{(1/2)\ e^{-t,strange,i/(\gamma t,strange,i1/2)}\}\ \{k_{,i}/100\}_2 + G_{strange,i}\{(1/2)\ e^{-t,strange,i/(\gamma t,strange,i1/2)}\}\{k_{,i}/100\}_1\{(1/2)\ e^{-t,strange,i/(\gamma t,strange,i1/2)}\}\{k_{,i}/100\}_2\{(1/2)\ e^{-t,strange,i/(\gamma t,strange,i1/2)}\}\{k_{,i}/100\}_3 + \ldots + G_{strange,i}\{(1/2)\ e^{-t,strange,i/(\gamma t,strange,i1/2)}\}\{k_{,i}/100\}_1\{(1/2)\ e^{-t,strange,i/(\gamma t,strange,i1/2)}\}\{k_{,i}/100\}_2 \ldots\{(1/2)\ e^{-t,strange,i/(\gamma t,strange,i1/2)}\}\{k_{,i}/100\}_{m,i}$$

Where the average yield energy per nucleon or nucleus conversion caused by transformation by particles if the ith species is E_i;

The total yield energy for a given QCD bomb is in first order:

$$\sum(i = 1;\ i = M)\ [(N_i)(E_i)] = \sum(i = 1;\ i = M)\ (E_i)\ \{G_{strange,i}\{e^{-t,strange,i/(Tstrange,i\gamma}\}\{k_{,i}/100\}_1 + G_{strange,i}\{e^{-t,strange,i/(Tstrange,i\gamma)}\}\{k_{,i}/100\}_1\{e^{-t,strange,i/(Tstrange,i\gamma)}\}\{k_{,i}/100\}_2 + G_{strange,i}\{e^{-t,strange,i/(Tstrange,i\gamma)}\}\{k/100\}_1\{e^{-t,strange,i/(Tstrange,i\gamma)}\}\{k_{,i}/100\}_2\{e^{-t,strange,i/(Tstrange,i\gamma)}\}\{k_{,i}/100\}_3 + \ldots + G_{strange,i}\{e^{-t,strange,i/(Tstrange,i\gamma)}\}\{k_{,i}/100\}_1\{e^{-t,strange,i/(Tstrange,i\gamma)}\}\{k_{,i}/100\}_2 \ldots\{e^{-t,strange,i/(Tstrange,i\gamma)}\}\{k_{,i}/100\}_{m,i}\}$$

$$= \{\sum(i = 1;\ i = M)\ (E_i)\ \{G_{strange,i}\{(1/2)\ e^{-t,strange,i/(\gamma t,strange,i1/2)}\}\ \{k_{,i}/100\}_1 + G_{strange,i}\{(1/2)\ e^{-t,strange,i/(\gamma t,strange,i1/2)}\}\{k_{,i}/100\}_1\{(1/2)\ e^{-t,strange,i/(\gamma t,strange,i1/2)}\}\ \{k_{,i}/100\}_2 + G_{strange,i}\{(1/2)\ e^{-t,strange,i/(\gamma t,strange,i1/2)}\}\{k_{,i}/100\}_1\{(1/2)\ e^{-t,strange,i/(\gamma t,strange,i1/2)}\}\{k_{,i}/100\}_2\{(1/2)\ e^{-t,strange,i/(\gamma t,strange,i1/2)}\}\{k_{,i}/100\}_3 + \ldots$$

$+ \quad G_{strange,i}\{(1/2) \quad e^{-t,strange,i/(\gamma t,strange,i1/2)}\}\{k_{,i}/100\}_1\{(1/2) \quad e^{-t,strange,i/(\gamma t,strange,i1/2)}\}\{k_{,i}/100\}_2 \ldots\{(1/2)\, e^{-t,strange,i/(\gamma t,strange,i1/2)}\}\{k_{,i}/100\}_{m,i}\}$

The time averaged explosive power of a given QCD bomb in the ship frame is in first order:

$$<d\{\sum(i = 1; i = M) [(N_i)(E_i)]\} /dt_{ship} > = <d\{\sum(i = 1; i = M) (E_i) \{G_{strange,i}\{e^{-t,strange,i/(Tstrange,i\gamma}\}\{k_{,i}/100\}_1 + G_{strange,i}\{e^{-t,strange,i/(Tstrange,i\gamma)}\}\{k_{,i}/100\}_1\{e^{-t,strange,i/(Tstrange,i\gamma)}\}\{k_{,i}/100\}_2 + G_{strange,i}\{e^{-t,strange,i\ /(Tstrange,i\ \gamma)}\}\{k/100\}_1\{e^{-t,strange,i\ /(Tstrange,i\ \gamma)}\}\{k_{,i}/100\}_2\{e^{-t,strange,i/(Tstrange,i\gamma)}\}\{k_{,i}/100\}_3 + \ldots + G_{strange,i}\{e^{-t,strange,i/(Tstrange,i\gamma)}\}\{k_{,i}/100\}_1\{e^{-t,strange,i/(Tstrange,i\gamma)}\}\{k_{,i}/100\}_2 \ldots\{e^{-t,strange,i/(Tstrange,i\gamma)}\}\{k_{,i}/100\}_{m,i}\}/dt_{ship} >$$

$$= <d\{\sum(i = 1; i = M) (E_i) \{G_{strange,i}\{(1/2)\, e^{-t,strange,i/(\gamma t,strange,i1/2)}\}\{k_{,i}/100\}_1 + G_{strange,i}\{(1/2)\, e^{-t,strange,i/(\gamma t,strange,i1/2)}\}\{k_{,i}/100\}_1\{(1/2)\, e^{-t,strange,i/(\gamma t,strange,i1/2)}\}\{k_{,i}/100\}_2 + G_{strange,i}\{(1/2)\, e^{-t,strange,i/(\gamma t,strange,i1/2)}\}\{k_{,i}/100\}_1\{(1/2)\, e^{-t,strange,i/(\gamma t,strange,i1/2)}\}\{k_{,i}/100\}_2\{(1/2)\, e^{-t,strange,i/(\gamma t,strange,i1/2)}\}\{k_{,i}/100\}_3 + \ldots + G_{strange,i}\{(1/2)\, e^{-t,strange,i/(\gamma t,strange,i1/2)}\}\{k_{,i}/100\}_1\{(1/2)\, e^{-t,strange,i/(\gamma t,strange,i1/2)}\}\{k_{,i}/100\}_2 \ldots\{(1/2)\, e^{-t,strange,i/(\gamma t,strange,i1/2)}\}\{k_{,i}/100\}_{m,i}\}/dt_{ship} >$$

We would definitely not want stable strangelets that are transmutative into-strange-matter, or ones that are sufficiently long lived so as to produce a reaction that could consume the entire planet, or propagate interstellar distances to reach other stars and planets where they could wreak havoc. Such strangelets may even convert interstellar gas and dust in a gradual propagating wavefront and literally gradually eat away at the universe or multiverse completely in a progressive reaction wavefront. However, to produce such strangelets or a black hole capable of doing this would require a galactic-sized accelerator—in other words, it is highly unlikely at best.

Regardless, strangelet-producing nuclear explosives might be used for the following:

1. Interstellar ramjet propulsion
2. Intergalactic ramjet propulsion
3. Hyperspatial rocket propulsion
4. Nuclear strange matter bomblet pulse drive
5. Nuclear strange-matter bomblet pellet runway propulsion

6. Mass fuel beam nuclear strange-matter bomblet propulsion
7. Einsteinian 4-D photon, electron, muon, tauon, positron, anti-muon, anti-tauon, proton, antiproton, ion, anti-ion, charged exotic meson, charged exotic baryon, electron neutrino, anti-electron neutrino, muon neutrino, antimuon neutrino, tauon neutrino, anti-tauon neutrino, gravity wave, and quantum-scale gravity wave or graviton rockets
8. Hyperspatial photon, electron, muon, tauon, positron, anti-muon, anti-tauon, proton, antiproton, ion, anti-ion, charged exotic meson, charged exotic baryon, electron neutrino, anti-electron neutrino, muon neutrino, antimuon neutrino, tauon neutrino, anti-tauon neutrino, gravity wave, and quantum scale gravity wave or graviton rockets. For rocket vehicle applications, the exothermic energy produced in strange-matter reactors can be used to energize thrust-stream generators.

Note that gravitons are hypothetical at this point. Their existence would require the discovery of the spin-2 boson, which is yet to be accomplished.

So for stable strangelet technology, only assured-to-die-out chain reactions must be produced; otherwise, we risk universal destruction. The ramifications of the technology as a power source, however, are too profound not to be mentioned. This is true not only from a philosophical perspective of enabling technology that is far superior in power than pure matter-antimatter conversion, but more to the point, a technology that can revolutionize commercial energy production for manned starship propulsion. Consider the case where the matter within any hyperspatial dimensions that are coupled to our universe would not have periodic table elements or baryons having the exact properties as those found within our ordinary 4-D Einsteinian space-time. For such hyperspatial matters that have quantum properties that are similar to the baryonic mass within our ordinary 4-D universe, such exothermic conversion may still be possible, thereby enabling the production of hyperspatial ramjets and hyperspatial photon, electron, muon, tauon, positron, anti-muon, anti-tauon, proton, antiproton, ion, anti-ion, charged exotic meson, charged exotic baryon, electron neutrino, anti-electron neutrino, muon neutrino, antimuon neutrino, tauon neutrino, anti-tauon neutrino, gravity wave, and quantum-scale gravity wave or graviton rockets.

There are several plausible candidates for mechanisms that can be used to perform nuclear fusion in order to produce the required temperatures and particle kinetic energies so as to enable the bulk production of metastable strangelets to provide for substantial but die-out forms of strange chain reactions. The devices are shaped-charge nuclear fusion and/or fission devices. To the best of our knowledge, none of the associated nuclear explosive types have yet been developed.

What can be accomplished with strange reactions might also work for charmonium catalysis and bottomonium catalysis. Toponium catalysis may prove possible in the long run; however, the production of toplets as yet have been unverified and may remain so for a long time. This is because toplets, which are bound states comprised of top quarks, would require the strong nuclear force to travel the distances to other quarks cocreated in particle accelerator collisions. All top quark-producing particle collisions to date have produced top quarks, which decay too quickly for the light-speed limited strong force to travel the distance between any cocreated top quarks thereby preventing any stabilizing bound states.

Chapter 12

Charm Matter Reactors and Bombs and Other Things Nice

This chapter includes a description of highly conjectural forms of QCD/QED reactors and bombs. Such reactors and bombs would be used to power highly relativistic starships by converting a large fuel mass at least partially into energy and/or exothermic conversion of interstellar ordinary baryonic matter into strange, charmed, and/or bottom quark-based matter.

Now that the nuclear and the eventual exotic QCD energy applications genie is permanently out of the bottle and exotic QCD energy applications loom on the horizon, how might such applications be used to reach the stars?

Below is a brief summary of thoughts on how exotic applications of nuclear and sub nuclear reactions can open up the cosmos for colonization and exploration by humanity. The technologies considered have likely already been pondered and studied to some degree; but if not, they definitely will be in the future.

Imagine a technology that could convert the elements in the periodic table atomic to other exotic forms of baryonic matter releasing prodigious quantities of energy in a mass-specific reaction energy that is higher than that released by nuclear fusion. It has been proposed that certain species of some stable charmlets that might be produced in particle accelerators may be able to convert periodic table atoms into charm matter. An unfortunate consequence of this may be a runaway reaction that could cause the planet to be exothermically converted into charm matter or matter comprised mostly of charm quarks with some additional species of quarks mixed in.

Stable catalytic charmlets may be ideal for powering future interstellar ramjet starships (ISRs). Imagine a large mass of stable charm matter for which interstellar and intergalactic matter would be funneled into the intake of a reaction chamber where the intake mass would be converted to quickly decaying charmlets, which would convert even more of the intake mass into energy through decay processes. The drag energy in the form of heat could be recycled to power photon, ion, electron, positron, muon, antimuon, tauon, antitauon, proton, antiproton, exotic charged meson, and/or exotic charged baryon rockets. The stable or meta-stable charmlets produced could be used as an accelerated thrust stream. A starship using this form of propulsion might accelerate and achieve high Lorentz factors and/or near light speeds in any hyper spatial dimension where periodic table atomic matter or other convertible baryonic matter is present.

Let us consider a scenario where a species of charmlet would not decay on average *n* percent of the time before making contact with an ordinary baryon or atomic nucleus and where the probability of the conversion of the baryon or atomic nucleus is 100 percent . The average probability of the initial charmlet converting an ordinary nucleon or atomic nucleus to another catalytic charmlet is n/100. After m time steps where, for which each time step, the probability of the average catalytic charmlet converting a baryon or nucleus it first contacts is n/100, the number of catalyzed baryons or nuclei is equal to in first order:

$$N = G_{charm}\{n/100\}_1 + G_{charm}\{n/100\}_1 \{n/100\}_2 + G_{charm}\{n/100\}_2 \{n/100\}_2\{n/100\}_3 + \ldots + G_{charm}\{n/100\}_1 \{n/100\}_2 \ldots\{ n/100\}_m$$

Where m is the number of time steps for the overall reaction extinction and G_{charm} is the initial population of charmlets.

Another way of presenting reaction wave propagation is to consider the average lifetime of the charmlets and the average distance traveled by the charmlets relative to the mean free path for charmlet and nucleon or atomic nuclei contact. For cases where the average charmlet would not decay with a probability of n/100 before completing the travel of its mean free path, the number of catalyzed baryons or nuclei is also equal to in first order:

$$N = G_{charm}\{n/100\}_1 + G_{charm}\{n/100\}_2 \{n/100\}_2 + G_{charm}\{n/100\}_2 \{n/100\}_2\{n/100\}_3 + \ldots + G_{charm}\{n/100\}_1 \{n/100\}_2 \ldots\{ n/100\}_m$$

Where m is the number of time steps for the overall reaction extinction.

Considering the length of time, t_m, traveled per mean free path where the average nondecay probability is n/100 for each mean path traveled, the velocity of the charmlet is L/t_m where the velocity of the charmlet is negligibly relativistic.

For cases where the charmlets produced are relativistic, the average distance traveled by such charmlets before it decays is given by:

$$D = [L/t_m](\gamma)\ (n)/(100) = [L/t_m]\{1/\{1 - [(v/C)^2]\}^{1/2}\}(n)/(100).$$

Now, applying the exponential decay formula:

$$N(t) = N_0\, e^{-t,\text{charm}/T\text{charm}} = N_0\, (1/2) e^{-t,\text{charm}/t,\text{charm}1/2}$$

Where T is the mean lifetime of a particle (such as an atom or subatomic particle), $t_{1/2}$ is the half-life of the particle, and N_0 is the starting population. $N(t)$ is the quantity that still remains and has not yet decayed after a time t.

Therefore the half-life is expressed as follows:

$$t_{1/2} = \ln(2)/\lambda = T \ln(2)$$

Here, λ is a positive number called the decay constant of the decaying quantity.

For a charmlet having a probability of nondecay along a mean path length of $P = n/100$, the number of charmlets out of a starting population of $N_0 = G_{\text{charm}}$ that contact a baryon or atomic nuclei is $N_0\{n/100\}$ where $\{n/100\} = e^{-t,\text{charm}/T\text{charm}} = (1/2)\, e^{-t,\text{charm}/t,\text{charm}1/2}$. Therefore, the number of catalyzed baryons becomes in first order:

$$N = G_{\text{charm}}\{e^{-t,\text{charm}/T\text{charm}}\}_1 + G_{\text{charm}}\{e^{-t,\text{charm}/T\text{charm}}\}\{e^{-t,\text{charm}/T\text{charm}}\}_2 + G_{\text{charm}}\, e^{-t,\text{charm}/T\text{charm}}\}\{e^{-t,\text{charm}/T\text{charm}}\}\{e^{-t,\text{charm}/T\text{charm}}\}_3 + \ldots + G_{\text{charm}}\{e^{-t,\text{charm}/T\text{charm}}\}\{e^{-t,\text{charm}/T\text{charm}}\}\ldots\{e^{-t,\text{charm}/T\text{charm}}\}_m$$

$$= G_{\text{charm}}\{(1/2)\, e^{-t,\text{charm}/t,\text{charm}1/2}\}_1 + G_{\text{charm}}\{(1/2)\, e^{-t,\text{charm}/t,\text{charm}1/2}\}\{(1/2)\, e^{-t,\text{charm}/t,\text{charm}1/2}\}_2 + G_{\text{charm}}\{(1/2)\, e^{-t,\text{charm}/t,\text{charm}1/2}\}\{(1/2)\, e^{-t,\text{charm}/t,\text{charm}1/2}\}\{(1/2)\, e^{-t,\text{charm}/t,\text{charm}1/2}\}_3 + \ldots + G_{\text{charm}}\{(1/2)\, e^{-t,\text{charm}/t,\text{charm}1/2}\}\{(1/2)\, e^{-t,\text{charm}/t,\text{charm}1/2}\}\ldots\{(1/2)\, e^{-t,\text{charm}/t,\text{charm}1/2}\}_m,$$

Where t is the travel time of the charmlets for the mean free path for charmlet to ordinary baryon contact.

Now, if $t_{1/2}$ and T are background stationary frames and the charmlets are relativistic, the number of catalyzed baryons becomes in first order:

$$N = G_{\text{charm}}\{e^{-t,\text{charm}/(T\text{charm}\gamma}\}_1 + G_{\text{charm}}\{e^{-t,\text{charm}/(T\text{charm}\gamma)}\}\{e^{-t,\text{charm}/(T\text{charm}\gamma)}\}_2 + G_{\text{charm}}\{e^{-t,\text{charm}/(T\text{charm}\gamma)}\}\{e^{-t,\text{charm}/(T\text{charm}\gamma)}\}\{e^{-t,\text{charm}/(T\text{charm}\gamma)}\}_3 + \ldots + G_{\text{charm}}\{e^{-t,\text{charm}/(T\text{charm}\gamma)}\}\{e^{-t,\text{charm}/(T\text{charm}\gamma)}\}\ldots\{e^{-t,\text{charm}/(T\text{charm}\gamma)}\}_m$$

$$= G_{\text{charm}}\{(1/2)\, e^{-t,\text{charm}/(\gamma t,\text{charm}1/2)}\}_1 + G_{\text{charm}}\{(1/2)\, e^{-t,\text{charm}/(\gamma t,\text{charm}1/2)}\}\{(1/2)\, e^{-t,\text{charm}/(\gamma t,\text{charm}1/2)}\}_2 + G_{\text{charm}}\{(1/2)\, e^{-t,\text{charm}/(\gamma t,\text{charm}1/2)}\}\{(1/2)\, e^{-t,\text{charm}/(\gamma t,\text{charm}1/2)}\}\{(1/2)\, e^{-t,\text{charm}/(\gamma t,\text{charm}1/2)}\}_3 + \ldots + G_{\text{charm}}\{(1/2)\, e^{-t,\text{charm}/(\gamma t,\text{charm}1/2)}\}\{(1/2)\, e^{-t,\text{charm}/(\gamma t,\text{charm}1/2)}\}\ldots\{(1/2)\, e^{-t,\text{charm}/(\gamma t,\text{charm}1/2)}\}_m$$

For cases where the average charmlet would not decay with a probability of n/100 after traveling the mean free path, and the probability of baryon or nuclei conversion upon contact is k/100, but where the charmlets are always destroyed when contacting the nuclei or converted into a non-charm-transmutative form with a probability of unity, the number of catalyzed baryons or nuclei is equal to in first order:

$$N = G_{charm}\{n/100\}\{k/100\}_1 + G_{charm}\{n/100\}\{k/100\}_1\{n/100\}\{k/100\}_2 + G_{charm}\{n/100\}\{k/100\}_1 \{n/100\}\{k/100\}_2\{n/100\}\{k/100\}_3 + \ldots + G_{charm}\{n/100\}\{k/100\}_1\{n/100\}\{k/100\}_2 \ldots\{n/100\}\{k/100\}_m,$$

Here, *m* is the number of time steps for the overall reaction extinction.

Considering the length of time, t_m, traveled per mean free path where the average nondecay probability is n/100 for each mean path traveled, the velocity of the charmlet is L/t_m, where the velocity of the charmlet is negligibly relativistic.

For cases where the charmlets produced are relativistic, the distance traveled by a given charmlet before it decays is given by:

$$[L/t_m](\gamma) = [L/t_m]\{1/\{1 - [(v/C)^2]\}^{1/2}\}$$

Once again, $N(t) = N_0\, e^{-t,charm/Tcharm} = N_0\,(1/2)e^{-t,charm/t,charm1/2}$, where T is the mean lifetime of a particle (such as an atom or subatomic particle), $t_{1/2}$ is the half-life of the particle, and N_0 is the starting population. $N(t)$ is the quantity that still remains and has not yet decayed after a time t; and half-life $t_{1/2} = \ln(2)/\lambda = T \ln(2)$, where λ is a positive number called the decay constant of the decaying quantity.

In the case of a charmlet species having a probability of nondecay along a mean path length of $P = n/100$, the number of charmlets out of a starting population of $N_0 = G_{charm}$ that contact a baryon or atomic nuclei is $N_0\{n/100\}$ where $\{n/100\} = e^{-t,charm/Tcharm} = (1/2)\, e^{-t,charm/t,charm1/2}$. Therefore, the number of catalyzed baryons becomes in first order:

$$N = G_{charm}\{e^{-t,charm/Tcharm}\}\{k/100\}_1 + G_{charm}\{e^{-t,charm/Tcharm}\}\{k/100\}_1\{e^{-t,charm/Tcharm}\}\{k/100\}_2 + G_{charm}\{e^{-t,charm/Tcharm}\}\{k/100\}_1\{e^{-t,charm/Tcharm}\}\{k/100\}_2\{e^{-t,charm/Tcharm}\}\{k/100\}_3 + \ldots + G_{charm}\{e^{-t,charm/Tcharm}\}\{k/100\}_1\{e^{-t,charm/Tcharm}\}\{k/100\}_2 \ldots\{e^{-t,charm/Tcharm}\}\{k/100\}_m$$

$$= G_{charm}\{(1/2)\ e^{-t,charm/t,charm1/2}\}\{k/100\}_1 + G_{charm}\{(1/2)\ e^{-t,charm/t,charm1/2}\}\{k/100\}_1\{(1/2)\ e^{-t,charm/t,charm1/2}\}\{k/100\}_2 + G_{charm}\{(1/2)\ e^{-t,charm/t,charm1/2}\}\{k/100\}_1\{(1/2)\ e^{-t,charm/t,charm1/2}\}\{k/100\}_2\ \{(1/2)\ e^{-t,charm/t,charm1/2}\}\{k/100\}_3 + \ldots + G_{charm}\{(1/2)\ e^{-t,charm/t,charm1/2}\}\{k/100\}_1\{(1/2)\ e^{-t,charm/t,charm1/2}\}\{k/100\}_2 \ldots\{(1/2)\ e^{-t,charm/t,charm1/2}\}\{k/100\}_m$$

If $t_{1/2}$ and T are background stationary frames and the charmlets are relativistic, the number of catalyzed baryons becomes in first order:

$$N = G_{charm}\{e^{-t,charm/(Tcharm\gamma}\}\{k/100\}_1 + G_{charm}\{e^{-t,charm/(Tcharm\gamma)}\}\{k/100\}_1\{e^{-t,charm/(Tcharm\gamma)}\}\{k/100\}_2 + G_{charm}\{e^{-t,charm/(Tcharm\gamma)}\}\{k/100\}_1\{e^{-t,charm/(Tcharm\gamma)}\}\{k/100\}_2\{e^{-t,charm/(Tcharm\gamma)}\}\{k/100\}_3 + \ldots + G_{charm}\{e^{-t,charm/(Tcharm\gamma)}\}\{k/100\}_1\{e^{-t,charm/(Tcharm\gamma)}\}\{k/100\}_2 \ldots\{e^{-t,charm/(Tcharm\gamma)}\}\{k/100\}_m$$

$$= G_{charm}\{(1/2)\ e^{-t,charm/(\gamma t,charm1/2)}\}\ \{k/100\}_1 + G_{charm}\{(1/2)\ e^{-t,charm/(\gamma t,charm1/2)}\}\{k/100\}_1\{(1/2)\ e^{-t,charm/(\gamma t,charm1/2)}\}\ \{k/100\}_2 + G_{charm}\{(1/2)\ e^{-t,charm/(\gamma t,charm1/2)}\}\{k/100\}_1\{(1/2)\ e^{-t,charm/(\gamma t,charm1/2)}\}\{k/100\}_2\{(1/2)\ e^{-t,charm/(\gamma t,charm1/2)}\}\{k/100\}_3 + \ldots + G_{charm}\{(1/2)\ e^{-t,charm/(\gamma t,charm1/2)}\}\{k/100\}_1\{(1/2)\ e^{-t,charm/(\gamma t,charm1/2)}\}\{k/100\}_2 \ldots\{(1/2)\ e^{-t,charm/(\gamma t,charm1/2)}\}\{k/100\}_m$$

Let us consider a scenario where a species, i, of charmlet would not decay on average n percent of the time before making contact with an ordinary baryon or atomic nucleus and where the probability of the conversion of the baryon or atomic nucleus is 100 percent. The average probability of the initial charmlet converting an ordinary nucleon or atomic nucleus to another catalytic charmlet is $n_{,i}/100$. After m time steps, where, for which each time step the probability of the average catalytic charmlet converting a baryon or nucleus first contacts is $n_{,i}/100$, the number of catalyzed baryons or nuclei is equal to in first order:

$$N_{,i} = G_{charm,i}\{n_{,i}/100\}_1 + G_{charm,i}\{n_i/100\}_1\ \{n_{,i}/100\}_2 + G_{charm,\ i}\{n_{,i}/100\}_2\ \{n_{,i}/100\}_2\{n_{,i}/100\}_3 + \ldots + G_{charm,i}\{n_{,i}/100\}_1\ \{n_{,i}/100\}_2 \ldots\{\ n_{,i}/100\}_{m,i}$$

Where $m_{,i}$ is the number of time steps for the overall reaction extinction and $G_{charm,i}$ is the initial population of charmlets.

So for M species of charmlets, where each species has its unique mean free path, initial population, and number of time steps for overall reaction extinctions, the number of catalyzed baryons or nuclei is equal to in first order:

$$\sum(i = 1; i = M)\ N_i = \sum(i = 1; i = M)\ \{G_{charm,i}\{n_{,i}/100\}_1 + G_{charm,i}\{n_{,i}/100\}_1\ \{n_{,i}/100\}_2 + G_{charm,\ i}\{n_{,i}/100\}_2\ \{n_{,i}/100\}_2\{n_{,i}/100\}_3 + \ldots + G_{charm,i}\{n_{,i}/100\}_1\ \{n_{,i}/100\}_2 \ldots \{n_{,i}/100\}_{m,i}\}$$

Where the average yield energy per nucleon or nucleus conversion caused by transformation by particles if the ith species is E_i.

The total yield energy for a given QCD bomb is in first order:

$$\sum(i = 1; i = M)\ [(N_i)(E_i)] = \sum(i = 1; i = M)\ (E_i)\ \{\{G_{charm,i}\{n_{,i}/100\}_1 + G_{charm,i}\{n_{,i}/100\}_1\ \{n_{,i}/100\}_2 + G_{charm,\ i}\{n_{,i}/100\}_2\ \{n_{,i}/100\}_2\{n_{,i}/100\}_3 + \ldots + G_{charm,i}\{n_{,i}/100\}_1\ \{n_{,i}/100\}_2 \ldots \{n_{,i}/100\}_{m,i}\}\}$$

The time averaged explosive power of a given QCD bomb in the ship frame is in first order:

$$<d\{\sum(i = 1; i = M)\ [(N_i)(E_i)]\}\ /dt_{ship} > = <d\{\sum(i = 1; i = M)\ (E_i)\{\{G_{charm,i}\{n_{,i}/100\}_1 + G_{charm,i}\{n_{,i}/100\}_1\ \{n_{,i}/100\}_2 + G_{charm,\ i}\{n_i/100\}_2\ \{n_{,i}/100\}_2\{n_{,i}/100\}_3 + \ldots + G_{charm,i}\{n_{,i}/100\}_1\ \{n_{,i}/100\}_2 \ldots \{n_{,i}/100\}_{m,i}\}\}/dt_{ship} >$$

Another way of presenting reaction wave propagation is to consider the average lifetime of the charmlets and the average distance traveled by the charmlets relative to the mean free path for charmlet and nucleon or atomic nuclei contact. For cases where the charmlet of the ith species of charmlets would nondecay with a probability of $n_{,i}/100$ before completing the travel of its mean free path, the number of catalyzed baryons or nuclei is also equal to in first order:

$$N_i = G_{charm,i}\{n_{,i}/100\}_1 + G_{charm,i}\{n_{,i}/100\}_1\ \{n_{,i}/100\}_2 + G_{charm,\ i}\{n_{,i}/100\}_2\ \{n_{,i}/100\}_2\{n_{,i}/100\}_3 + \ldots + G_{charm,i}\{n_{,i}/100\}_1\ \{n_{,i}/100\}_2 \ldots \{n_{,i}/100\}_{m,i}$$

Where $m_{,i}$ is the number of time steps for the overall reaction extinction for the ith species of charmlet.

So, for M species of charmlets, where each species has its unique mean free path, initial population, and number of time steps for overall reaction extinctions, the number of catalyzed baryons or nuclei is equal to in first order:

$$\sum(i = 1; i = M)\ N_i = \sum(i = 1; i = M)\ \{G_{charm,i}\{n_{,i}/100\}_1 + G_{charm,i}\{n_{,i}/100\}_1\ \{n_{,i}/100\}_2 + G_{charm,\ i}\{n_{,i}/100\}_2\ \{n_{,i}/100\}_2\{n_{,i}/100\}_3 + \ldots + G_{charm,i}\{n_{,i}/100\}_1\ \{n_{,i}/100\}_2 \ldots \{n_{,i}/100\}_{m,i}\}$$

Where the average yield energy per nucleon or nucleus conversion caused by transformation by particles if the ith species is E_i:

The total yield energy for a given QCD bomb is in first order:

$$\sum(i = 1; i = M) [(N_i)(E_i)] = \sum(i = 1; i = M) (E_i)\{\{G_{charm,i}\{n_{,i}/100\}_1 + G_{charm,i}\{n_{,i}/100\}_1 \{n_{,i}/100\}_2 + G_{charm,\ i}\{n_{,i}/100\}_2 \{n_{,i}/100\}_2\{n_{,i}/100\}_3 + \ldots + G_{charm,i}\{n_{,i}/100\}_1 \{n_{,i}/100\}_2 \ldots\{ n_{,i}/100\}_{m,i}\}\}$$

The time averaged explosive power of a given QCD bomb in the ship frame is in first order:

$$<d\{\sum(i = 1; i = M) [(N_i)(E_i)]\} /dt_{ship} > = <d\{\sum(i = 1; i = M) (E_i)\{\{G_{charm,i}\{n_{,i}/100\}_1 + G_{charm,i}\{n_{,i}/100\}_1 \{n_{,i}/100\}_2 + G_{charm,\ i}\{n_{,i}/100\}_2 \{n_{,i}/100\}_2\{n_{,i}/100\}_3 + \ldots + G_{charm,i}\{n_{,i}/100\}_1 \{n_{,i}/100\}_2 \ldots\{ n_{,i}/100\}_{m,i}\}\}\}/dt_{ship} >$$

Considering the length of time, $t_{m,i}$, traveled per mean free path where the average nondecay probability is $n_{,i}/100$ for each mean path traveled, the velocity of the charmlet is $L_i/t_{m.i}$ where the velocity of the charmlet is negligibly relativistic for the ith species of charmlets.

For cases where the ith species of charmlets produced are relativistic, the average distance traveled by such charmlets before it decays is given by:

$$D_i = [L_i/t_{m,i}](\gamma_{,i}) (n_{,i})/(100) = [L_i/t_{m,i}]\{1/\{1 - [(v_{,i}/C)^2]\}^{1/2}\}(n_{,i})/(100)$$

Now, applying the exponetial decay formula:

$$N_i(t_i) = N_{0i}\, e^{-t,charm.i/Tcharm,i} = N_{0i}\, (1/2)e^{-t,charm,i/t,charm,i1/2}$$

Where T_i is the mean lifetime of a particle (such as an atom or subatomic particle), $t_{i1/2}$ is the half-life of the particle, and N_{0i} is the starting population. $N_i(t_i)$ is the quantity that still remains and has not yet decayed after a time t_i.

Therefore the half-life is expressed as follows:

$$t_{i1/2} = \ln(2)/\lambda_i = T_i \ln (2)$$

Here, λ_i is a positive number called the decay constant of the decaying quantity.

For a charmlet of the ith charmlet species having a probability of non-decay along a mean path length of P_i = $n_{,i}/100$, the number of charmlets out of a starting population of N_{0i} = $G_{charm,i}$ that contact a baryon or atomic nuclei is $N_{0i}\{n_{,i}/100\}$ where $\{n_{,i}/100\} = e^{-t,charm,i/Tcharm,i} = (1/2)\, e^{-t,charm,i/t,charm,i1/2}$. Therefore, the number of catalyzed baryons becomes in first order:

$$N_i = G_{charm,i}\{e^{-t,charm,i/Tcharm,i}\}_1 + G_{charm,i}\{e^{-t,charm,i/Tcharm,i}\}\{e^{-t,charm,i/Tcharm,i}\}_2 + G_{charm,i}\, e^{-t,charm,i/Tcharm,i}\}\{e^{-t,charm,i/Tcharm,i}\}\{e^{-t,charm,i/Tcharm,i}\}_3 + \ldots + G_{charm,i}\{e^{-t,charm,i/Tcharm,i}\}\{e^{-t,charm,i/Tcharm,i}\}\ldots\{e^{-t,charm,i/Tcharm,i}\}_{m,i}$$

$$= G_{charm,i}\{(1/2)\, e^{-t,charm,i/t,charm,i1/2}\}_1 + G_{charm,i}\{(1/2)\, e^{-t,charm,i/t,charm,i1/2}\}\{(1/2)\, e^{-t,charm,i/t,charm,i1/2}\}_2 + G_{charm,i}\{(1/2)\, e^{-t,charm,i/t,charm,i1/2}\}\{(1/2)\, e^{-t,charm,i/t,charm,i1/2}\}\{(1/2)\, e^{-t,charm,i/t,charm,i1/2}\}_3 + \ldots + G_{charm,i}\{(1/2)\, e^{-t,charm,i/t,charm,i1/2}\}\{(1/2)\, e^{-t,charm,i/t,charm,i1/2}\}\ldots\{(1/2)\, e^{-t,charm,i/t,charm,i1/2}\}_{m,i},$$

Where t_i is the travel time of the ith species of charmlets for the mean free path for charmlet to ordinary baryon contact.

Where the average yield energy per nucleon or nucleus conversion caused by transformation by particles if the ith species is E_i.

The total yield energy for a given QCD bomb is in first order:

$$\sum(i = 1; i = M)\, [(N_i)(E_i)] = \sum(i = 1; i = M)\, (E_i)\, \{G_{charm,i}\{e^{-t,charm,i/Tcharm,i}\}_1 + G_{charm,i}\{e^{-t,charm,i/Tcharm,i}\}\{e^{-t,charm,i/Tcharm,i}\}_2 + G_{charm,i}\, e^{-t,charm,i/Tcharm,i}\}\{e^{-t,charm,i/Tcharm,i}\}\{e^{-t,charm,i/Tcharm,i}\}_3 + \ldots + G_{charm,i}\{e^{-t,charm,i/Tcharm,i}\}\{e^{-t,charm,i/Tcharm,i}\}\ldots\{e^{-t,charm,i/Tcharm,i}\}_{m,i}\}$$

$$= \sum(i = 1; i = M)\, (E_i)\{G_{charm,i}\{(1/2)\, e^{-t,charm,i/t,charm,i1/2}\}_1 + G_{charm,i}\{(1/2)\, e^{-t,charm,i/t,charm,i1/2}\}\{(1/2)\, e^{-t,charm,i/t,charm,i1/2}\}_2 + G_{charm,i}\{(1/2)\, e^{-t,charm,i/t,charm,i1/2}\}\{(1/2)\, e^{-t,charm,i/t,charm,i1/2}\}\{(1/2)\, e^{-t,charm,i/t,charm,i1/2}\}_3 + \ldots + G_{charm,i}\{(1/2)\, e^{-t,charm,i/t,charm,i1/2}\}\{(1/2)\, e^{-t,charm,i/t,charm,i1/2}\}\ldots\{(1/2)\, e^{-t,charm,i/t,charm,i1/2}\}_{m,i}\},$$

The time averaged explosive power of a given QCD bomb in the ship frame is in first order:

$$<d\{\sum(i = 1; i = M)\, [(N_i)(E_i)]\}/dt_{ship}> = <d\{\sum(i = 1; i = M)\, (E_i)\, \{G_{charm,i}\{e^{-t,charm,i/Tcharm,i}\}_1 + G_{charm,i}\{e^{-t,charm,i/Tcharm,i}\}\{e^{-t,charm,i/Tcharm,i}\}_2 + G_{charm,i}\, e^{-t,charm,i/Tcharm,i}\}\{e^{-t,charm,i/Tcharm,i}\}\{e^{-t,charm,i/Tcharm,i}\}_3 + \ldots +$$

$G_{charm,i}\{ e^{-t,charm,i/Tcharm,i}\} \{ e^{-t,charm,i/Tcharm,i}\} \ldots\{ e^{-t,charm,i/Tcharm,i}\}_{m,i}\}\}/dt_{ship}>$

$= <d\{\sum(i = 1; i = M) (E_i)\{G_{charm,i}\{(1/2) e^{-t,charm,i/t,charm,i1/2}\}_1 + G_{charm,i}\{(1/2) e^{-t,charm,i/t,charm,i1/2}\} \{(1/2) e^{-t,charm,i/t,charm,i1/2}\}_2 + G_{charm,i}\{(1/2) e^{-t,charm,i/t,charm,i1/2}\} \{(1/2) e^{-t,charm,i/t,charm,i1/2}\} \{(1/2) e^{-t,charm,i/t,charm,i1/2}\}_3 + \ldots + G_{charm,i}\{(1/2) e^{-t,charm,i/t,charm,i1/2}\} \{(1/2) e^{-t,charm,i/t,charm,i1/2}\} \ldots\{(1/2) e^{-t,charm,i/t,charm,i1/2}\}_{m,i}\}\}/dt_{ship}>$

Now, if $t_{i1/2}$ and T_i are background stationary frames and the ith species of charmlets are relativistic, the number of catalyzed baryons becomes in first order:

$N = G_{charm,i}\{e^{-t,charm,i/(Tcharm,i\gamma}\}_1 + G_{charm,i}\{e^{-t,charm,i/(Tcharm,i\gamma)}\}\{ e^{-t,charm,i/(Tcharm,i\gamma)}\}_2 + G_{charm,i}\{ e^{-t,charm,i/(Tcharm,i\gamma)}\}\{ e^{-t,charm,i/(Tcharm,i\gamma)}\}\{ e^{-t,charm,i/(Tcharm,i\gamma)}\}_3 + \ldots + G_{charm,i}\{ e^{-t,charm,i/(Tcharm,i\gamma)}\}\{ e^{-t,charm,i/(Tcharm,i\gamma)}\} \ldots\{ e^{-t,charm,i/(Tcharm,i\gamma)}\}_{m,i}$

$= G_{charm,i}\{(1/2) e^{-t,charm,i/(\gamma t,charm,i1/2)}\}_1 + G_{charm,i}\{(1/2) e^{-t,charm,i/(\gamma t,charm,i1/2)}\}\{(1/2) e^{-t,charm,i/(\gamma t,charm,i1/2)}\}_2 + G_{charm,i}\{(1/2) e^{-t,charm,i/(\gamma t,charm,i1/2)}\}\{(1/2) e^{-t,charm,i/(\gamma t,charm,i1/2)}\}\{(1/2) e^{-t,charm,i/(\gamma t,charm,i1/2)}\}_3 + \ldots + G_{charm,i}\{(1/2) e^{-t,charm,i/(\gamma t,charm,i1/2)}\}\{(1/2) e^{-t,charm,i/(\gamma t,charm,i1/2)}\} \ldots\{(1/2) e^{-t,charm,i/(\gamma t,charm,i1/2)}\}_{m,i}$

Where the average yield energy per nucleon or nucleus conversion caused by transformation by particles if the ith species is E_i.

The total yield energy for a given QCD bomb is in first order:

$\sum(i = 1; i = M) [(N_i)(E_i)] = \sum(i = 1; i = M) (E_i) \{G_{charm,i}\{e^{-t,charm,i/(Tcharm,i\gamma}\}_1 + G_{charm,i}\{e^{-t,charm,i/(Tcharm,i\gamma)}\}\{ e^{-t,charm,i/(Tcharm,i\gamma)}\}_2 + G_{charm,i}\{ e^{-t,charm,i/(Tcharm,i\gamma)}\}\{ e^{-t,charm,i/(Tcharm,i\gamma)}\}\{ e^{-t,charm,i/(Tcharm,i\gamma)}\}_3 + \ldots + G_{charm,i}\{ e^{-t,charm,i/(Tcharm,i\gamma)}\}\{ e^{-t,charm,i/(Tcharm,i\gamma)}\} \ldots\{ e^{-t,charm,i/(Tcharm,i\gamma)}\}_{m,i}\}$

$= \sum(i = 1; i = M) (E_{,i}) \{G_{charm,i}\{(1/2) e^{-t,charm,i/(\gamma t,charm,i1/2)}\}_1 + G_{charm,i}\{(1/2) e^{-t,charm,i/(\gamma t,charm,i1/2)}\}\{(1/2) e^{-t,charm,i/(\gamma t,charm,i1/2)}\}_2 + G_{charm,i}\{(1/2) e^{-t,charm,i/(\gamma t,charm,i1/2)}\}\{(1/2) e^{-t,charm,i/(\gamma t,charm,i1/2)}\}\{(1/2) e^{-t,charm,i/(\gamma t,charm,i1/2)}\}_3 + \ldots + G_{charm,i}\{(1/2) e^{-t,charm,i/(\gamma t,charm,i1/2)}\}\{(1/2) e^{-t,charm,i/(\gamma t,charm,i1/2)}\} \ldots\{(1/2) e^{-t,charm,i/(\gamma t,charm,i1/2)}\}_{m,i}\}$

The time averaged explosive power of a given QCD bomb in the ship frame is in first order:

$$<d\{\sum(i = 1; i = M) [(N_i)(E_i)]\} /dt_{ship} > = <d\{\sum(i = 1; i = M) (E_i) \{G_{charm,i}\{e^{-t,charm,i/(Tcharm,i\gamma}\}_1 + G_{charm,i}\{e^{-t,charm,i/(Tcharm,i\gamma)}\}\{ e^{-t,charm,i/(Tcharm,i\gamma)}\}_2 + G_{charm,i}\{ e^{-t,charm,i/(Tcharm,i\gamma)}\}\{ e^{-t,charm,i/(Tcharm,i\gamma)}\}\{ e^{-t,charm,i/(Tcharm,i\gamma)}\}_3 + \ldots + G_{charm,i}\{ e^{-t,charm,i/(Tcharm,i\gamma)}\}\{ e^{-t,charm,i/(Tcharm,i\gamma)}\} \ldots\{ e^{-t,charm,i/(Tcharm,i\gamma)}\}_{m,i}\}\}/dt_{ship} >$$

$$= <d\{\sum(i = 1; i = M) (E_i) \{G_{charm,i}\{(1/2) e^{-t,charm,i/(\gamma t,charm,i1/2)}\}_1 + G_{charm,i}\{(1/2) e^{-t,charm,i/(\gamma t,charm,i1/2)}\}\{(1/2) e^{-t,charm,i/(\gamma t,charm,i1/2)}\}_2 + G_{charm,i}\{(1/2) e^{-t,charm,i/(\gamma t,charm,i1/2)}\}\{(1/2) e^{-t,charm,i/(\gamma t,charm,i1/2)}\}\{(1/2) e^{-t,charm,i/(\gamma t,charm,i1/2)}\}_3 + \ldots + G_{charm,i}\{(1/2) e^{-t,charm,i/(\gamma t,charm,i1/2)}\}\{(1/2) e^{-t,charm,i/(\gamma t,charm,i1/2)}\} \ldots\{(1/2) e^{-t,charm,i/(\gamma t,charm,i1/2)}\}_{m,i}\}\}/dt_{ship} >$$

For cases where the average charmlet of the ith species would not decay with a probability of $n_{,i}/100$ after traveling the mean free path, and the probability of baryon or nuclei conversion upon contact is $k_{,i}/100$, but where the charmlets are always destroyed when contacting the nuclei or converted into a non-charm-transmutative form with a probability of unity, the number of catalyzed baryons or nuclei is equal to in first order:

$$N_i = G_{charm,i}\{n_{,i}/100\}\{k_i/100\}_1 + G_{charm,i}\{n_{,i}/100\}\{k_{,i}/100\}_1\{n_{,i}/100\}\{k_{,i}/100\}_2 + G_{charm,i}\{n_{,i}/100\}\{k_{,i}/100\}_1 \{n_{,i}/100\}\{k_{,i}/100\}_2\{n_{,i}/100\}\{k_{,i}/100\}_3 + \ldots + G_{charm,i}\{n_{,i}/100\}\{k_{,i}/100\}_1\{n_{,i}/100\}\{k_{,i}/100\}_2 \ldots\{ n_{,i}/100\}\{k_{,i}/100\}_{m,i},$$

Here, *m* is the number of time steps for the overall reaction extinction.

Where the average yield energy per nucleon or nucleus conversion caused by transformation by particles if the ith species is E_i.

The total yield energy for a given QCD bomb is in first order:

$$\sum(i = 1; i = M) [(N_i)(E_i)] = \sum(i = 1; i = M) (E_i) \{G_{charm,i}\{n_{,i}/100\}\{k_{,i}/100\}_1 + G_{charm,i}\{n_{,i}/100\}\{k_{,i}/100\}_1\{n_{,i}/100\}\{k_{,i}/100\}_2 + G_{charm,i}\{n_{,i}/100\}\{k_{,i}/100\}_1 \{n_{,i}/100\}\{k_{,i}/100\}_2\{n_{,i}/100\}\{k_{,i}/100\}_3 + \ldots + G_{charm,i}\{n_{,i}/100\}\{k_{,i}/100\}_1\{n_{,i}/100\}\{k_{,i}/100\}_2 \ldots\{ n_{,i}/100\}\{k_{,i}/100\}_{m,i}\},$$

The time averaged explosive power of a given QCD bomb in the ship frame is in first order:

$$<d\{\sum(i = 1; i = M) [(N_i)(E_i)]\} /dt_{ship} > = <d\{\sum(i = 1; i = M) (E_i) \{G_{charm,i}\{n_{,i}/100\}\{k_{,i}/100\}_1 + G_{charm,i}\{n_{,i}/100\}\{k_{,i}/100\}_1\{n_{,i}/100\}\{k_{,i}/100\}_2 + G_{charm,i}\{n_{,i}/100\}\{k_{,i}/100\}_1 \{n_{,i}/100\}\{k_{,i}/100\}_2\{n_{,i}/100\}\{k_{,i}/100\}_3 + \ldots + G_{charm,i}\{n_{,i}/100\}\{k_{,i}/100\}_1\{n_{,i}/100\}\{k_{,i}/100\}_2 \ldots\{ n_{,i}/100\}\{k_{,i}/100\}_m\}\}/dt_{ship} >$$

Considering the length of time, $t_{m,i}$, traveled per mean free path where the average nondecay probability is n/100 for each mean path traveled, the velocity of the ith species of charmlet is $L_i/t_{m,i}$, where the velocity of the charmlet is negligibly relativistic.

For cases where the ith species of charmlets produced are relativistic, the distance traveled by a given charmlet before it decays is given by:

$$[L_i/t_{mi}](\gamma) = [L_i/t_{m,ii}]\{1/\{1 - [(v_{,i}/C)^2]\}^{1/2}\}.$$

Once again, $N_i\ (t_i) = N_{0i}\, e^{-t,charm,i/Tcharm,i} = N_{0i}\ (1/2)e^{-t,charm,i/t,charm,i1/2}$, where T_i is the mean lifetime of a particle (such as an atom or subatomic particle), $t_{1/2}$ is the half-life of the particle, and N_{0i} is the starting population. $N_i\ (t_i)$ is the quantity that still remains and has not yet decayed after a time t; and half-life $t_{1/2} = \ln(2)/\lambda_i = T_i \ln(2)$, where λ_i is a positive number called the decay constant of the decaying quantity.

In the case of a charmlet species having a probability of nondecay along a mean path length of $P_i = n_i/100$, the number of charmlets out of a starting population of $N_{0i} = G_{charm,i}$ that contact a baryon or atomic nuclei is $N_{0i}\{n_i/100\}$ where $\{n_{,i}/100\} = e^{-t,charm,i/Tcharm,i} = (1/2)\ e^{-t,charm,i/t,charm,i1/2}$. Therefore, the number of catalyzed baryons becomes in first order:

$$N_i = G_{charmi}\{e^{-t,charm,i/Tcharm,i}\}\{k_{,i}/100\}_1 + G_{charm,i}\{e^{-t,charm,i/Tcharm,i}\}\{k_{,i}/100\}_1\{e^{-t,charm,i/Tcharm,i}\}\{k_{,i}/100\}_2 + G_{charm,i}\ \{e^{-t,charm,i/Tcharm,i}\}\{k_{,i}/100\}_1\ \{e^{-t,charm,i/Tcharm,i}\}\{k_{,i}/100\}_2\ \{e^{-t,charm,i/Tcharm,i}\}\{k_{,i}/100\}_3 + \ldots + G_{charm,i}\{e^{-t,charm,i/Tcharm,i}\}\{k_{,i}/100\}_1\{e^{-t,charm,i/Tcharm,i}\}\{k_{,i}/100\}_2 \ldots\{e^{-t,charm,i/Tcharm,i}\}\{k_{,i}/100\}_{m,i}$$

$$= G_{charm,i}\{(1/2)\ e^{-t,charm,i/t,charm,i1/2}\}\{k_{,i}/100\}_1 + G_{charm,i}\{(1/2)\ e^{-t,charm,i/t,charm,i1/2}\}\{k_{,i}/100\}_1\{(1/2)\ e^{-t,charm,i/t,charm,i1/2}\}\{k_{,i}/100\}_2 + G_{charm,i}\{(1/2)\ e^{-t,charm,i/t,charm,i1/2}\}\{k_{,i}/100\}_1\{(1/2)\ e^{-t,charm,i/t,charm,i1/2}\}\{k_{,i}/100\}_2\ \{(1/2)\ e^{-t,charm,i/t,charm,i1/2}\}\{k_{,i}/100\}_3 + \ldots + G_{charm,i}\{(1/2)\ e^{-}$$

$$^{t,charm,i/t,charm,i1/2}\}\{k_{,i}/100\}_1\{(1/2)\, e^{-t,charm,i/t,charm,i1/2}\}\{k_{,i}/100\}_2 \ldots\{(1/2)\, e^{-t,charm,i/t,charm,i1/2}\}\{k_{,i}/100\}_{m,i}$$

Where the average yield energy per nucleon or nucleus conversion caused by transformation by particles if the ith species is E_i.

The total yield energy for a given QCD bomb is in first order:

$$\sum(i = 1;\ i = M)\ [(N_i)(E_i)] = \sum(i = 1;\ i = M)\ (E_i)\ \{G_{charm,i}\{e^{-t,charm,i/Tcharm,i}\}\{k_{,i}/100\}_1 + G_{charm,i}\{e^{-t,charm,i/Tcharm,i}\}\{k_{,i}/100\}_1\{e^{-t,charm,i/Tcharm,i}\}\{k_{,i}/100\}_2 + G_{charm,i}\ \{e^{-t,charm,i/Tcharm,i}\}\{k_{,i}/100\}_1\ \{e^{-t,charm,i/Tcharm,i}\}\{k_{,i}/100\}_2\ \{\ e^{-t,charm,i/Tcharm,i}\}\{k_{,i}/100\}_3 + \ldots + G_{charm,i}\{\ e^{-t,charm,i/Tcharm,i}\}\{k_{,i}/100\}_1\{\ e^{-t,charm,i/Tcharm,i}\}\{k_{,i}/100\}_2\ \ldots\{\ e^{-t,charm,i/Tcharm,i}\}\{k_{,i}/100\}_{m,i}\}$$

$$= \sum(i = 1;\ i = M)\ (E_i)\ \{G_{charm,i}\{(1/2)\ e^{-t,charm,i/t,charm,i1/2}\}\{k_{,i}/100\}_1 + G_{charm,i}\{(1/2)\ e^{-t,charm,i/t,charm,i1/2}\}\{k_{,i}/100\}_1\{(1/2)\ e^{-t,charm,i/t,charm,i1/2}\}\{k_{,i}/100\}_2 + G_{charm,i}\{(1/2)\ e^{-t,charm,i/t,charm,i1/2}\}\{k_{,i}/100\}_1\{(1/2)\ e^{-t,charm,i/t,charm,i1/2}\}\{k_{,i}/100\}_2\ \{(1/2)\ e^{-t,charm,i/t,charm,i1/2}\}\{k_{,i}/100\}_3 + \ldots + G_{charm,i}\{(1/2)\ e^{-t,charm,i/t,charm,i1/2}\}\{k_{,i}/100\}_1\{(1/2)\ e^{-t,charm,i/t,charm,i1/2}\}\{k_{,i}/100\}_2 \ldots\{(1/2)\ e^{-t,charm,i/t,charm,i1/2}\}\{k_{,i}/100\}_{m,i}\}$$

The time averaged explosive power of a given QCD bomb in the ship frame is in first order:

$$<d\{\sum(i = 1;\ i = M)\ [(N_i)(E_i)]\}\ /dt_{ship}> = <d\{\sum(i = 1;\ i = M)\ (E_i)\ \{G_{charm,i}\{e^{-t,charm,i/Tcharm,i}\}\{k_{,i}/100\}_1 + G_{charm,i}\{e^{-t,charm,i/Tcharm,i}\}\{k_{,i}/100\}_1\{e^{-t,charm,i/Tcharm,i}\}\{k_{,i}/100\}_2 + G_{charm,i}\ \{e^{-t,charm,i/Tcharm,i}\}\{k_{,i}/100\}_1\ \{e^{-t,charm,i/Tcharm,i}\}\{k_{,i}/100\}_2\ \{\ e^{-t,charm,i/Tcharm,i}\}\{k_{,i}/100\}_3 + \ldots + G_{charm,i}\{\ e^{-t,charm,i/Tcharm,i}\}\{k_{,i}/100\}_1\{\ e^{-t,charm,i/Tcharm,i}\}\{k_{,i}/100\}_2\ \ldots\{\ e^{-t,charm,i/Tcharm,i}\}\{k_{,i}/100\}_{m,i}\}\}/dt_{ship}>$$

$$= <d\{\sum(i = 1;\ i = M)\ (E_i)\ \{G_{charm,i}\{(1/2)\ e^{-t,charm,i/t,charm,i1/2}\}\{k_{,i}/100\}_1 + G_{charm,i}\{(1/2)\ e^{-t,charm,i/t,charm,i1/2}\}\{k_{,i}/100\}_1\{(1/2)\ e^{-t,charm,i/t,charm,i1/2}\}\{k_{,i}/100\}_2 + G_{charm,i}\{(1/2)\ e^{-t,charm,i/t,charm,i1/2}\}\{k_{,i}/100\}_1\{(1/2)\ e^{-t,charm,i/t,charm,i1/2}\}\{k_{,i}/100\}_2\ \{(1/2)\ e^{-t,charm,i/t,charm,i1/2}\}\{k_{,i}/100\}_3 + \ldots + G_{charm,i}\{(1/2)\ e^{-t,charm,i/t,charm,i1/2}\}\{k_{,i}/100\}_1\{(1/2)\ e^{-t,charm,i/t,charm,i1/2}\}\{k_{,i}/100\}_2 \ldots\{(1/2)\ e^{-t,charm,i/t,charm,i1/2}\}\{k_{,i}/100\}_{m,i}\}\}/dt_{ship}>$$

If $t_{i1/2}$ and T_i are background stationary frames and the charmlets are relativistic, the number of catalyzed baryons becomes in first order:

$$N_i = G_{charm,i}\{e^{-t,charm,i/(Tcharm,i\gamma}\}\{k_{,i}/100\}_1 + G_{charm,i}\{e^{-t,charm,i/(Tcharm,i\gamma)}\}\{k_{,i}/100\}_1\{e^{-t,charm,i/(Tcharm,i\gamma)}\}\{k_{,i}/100\}_2 + G_{charm,i}\{e^{-t,charm,i/(Tcharm,i\gamma)}\}\{k/100\}_1\{e^{-t,charm,i\ /(Tcharm,i\ \gamma)}\}\{k_{,i}/100\}_2\{e^{-t,charm,i/(Tcharm,i\gamma)}\}\{k_{,i}/100\}_3 + \ldots + G_{charm,i}\{e^{-t,charm,i/(Tcharm,i\gamma)}\}\{k_{,i}/100\}_1\{e^{-t,charm,i/(Tcharm,i\gamma)}\}\{k_{,i}/100\}_2 \ldots\{e^{-t,charm,i/(Tcharm,i\gamma)}\}\{k_{,i}/100\}_{m,i}$$

$$= G_{charm,i}\{(1/2)\ e^{-t,charm,i/(\gamma t,charm,i1/2)}\}\{k_{,i}/100\}_1 + G_{charm,i}\{(1/2)\ e^{-t,charm,i/(\gamma t,charm,i1/2)}\}\{k_{,i}/100\}_1\{(1/2)\ e^{-t,charm,i/(\gamma t,charm,i1/2)}\}\{k_{,i}/100\}_2 + G_{charm,i}\{(1/2)\ e^{-t,charm,i/(\gamma t,charm,i1/2)}\}\{k_{,i}/100\}_1\{(1/2)\ e^{-t,charm,i/(\gamma t,charm,i1/2)}\}\{k_{,i}/100\}_2\{(1/2)\ e^{-t,charm,i/(\gamma t,charm,i1/2)}\}\{k_{,i}/100\}_3 + \ldots + G_{charm,i}\{(1/2)\ e^{-t,charm,i/(\gamma t,charm,i1/2)}\}\{k_{,i}/100\}_1\{(1/2)\ e^{-t,charm,i/(\gamma t,charm,i1/2)}\}\{k_{,i}/100\}_2 \ldots\{(1/2)\ e^{-t,charm,i/(\gamma t,charm,i1/2)}\}\{k_{,i}/100\}_{m,i}$$

Where the average yield energy per nucleon or nucleus conversion caused by transformation by particles in the ith species is E_i.

The total yield energy for a given QCD bomb is in first order:

$$\sum(i = 1;\ i = M)\ [(N_i)(E_i)] = \sum(i = 1;\ i = M)\ (E_i)\ \{G_{charm,i}\{e^{-t,charm,i/(Tcharm,i\gamma}\}\{k_{,i}/100\}_1 + G_{charm,i}\{e^{-t,charm,i/(Tcharm,i\gamma)}\}\{k_{,i}/100\}_1\{e^{-t,charm,i/(Tcharm,i\gamma)}\}\{k_{,i}/100\}_2 + G_{charm,i}\{e^{-t,charm,i\ /(Tcharm,i\ \gamma)}\}\{k_i/100\}_1\{e^{-t,charm,i\ /(Tcharm,i\ \gamma)}\}\{k_{,i}/100\}_2\{e^{-t,charm,i/(Tcharm,i\gamma)}\}\{k_{,i}/100\}_3 + \ldots + G_{charm,i}\{e^{-t,charm,i/(Tcharm,i\gamma)}\}\{k_{,i}/100\}_1\{e^{-t,charm,i/(Tcharm,i\gamma)}\}\{k_{,i}/100\}_2 \ldots\{e^{-t,charm,i/(Tcharm,i\gamma)}\}\{k_{,i}/100\}_{m,i}\}$$

$$= \{\sum(i = 1;\ i = M)\ (E_i)\ \{G_{charm,i}\{(1/2)\ e^{-t,charm,i/(\gamma t,charm,i1/2)}\}\{k_{,i}/100\}_1 + G_{charm,i}\{(1/2)\ e^{-t,charm,i/(\gamma t,charm,i1/2)}\}\{k_{,i}/100\}_1\{(1/2)\ e^{-t,charm,i/(\gamma t,charm,i1/2)}\}\{k_{,i}/100\}_2 + G_{charm,i}\{(1/2)\ e^{-t,charm,i/(\gamma t,charm,i1/2)}\}\{k_{,i}/100\}_1\{(1/2)\ e^{-t,charm,i/(\gamma t,charm,i1/2)}\}\{k_{,i}/100\}_2\{(1/2)\ e^{-t,charm,i/(\gamma t,charm,i1/2)}\}\{k_{,i}/100\}_3 + \ldots + G_{charm,i}\{(1/2)\ e^{-t,charm,i/(\gamma t,charm,i1/2)}\}\{k_{,i}/100\}_1\{(1/2)\ e^{-t,charm,i/(\gamma t,charm,i1/2)}\}\{k_{,i}/100\}_2 \ldots\{(1/2)\ e^{-t,charm,i/(\gamma t,charm,i1/2)}\}\{k_{,i}/100\}_{m,i}\}$$

The time averaged explosive power of a given QCD bomb in the ship frame is in first order:

$$<d\{\sum(i = 1;\ i = M)\ [(N_i)(E_i)]\}\ /dt_{ship}> = <d\{\sum(i = 1;\ i = M)\ (E_i)\ \{G_{charm,i}\{e^{-t,charm,i/(Tcharm,i\gamma}\}\{k_{,i}/100\}_1 + G_{charm,i}\{e^{-t,charm,i/(Tcharm,i\gamma)}\}\{k_{,i}/100\}_1\{e^{-t,charm,i/(Tcharm,i\gamma)}\}\{k_{,i}/100\}_2 + G_{charm,i}\{e^{-t,charm,i\ /(Tcharm,i\ \gamma)}\}\{k_i/100\}_1\{e^{-t,charm,i\ /(Tcharm,i\ \gamma)}\}\{k_{,i}/100\}_2\{e^{-t,charm,i/(Tcharm,i\gamma)}\}\{k_{,i}/100\}_3 + \ldots + G_{charm,i}\{e^{-t,charm,i/(Tcharm,i\gamma)}\}\{k_{,i}/100\}_1\{e^{-t,charm,i/(Tcharm,i\gamma)}\}\{k_{,i}/100\}_2 \ldots\{e^{-t,charm,i/(Tcharm,i\gamma)}\}\{k_{,i}/100\}_{m,i}\}/dt_{ship}>$$

$$= \langle d\{\sum(i = 1; i = M) (E_i) \{G_{charm,i}\{(1/2)\, e^{-t,charm,i/(\gamma t,charm,i1/2)}\} \{k_{,i}/100\}_1 + G_{charm,i}\{(1/2)\, e^{-t,charm,i/(\gamma t,charm,i1/2)}\} \{k_{,i}/100\}_1 \{(1/2)\, e^{-t,charm,i/(\gamma t,charm,i1/2)}\} \{k_{,i}/100\}_2 + G_{charm,i}\{(1/2)\, e^{-t,charm,i/(\gamma t,charm,i1/2)}\} \{k_{,i}/100\}_1 \{(1/2)\, e^{-t,charm,i/(\gamma t,charm,i1/2)}\} \{k_{,i}/100\}_2 \{(1/2)\, e^{-t,charm,i/(\gamma t,charm,i1/2)}\} \{k_{,i}/100\}_3 + \ldots + G_{charm,i}\{(1/2)\, e^{-t,charm,i/(\gamma t,charm,i1/2)}\} \{k_{,i}/100\}_1 \{(1/2)\, e^{-t,charm,i/(\gamma t,charm,i1/2)}\} \{k_{,i}/100\}_2 \ldots \{(1/2)\, e^{-t,charm,i/(\gamma t,charm,i1/2)}\} \{k_{,i}/100\}_{m,i}\}/dt_{ship} \rangle$$

We would definitely not want stable charmlets that are transmutative into-charm-matter or ones that are sufficiently long lived so as to produce a reaction that could consume the entire planet, or propagate interstellar distances to reach other stars and planets where they could wreak havoc. Such charmlets may even convert interstellar gas and dust in a gradual propagating wave front and literally gradually eat away at the universe or multiverse completely in a progressive reaction wave front. However, to produce such charmlets or a black hole capable of doing this would require a galactic-sized accelerator—in other words, it is highly unlikely at best.

Regardless, charmlet-producing nuclear explosives might be used for the following:

1. Interstellar ramjet propulsion
2. Intergalactic ramjet propulsion
3. Hyper spatial rocket propulsion
4. Nuclear charm matter bomblet pulse drive
5. Nuclear charm matter bomblet pellet runway propulsion
6. Mass fuel beam nuclear charm matter bomblet propulsion
7. Einsteinian 4-D photon, electron, muon, tauon, positron, anti-muon, anti-tauon, proton, antiproton, ion, anti-ion, charged exotic meson, charged exotic baryon, electron neutrino, anti-electron neutrino, muon neutrino, antimuon neutrino, tauon neutrino, anti-tauon neutrino, gravity wave, and quantum scale gravity wave or graviton rockets
8. Hyper spatial photon, electron, muon, tauon, positron, anti-muon, anti-tauon, proton, antiproton, ion, anti-ion, charged exotic meson, charged exotic baryon, electron neutrino, anti-electron neutrino, muon neutrino, antimuon neutrino, tauon neutrino, anti-tauon neutrino, gravity wave, and quantum-scale gravity wave or graviton rockets. For rocket vehicle applications, the exothermic energy produced in charm matter reactors can be used to energize thrust-stream generators.

Note that gravitons are hypothetical at this point. Their existence would require the discovery of the spin-2 boson, which is yet to be accomplished.

So for stable charmlet technology, only assured-to-die-out chain reactions must be produced; otherwise, we risk universal destruction. The ramifications of the technology as a power source, however, are too profound not to be mentioned. This is true, not only from a philosophical perspective of enabling technology that is far superior in power than pure matter-antimatter conversion, but, more to the point, a technology that can revolutionize commercial energy production for manned starship propulsion. Consider the case where the matter within any hyper spatial dimensions that are coupled to our universe would not have periodic table elements or baryons having the exact properties as those found within our ordinary 4-D Einsteinian space-time. For such hyper spatial matters that have quantum properties that are similar to the baryonic mass within our ordinary 4-D universe, such exothermic conversion may still be possible, thereby enabling the production of hyper spatial ramjets and hyper spatial photon, electron, muon, tauon, positron, anti-muon, anti-tauon, proton, antiproton, ion, anti-ion, charged exotic meson, charged exotic baryon, electron neutrino, anti-electron neutrino, muon neutrino, antimuon neutrino, tauon neutrino, anti-tauon neutrino, gravity wave, and quantum-scale gravity wave or graviton rockets.

There are several plausible candidates for mechanisms that can be used to perform nuclear fusion in order to produce the required temperatures and particle kinetic energies so as to enable the bulk production of metastable charmlets to provide for substantial but die-out forms of charm chain reactions. The devices are shaped charge nuclear fusion and/or fission devices. To the best of our knowledge, none of the associated nuclear explosive types have yet been developed.

What can be accomplished with charm reactions might also work for charmonium catalysis and bottomonium catalysis. Toponium catalysis may prove possible in the long run; however, the production of toplets as yet have been unverified and may remain so for a long time. This is because toplets, which are bound states comprised of top quarks, would require the strong nuclear force to travel the distances to other quarks cocreated in particle accelerator collisions. All top quark-producing particle collisions to date have produced top quarks that decay too quickly for the light-speed-limited strong force to travel the distance between any cocreated top quarks, thereby preventing any stabilizing bound states.

Chapter 13

Bottom Matter Reactors and Bombs and Other Things Nice

This chapter includes a description of highly conjectural forms of QCD/QED reactors and bombs. Such reactors and bombs would be used to power highly relativistic starships by converting a large fuel mass at least partially into energy and/or exothermic conversion of interstellar ordinary baryonic matter into strange, charmed, and/or bottom quark-based matter. Herein, attractive use of Times New Roman font enlarged to 14 point size made for clarity.

Now that the nuclear and the eventual exotic QCD energy applications genie is permanently out of the bottle and exotic QCD energy applications loom on the horizon, how might such applications be used to reach the stars?

Below is a brief summary of thoughts on how exotic applications of nuclear and subnuclear reactions can open up the cosmos for colonization and exploration by humanity. The technologies considered have likely already been pondered and studied to some degree; but if not, they definitely will be in the future.

Imagine a technology that could convert the elements in the periodic table atomic to other exotic forms of baryonic matter releasing prodigious quantities of energy in a mass-specific reaction energy that is higher than that released by nuclear fusion. It has been proposed that certain species of some stable bottomlets that might be produced in particle accelerators may be able to convert periodic table atoms into bottom matter. An unfortunate consequence of this may be a runaway reaction that could cause the planet to be exothermically converted into bottom matter or matter comprised mostly of bottom quarks with some additional species of quarks mixed in.

Stable catalytic bottomlets may be ideal for powering future interstellar ramjet starships (ISRs). Imagine a large mass of stable bottom matter for which interstellar and intergalactic matter would be funneled into the intake of a reaction chamber where the intake mass would be converted to quickly decaying bottomlets that would convert even more of the intake mass into energy through decay processes. The drag energy in the form of heat could be recycled to power photon, ion, electron, positron, muon, antimuon, tauon, antitauon, proton, antiproton, exotic charged meson, and/or exotic charged baryon rockets. The stable or metastable bottomlets produced could be used as an accelerated thrust stream. A starship using this form of propulsion might accelerate and achieve high Lorentz factors and/or near light speeds in any hyperspatial dimension where periodic table atomic matter or other convertible baryonic matter is present.

Let us consider a scenario where a species of bottomlet would not decay on average n percent of the time before making contact with an ordinary baryon or atomic nucleus and where the probability of the conversion of the baryon or atomic nucleus is 100 percent. The average probability of the initial bottomlet converting an ordinary nucleon or atomic nucleus to another catalytic bottomlet is n/100. After m time steps where, for which each time step the probability of the average catalytic bottomlet converting a baryon or nucleus it first contacts is n/100, the number of catalyzed baryons or nuclei is equal to in first order:

$$N = G_{bottom}\{n/100\}_1 + G_{bottom}\{n/100\}_1 \{n/100\}_2 + G_{bottom}\{n/100\}_2 \{n/100\}_2\{n/100\}_3 + \ldots + G_{bottom}\{n/100\}_1 \{n/100\}_2 \ldots\{ n/100\}_m$$

Where m is the number of time steps for the overall reaction extinction and G_{bottom} is the initial population of bottomlets.

Another way of presenting reaction-wave propagation is to consider the average lifetime of the bottomlets and the average distance traveled by the bottomlets relative to the mean free path for bottomlet and nucleon or atomic nuclei contact. For cases where the average bottomlet would not decay with a probability of n/100 before completing the travel of its mean free path, the number of catalyzed baryons or nuclei is also equal to in first order:

$$N = G_{bottom}\{n/100\}_1 + G_{bottom}\{n/100\}_2 \{n/100\}_2 + G_{bottom}\{n/100\}_2 \{n/100\}_2\{n/100\}_3 + \ldots + G_{bottom}\{n/100\}_1 \{n/100\}_2 \ldots\{ n/100\}_m$$

Where m is the number of time steps for the overall reaction extinction.

Considering the length of time, t_m, traveled per mean free path where the average nondecay probability is n/100 for each mean path traveled, the velocity of the bottomlet is L/t_m, where the velocity of the bottomlet is negligibly relativistic.

For cases where the bottomlets produced are relativistic, the average distance traveled by such bottomlets before it decays is given by:

$$D = [L/t_m](\gamma)\ (n)/(100) = [L/t_m]\{1/\{1 - [(v/C)^2]\}^{1/2}\}(n)/(100)$$

Now, applying the exponetial decay formula:

$$N(t) = N_0\, e^{-t,bottom/Tbottom} = N_0\, (1/2) e^{-t,bottom/t,bottom1/2}$$

Where T is the mean lifetime of a particle (such as an atom or subatomic particle), $t_{1/2}$ is the half-life of the particle, and N_0 is the starting population. $N(t)$ is the quantity that still remains and has not yet decayed after a time t.

Therefore the half-life is expressed as follows:

$$t_{1/2} = \ln(2)/\lambda = T \ln(2)$$

Here, λ is a positive number called the decay constant of the decaying quantity.

For a bottomlet having a probability of nondecay along a mean path length of P = n/100, the number of bottomlets out of a starting population of $N_0 = G_{bottom}$ that contact a baryon or atomic nuclei is $N_0\{n/100\}$ where $\{n/100\} = e^{-t,bottom/Tbottom} = (1/2)\, e^{-t,bottom/t,bottom1/2}$.

Therefore, the number of catalyzed baryons becomes:

$$N = G_{bottom}\{e^{-t,bottom/Tbottom}\}_1 + G_{bottom}\{e^{-t,bottom/Tbottom}\}\{e^{-t,bottom/Tbottom}\}_2 + G_{bottom}\, e^{-t,bottom/Tbottom}\}\{e^{-t,bottom/Tbottom}\}\{e^{-t,bottom/Tbottom}\}_3 + \ldots + G_{bottom}\{e^{-t,bottom/Tbottom}\}\{e^{-t,bottom/Tbottom}\}\ldots\{e^{-t,bottom/Tbottom}\}_m$$

$$= G_{bottom}\{(1/2)\, e^{-t,bottom/t,bottom1/2}\}_1 + G_{bottom}\{(1/2)\, e^{-t,bottom/t,bottom1/2}\}\{(1/2)\, e^{-t,bottom/t,bottom1/2}\}_2 + G_{bottom}\{(1/2)\, e^{-t,bottom/t,bottom1/2}\}\{(1/2)\, e^{-t,bottom/t,bottom1/2}\}\{(1/2)\, e^{-t,bottom/t,bottom1/2}\}_3 + \ldots + G_{bottom}\{(1/2)\, e^{-t,bottom/t,bottom1/2}\}\{(1/2)\, e^{-t,bottom/t,bottom1/2}\}\ldots\{(1/2)\, e^{-t,bottom/t,bottom1/2}\}_m,$$

Where t is the travel time of the bottomlets for the mean free path for bottomlet to ordinary baryon contact.

Now, if $t_{1/2}$ and T are background stationary frames and the bottomlets are relativistic, the number of catalyzed baryons becomes in first order:

$$N = G_{bottom}\{e^{-t,bottom/(Tbottom\gamma}\}_1 + G_{bottom}\{e^{-t,bottom/(Tbottom\gamma)}\}\{e^{-t,bottom/(Tbottom\gamma)}\}_2 + G_{bottom}\{e^{-t,bottom/(Tbottom\gamma)}\}\{e^{-t,bottom/(Tbottom\gamma)}\}\{e^{-t,bottom/(Tbottom\gamma)}\}_3 + \ldots + G_{bottom}\{e^{-t,bottom/(Tbottom\gamma)}\}\{e^{-t,bottom/(Tbottom\gamma)}\}\ldots\{e^{-t,bottom/(Tbottom\gamma)}\}_m$$

$$= G_{bottom}\{(1/2)\, e^{-t,bottom/(\gamma t,bottom1/2)}\}_1 + G_{bottom}\{(1/2)\, e^{-t,bottom/(\gamma t,bottom1/2)}\}\{(1/2)\, e^{-t,bottom/(\gamma t,bottom1/2)}\}_2 + G_{bottom}\{(1/2)\, e^{-t,bottom/(\gamma t,bottom1/2)}\}\{(1/2)\, e^{-t,bottom/(\gamma t,bottom1/2)}\}\{(1/2)\, e^{-t,bottom/(\gamma t,bottom1/2)}\}_3 + \ldots$$

$+ G_{bottom}\{(1/2)\ e^{-t,bottom/(\gamma t,bottom1/2)}\}\{(1/2)\ e^{-t,bottom/(\gamma t,bottom1/2)}\} \ldots \{(1/2)\ e^{-t,bottom/(\gamma t,bottom1/2)}\}_m$

For cases where the average bottomlet would not decay with a probability of n/100 after traveling the mean free path, and the probability of baryon or nuclei conversion upon contact is k/100, but where the bottomlets are always destroyed when contacting the nuclei or converted into a non-bottom-transmutative form with a probability of unity, the number of catalyzed baryons or nuclei is equal to in first order:

$$N = G_{bottom}\{n/100\}\{k/100\}_1 + G_{bottom}\{n/100\}\{k/100\}_1\{n/100\}\{k/100\}_2 + G_{bottom}\{n/100\}\{k/100\}_1\{n/100\}\{k/100\}_2\{n/100\}\{k/100\}_3 + \ldots + G_{bottom}\{n/100\}\{k/100\}_1\{n/100\}\{k/100\}_2 \ldots \{n/100\}\{k/100\}_m,$$

Here, *m* is the number of time steps for the overall reaction extinction.

Considering the length of time, t_m, traveled per mean free path where the average nondecay probability is n/100 for each mean path traveled, the velocity of the bottomlet is L/t_m, where the velocity of the bottomlet is negligibly relativistic.

For cases where the bottomlets produced are relativistic, the distance traveled by a given bottomlet before it decays is given in first order:

$$[L/t_m](\gamma) = [L/t_m]\{1/\{1 - [(v/C)^2]\}^{1/2}\}$$

Once again, $N(t) = N_0\ e^{-t,bottom/Tbottom} = N_0\ (1/2)e^{-t,bottom/t,bottom1/2}$, where T is the mean lifetime of a particle (such as an atom or subatomic particle), $t_{1/2}$ is the half-life of the particle, and N_0 is the starting population. $N(t)$ is the quantity that still remains and has not yet decayed after a time t, and half-life $t_{1/2} = \ln(2)/\lambda = T \ln(2)$, where λ is a positive number called the decay constant of the decaying quantity.

In the case of a bottomlet species having a probability of nondecay along a mean path length of $P = n/100$, the number of bottomlets out of a starting population of $N_0 = G_{bottom}$ that contact a baryon or atomic nuclei is $N_0\{n/100\}$ where $\{n/100\} = e^{-t,bottom/Tbottom} = (1/2)\ e^{-t,bottom/t,bottom1/2}$. Therefore, the number of catalyzed baryons becomes in first order:

$N = G_{bottom}\{e^{-t,bottom/Tbottom}\}\{k/100\}_1 + G_{bottom}\{e^{-t,bottom/Tbottom}\}\{k/100\}_1\{e^{-t,bottom/Tbottom}\}\{k/100\}_2 + G_{bottom}\{e^{-t,bottom/Tbottom}\}\{k/100\}_1\{e^{-t,bottom/Tbottom}\}\{k/100\}_2\{e^{-t,bottom/Tbottom}\}\{k/100\}_3 + \ldots + G_{bottom}\{e^{-}$

$^{t,bottom/Tbottom}\}\{k/100\}_1\{ e^{-t,bottom/Tbottom}\}\{k/100\}_2 \ldots\{ e^{-t,bottom/Tbottom}\}\{k/100\}_m$

$$= G_{bottom}\{(1/2)\, e^{-t,bottom/t,bottom1/2}\}\{k/100\}_1 + G_{bottom}\{(1/2)\, e^{-t,bottom/t,bottom1/2}\}\{k/100\}_1\{(1/2)\, e^{-t,bottom/t,bottom1/2}\}\{k/100\}_2 + G_{bottom}\{(1/2)\, e^{-t,bottom/t,bottom1/2}\}\{k/100\}_1\{(1/2)\, e^{-t,bottom/t,bottom1/2}\}\{k/100\}_2\ \{(1/2)\, e^{-t,bottom/t,bottom1/2}\}\{k/100\}_3 + \ldots + G_{bottom}\{(1/2)\, e^{-t,bottom/t,bottom1/2}\}\{k/100\}_1\{(1/2)\, e^{-t,bottom/t,bottom1/2}\}\{k/100\}_2 \ldots\{(1/2)\, e^{-t,bottom/t,bottom1/2}\}\{k/100\}_m$$

If $t_{1/2}$ and T are background stationary frames and the bottomlets are relativistic, the number of catalyzed baryons becomes in first order:

$$N = G_{bottom}\{e^{-t,bottom/(Tbottom\gamma}\}\{k/100\}_1 + G_{bottom}\{e^{-t,bottom/(Tbottom\gamma)}\}\{k/100\}_1\{ e^{-t,bottom/(Tbottom\gamma)}\}\{k/100\}_2 + G_{bottom}\{ e^{-t,bottom/(Tbottom\gamma)}\}\{k/100\}_1\{ e^{-t,bottom/(Tbottom\gamma)}\}\{k/100\}_2\{ e^{-t,bottom/(Tbottom\gamma)}\}\{k/100\}_3 + \ldots + G_{bottom}\{ e^{-t,bottom/(Tbottom\gamma)}\}\{k/100\}_1\{ e^{-t,bottom/(Tbottom\gamma)}\}\{k/100\}_2 \ldots\{ e^{-t,bottom/(Tbottom\gamma)}\}\{k/100\}_m$$

$$= G_{bottom}\{(1/2)\, e^{-t,bottom/(\gamma t,bottom1/2)}\}\{k/100\}_1 + G_{bottom}\{(1/2)\, e^{-t,bottom/(\gamma t,bottom1/2)}\}\{k/100\}_1\{(1/2)\, e^{-t,bottom/(\gamma t,bottom1/2)}\}\{k/100\}_2 + G_{bottom}\{(1/2)\, e^{-t,bottom/(\gamma t,bottom1/2)}\}\{k/100\}_1\{(1/2)\, e^{-t,bottom/(\gamma t,bottom1/2)}\}\{k/100\}_2\{(1/2)\, e^{-t,bottom/(\gamma t,bottom1/2)}\}\{k/100\}_3 + \ldots + G_{bottom}\{(1/2)\, e^{-t,bottom/(\gamma t,bottom1/2)}\}\{k/100\}_1\{(1/2)\, e^{-t,bottom/(\gamma t,bottom1/2)}\}\{k/100\}_2 \ldots\{(1/2)\, e^{-t,bottom/(\gamma t,bottom1/2)}\}\{k/100\}_m$$

Let us consider a scenario where a species, i, of bottomlet would not decay on average n percent of the time before making contact with an ordinary baryon or atomic nucleus and where the probability of the conversion of the baryon or atomic nucleus is 100 percent. The average probability of the initial bottomlet converting an ordinary nucleon or atomic nucleus to another catalytic bottomlet is $n_{,i}/100$. After m time steps, where, for which each time step the probability of the average catalytic bottomlet converting a baryon or nucleus it first contacts is $n_{,i}/100$, the number of catalyzed baryons or nuclei is equal to in first order:

$$N_{,i} = G_{bottom,i}\{n_{,i}/100\}_1 + G_{bottom,i}\{n_i/100\}_1\ \{n_{,i}/100\}_2 + G_{bottom,\ i}\{n_{,i}/100\}_2\ \{n_{,i}/100\}_2\{n_{,i}/100\}_3 + \ldots + G_{bottom,i}\{n_{,i}/100\}_1\ \{n_{,i}/100\}_2 \ldots\{\ n_{,i}/100\}_{m,i}$$

Where $m_{,i}$ is the number of time steps for the overall reaction extinction and $G_{bottom,i}$ is the initial population of bottomlets.

So for M species of bottomlets, where each species has its unique mean free path, initial population, and number of time steps for overall reaction extinctions, the number of catalyzed baryons or nuclei is equal to in first order:

$$\sum(i = 1; i = M)\ N_i = \sum(i = 1; i = M)\ \{G_{bottom,i}\{n_{,i}/100\}_1 + G_{bottom,i}\{n_{,i}/100\}_1 \{n_{,i}/100\}_2 + G_{bottom,\ i}\{n_{,i}/100\}_2\ \{n_{,i}/100\}_2\{n_{,i}/100\}_3 + \ldots + G_{bottom,i}\{n_{,i}/100\}_1\ \{n_{,i}/100\}_2 \ldots\{ n_{,i}/100\}_{m,i}\}$$

Where the average yield energy per nucleon or nucleus conversion caused by transformation by particles if the ith species is E_i.

The total yield energy for a given QCD bomb is in first order:

$$\sum(i = 1; i = M)\ [(N_i)(E_i)] = \sum(i = 1; i = M)\ (E_i)\ \{\{G_{bottom,i}\{n_{,i}/100\}_1 + G_{bottom,i}\{n_{,i}/100\}_1\ \{n_{,i}/100\}_2 + G_{bottom,\ i}\{n_{,i}/100\}_2\ \{n_{,i}/100\}_2\{n_{,i}/100\}_3 + \ldots + G_{bottom,i}\{n_{,i}/100\}_1\ \{n_{,i}/100\}_2 \ldots\{ n_{,i}/100\}_{m,i}\}\}$$

The time averaged explosive power of a given QCD bomb in the ship frame is in first order:

$$<d\{\sum(i = 1; i = M)\ [(N_i)(E_i)]\}\ /dt_{ship} > = <d\{\sum(i = 1; i = M)\ (E_i)\{\{G_{bottom,i}\{n_{,i}/100\}_1 + G_{bottom,i}\{n_{,i}/100\}_1\ \{n_{,i}/100\}_2 + G_{bottom,\ i}\{n_i/100\}_2\ \{n_{,i}/100\}_2\{n_{,i}/100\}_3 + \ldots + G_{bottom,i}\{n_{,i}/100\}_1\ \{n_{,i}/100\}_2 \ldots\{ n_{,i}/100\}_{m,i}\}\}/dt_{ship}>$$

Another way of presenting reaction-wave propagation is to consider the average lifetime of the bottomlets and the average distance traveled by the bottomlets relative to the mean free path for bottomlet and nucleon or atomic nuclei contact. For cases where the bottomlet of the ith species of bottomlets would not decay with a probability of $n_{,i}/100$ before completing the travel of its mean free path, the number of catalyzed baryons or nuclei is also equal to in first order:

$$N_i = G_{bottom,i}\{n_{,i}/100\}_1 + G_{bottom,i}\{n_{,i}/100\}_1\ \{n_{,i}/100\}_2 + G_{bottom,\ i}\{n_{,i}/100\}_2\ \{n_{,i}/100\}_2\{n_{,i}/100\}_3 + \ldots + G_{bottom,i}\{n_{,i}/100\}_1\ \{n_{,i}/100\}_2 \ldots\{ n_{,i}/100\}_{m,i}$$

Where $m_{,i}$ is the number of time steps for the overall reaction extinction for the ith species of bottomlet.

So for M species of bottomlets, where each species has its unique mean free path, initial population, and number of time steps for overall reaction extinctions, the number of catalyzed baryons or nuclei is equal to in first order:

$$\sum(i = 1; i = M) N_i = \sum(i = 1; i = M) \{G_{bottom,i}\{n_{,i}/100\}_1 + G_{bottom,i}\{n_{,i}/100\}_1 \{n_{,i}/100\}_2 + G_{bottom,i}\{n_{,i}/100\}_2 \{n_{,i}/100\}_2\{n_{,i}/100\}_3 + \ldots + G_{bottom,i}\{n_{,i}/100\}_1 \{n_{,i}/100\}_2 \ldots\{ n_{,i}/100\}_{m,i}\}$$

Where the average yield energy per nucleon or nucleus conversion caused by transformation by particles if the ith species is E_i.

The total yield energy for a given QCD bomb is in first order:

$$\sum(i = 1; i = M) [(N_i)(E_i)] = \sum(i = 1; i = M) (E_i)\{\{G_{bottom,i}\{n_{,i}/100\}_1 + G_{bottom,i}\{n_{,i}/100\}_1 \{n_{,i}/100\}_2 + G_{bottom,i}\{n_{,i}/100\}_2 \{n_{,i}/100\}_2\{n_{,i}/100\}_3 + \ldots + G_{bottom,i}\{n_{,i}/100\}_1 \{n_{,i}/100\}_2 \ldots\{ n_{,i}/100\}_{m,i}\}\}$$

The time averaged explosive power of a given QCD bomb in the ship frame is in first order:

$$<d\{\sum(i = 1; i = M) [(N_i)(E_i)]\} /dt_{ship} > = <d\{\sum(i = 1; i = M) (E_i)\{\{G_{bottom,i}\{n_{,i}/100\}_1 + G_{bottom,i}\{n_{,i}/100\}_1 \{n_{,i}/100\}_2 + G_{bottom,i}\{n_{,i}/100\}_2 \{n_{,i}/100\}_2\{n_{,i}/100\}_3 + \ldots + G_{bottom,i}\{n_{,i}/100\}_1 \{n_{,i}/100\}_2 \ldots\{ n_{,i}/100\}_{m,i}\}\}\}/dt_{ship}>$$

Considering the length of time, $t_{m,i}$, traveled per mean free path where the average nondecay probability is $n_{,i}/100$ for each mean path traveled, the velocity of the bottomlet is $L_i/t_{m.i}$, where the velocity of the bottomlet is negligibly relativistic for the ith species of bottomlets.

For cases where the ith species of bottomlets produced are relativistic, the average distance traveled by such bottomlets before it decays is given by:

$$D_i = [L_i/t_{m,i}](\gamma_{,i}) (n_{,i})/(100) = [L_i/t_{m,i}]\{1/\{1 - [(v_{,i}/C)^2]\}^{1/2}\}(n_{,i})/(100).$$

Now, applying the exponetial decay formula:

$$N_i(t_i) = N_{0i}\, e^{-t,bottom.i/Tbottom,i} = N_{0i}\, (1/2)e^{-t,bottom,i/t,bottom,i1/2}$$

Where T_i is the mean lifetime of a particle (such as an atom or subatomic particle), $t_{i1/2}$ is the half-life of the particle, and N_{0i} is the starting population. $N_i(t_i)$ is the quantity that still remains and has not yet decayed after a time t_i.

Therefore the half-life is expressed as in first order:

$t_{i1/2} = \ln(2)/\lambda_i = T_i \ln(2)$

Here λ_i is a positive number called the decay constant of the decaying quantity.

For a bottomlet of the ith bottomlet species having a probability of nondecay along a mean path length of $P_i = n_{,i}/100$, the number of bottomlets out of a starting population of $N_{0i} = G_{bottom,i}$ that contact a baryon or atomic nuclei is $N_{0i}\{n_{,i}/100\}$ where $\{n_{,i}/100\} = e^{-t,bottom,i/Tbottom,i} = (1/2)\, e^{-t,bottom,i/t,bottom,i1/2}$. Therefore, the number of catalyzed baryons becomes in first order:

$$N_i = G_{bottom,i}\{e^{-t,bottom,i/Tbottom,i}\}_1 + G_{bottom,i}\{e^{-t,bottom,i/Tbottom,i}\}\{e^{-t,bottom,i/Tbottom,i}\}_2 + G_{bottom,i}\, e^{-t,bottom,i/Tbottom,i}\}\{e^{-t,bottom,i/Tbottom,i}\}\{e^{-t,bottom,i/Tbottom,i}\}_3 + \ldots + G_{bottom,i}\{e^{-t,bottom,i/Tbottom,i}\}\{e^{-t,bottom,i/Tbottom,i}\}\ldots\{e^{-t,bottom,i/Tbottom,i}\}_{m,i}$$

$$= G_{bottom,i}\{(1/2)\, e^{-t,bottom,i/t,bottom,i1/2}\}_1 + G_{bottom,i}\{(1/2)\, e^{-t,bottom,i/t,bottom,i1/2}\}\{(1/2)\, e^{-t,bottom,i/t,bottom,i1/2}\}_2 + G_{bottom,i}\{(1/2)\, e^{-t,bottom,i/t,bottom,i1/2}\}\{(1/2)\, e^{-t,bottom,i/t,bottom,i1/2}\}\{(1/2)\, e^{-t,bottom,i/t,bottom,i1/2}\}_3 + \ldots + G_{bottom,i}\{(1/2)\, e^{-t,bottom,i/t,bottom,i1/2}\}\{(1/2)\, e^{-t,bottom,i/t,bottom,i1/2}\}\ldots\{(1/2)\, e^{-t,bottom,i/t,bottom,i1/2}\}_{m,i},$$

Where t_i is the travel time of the ith species of bottomlets for the mean free path for bottomlet to ordinary baryon contact.

Where the average yield energy per nucleon or nucleus conversion caused by transformation by particles if the ith species is E_i.

The total yield energy for a given QCD bomb is in first order:

$$\sum(i = 1;\, i = M)\,[(N_i)(E_i)] = \sum(i = 1;\, i = M)\,(E_i)\,\{G_{bottom,i}\{e^{-t,bottom,i/Tbottom,i}\}_1 + G_{bottom,i}\{e^{-t,bottom,i/Tbottom,i}\}\{e^{-t,bottom,i/Tbottom,i}\}_2 + G_{bottom,i}\, e^{-t,bottom,i/Tbottom,i}\}\{e^{-t,bottom,i/Tbottom,i}\}\{e^{-t,bottom,i/Tbottom,i}\}_3 + \ldots + G_{bottom,i}\{e^{-t,bottom,i/Tbottom,i}\}\{e^{-t,bottom,i/Tbottom,i}\}\ldots\{e^{-t,bottom,i/Tbottom,i}\}_{m,i}\}$$

$$= \sum(i = 1;\, i = M)\,(E_i)\{G_{bottom,i}\{(1/2)\, e^{-t,bottom,i/t,bottom,i1/2}\}_1 + G_{bottom,i}\{(1/2)\, e^{-t,bottom,i/t,bottom,i1/2}\}\{(1/2)\, e^{-t,bottom,i/t,bottom,i1/2}\}_2 + G_{bottom,i}\{(1/2)\, e^{-t,bottom,i/t,bottom,i1/2}\}\{(1/2)\, e^{-t,bottom,i/t,bottom,i1/2}\}\{(1/2)\, e^{-t,bottom,i/t,bottom,i1/2}\}_3 + \ldots + G_{bottom,i}\{(1/2)\, e^{-t,bottom,i/t,bottom,i1/2}\}\{(1/2)\, e^{-t,bottom,i/t,bottom,i1/2}\}\ldots\{(1/2)\, e^{-t,bottom,i/t,bottom,i1/2}\}_{m,i}\},$$

The time averaged explosive power of a given QCD bomb in the ship frame is in first order:

$$<d\{\sum(i = 1; i = M)\ [(N_i)(E_i)]\}/dt_{ship}> = <d\{\sum(i = 1; i = M)\ (E_i)\ \{G_{bottom,i}\{e^{-t,bottom,i/Tbottom,i}\}_1 + G_{bottom,i}\{e^{-t,bottom,i/Tbottom,i}\}\ \{e^{-t,bottom,i/Tbottom,i}\}_2 + G_{bottom,i}\ e^{-t,bottom,i/Tbottom,i}\}\ \{e^{-t,bottom,i/Tbottom,i}\}\ \{e^{-t,bottom,i/Tbottom,i}\}_3 + \ldots + G_{bottom,i}\{e^{-t,bottom,i/Tbottom,i}\}\ \{e^{-t,bottom,i/Tbottom,i}\}\ \ldots\{e^{-t,bottom,i/Tbottom,i}\}_{m,i}\}/dt_{ship}>$$

$$= <d\{\sum(i = 1; i = M)\ (E_i)\{G_{bottom,i}\{(1/2)\ e^{-t,bottom,i/t,bottom,i1/2}\}_1 + G_{bottom,i}\{(1/2)\ e^{-t,bottom,i/t,bottom,i1/2}\}\ \{(1/2)\ e^{-t,bottom,i/t,bottom,i1/2}\}_2 + G_{bottom,i}\{(1/2)\ e^{-t,bottom,i/t,bottom,i1/2}\}\ \{(1/2)\ e^{-t,bottom,i/t,bottom,i1/2}\}\ \{(1/2)\ e^{-t,bottom,i/t,bottom,i1/2}\}_3 + \ldots + G_{bottom,i}\{(1/2)\ e^{-t,bottom,i/t,bottom,i1/2}\}\ \{(1/2)\ e^{-t,bottom,i/t,bottom,i1/2}\}\ \ldots\{(1/2)\ e^{-t,bottom,i/t,bottom,i1/2}\}_{m,i}\}\}/dt_{ship}>$$

Now if $t_{i1/2}$ and T_i are background stationary frames and the ith species of bottomlets are relativistic, the number of catalyzed baryons becomes in first order:

$$N = G_{bottom,i}\{e^{-t,bottom,i/(Tbottom,i\gamma}\}_1 + G_{bottom,i}\{e^{-t,bottom,i/(Tbottom,i\gamma)}\}\{e^{-t,bottom,i/(Tbottom,i\gamma)}\}_2 + G_{bottom,i}\{e^{-t,bottom,i/(Tbottom,i\gamma)}\}\{e^{-t,bottom,i/(Tbottom,i\gamma)}\}\{e^{-t,bottom,i/(Tbottom,i\gamma)}\}_3 + \ldots + G_{bottom,i}\{e^{-t,bottom,i/(Tbottom,i\gamma)}\}\{e^{-t,bottom,i/(Tbottom,i\gamma)}\}\ldots\{e^{-t,bottom,i/(Tbottom,i\gamma)}\}_{m,i}$$

$$= G_{bottom,i}\{(1/2)\ e^{-t,bottom,i/(\gamma t,bottom,i1/2)}\}_1 + G_{bottom,i}\{(1/2)\ e^{-t,bottom,i/(\gamma t,bottom,i1/2)}\}\{(1/2)\ e^{-t,bottom,i/(\gamma t,bottom,i1/2)}\}_2 + G_{bottom,i}\{(1/2)\ e^{-t,bottom,i/(\gamma t,bottom,i1/2)}\}\{(1/2)\ e^{-t,bottom,i/(\gamma t,bottom,i1/2)}\}\{(1/2)\ e^{-t,bottom,i/(\gamma t,bottom,i1/2)}\}_3 + \ldots + G_{bottom,i}\{(1/2)\ e^{-t,bottom,i/(\gamma t,bottom,i1/2)}\}\{(1/2)\ e^{-t,bottom,i/(\gamma t,bottom,i1/2)}\}\ldots\{(1/2)\ e^{-t,bottom,i/(\gamma t,bottom,i1/2)}\}_{m,i}$$

Where the average yield energy per nucleon or nucleus conversion caused by transformation by particles if the ith species is E_i.

The total yield energy for a given QCD bomb is in first order:

$$\sum(i = 1; i = M)\ [(N_i)(E_i)] = \sum(i = 1; i = M)\ (E_i)\ \{G_{bottom,i}\{e^{-t,bottom,i/(Tbottom,i\gamma}\}_1 + G_{bottom,i}\{e^{-t,bottom,i/(Tbottom,i\gamma)}\}\{e^{-t,bottom,i/(Tbottom,i\gamma)}\}_2 + G_{bottom,i}\{e^{-t,bottom,i/(Tbottom,i\gamma)}\}\{e^{-t,bottom,i/(Tbottom,i\gamma)}\}\{e^{-t,bottom,i/(Tbottom,i\gamma)}\}_3 + \ldots + G_{bottom,i}\{e^{-t,bottom,i/(Tbottom,i\gamma)}\}\{e^{-t,bottom,i/(Tbottom,i\gamma)}\}\ldots\{e^{-t,bottom,i/(Tbottom,i\gamma)}\}_{m,i}\}$$

$$= \sum(i = 1;\ i = M)\ (E_{,i})\ \{G_{bottom,i}\{(1/2)\ e^{-t,bottom,i/(\gamma t,bottom,i1/2)}\}_1 + G_{bottom,i}\{(1/2)\ e^{-t,bottom,i/(\gamma t,bottom,i1/2)}\}\{(1/2)\ e^{-t,bottom,i/(\gamma t,bottom,i1/2)}\}_2 + G_{bottom,i}\{(1/2)\ e^{-t,bottom,i/(\gamma t,bottom,i1/2)}\}\{(1/2)\ e^{-t,bottom,i/(\gamma t,bottom,i1/2)}\}\{(1/2)\ e^{-t,bottom,i/(\gamma t,bottom,i1/2)}\}_3 + \ldots + G_{bottom,i}\{(1/2)\ e^{-t,bottom,i/(\gamma t,bottom,i1/2)}\}\{(1/2)\ e^{-t,bottom,i/(\gamma t,bottom,i1/2)}\}\ldots\{(1/2)\ e^{-t,bottom,i/(\gamma t,bottom,i1/2)}\}_{m,i}\}$$

The time averaged explosive power of a given QCD bomb in the ship frame is in first order:

$$<d\{\sum(i = 1;\ i = M)\ [(N_i)(E_i)]\}\ /dt_{ship}> = <d\{\sum(i = 1;\ i = M)\ (E_i)\ \{G_{bottom,i}\{e^{-t,bottom,i/(Tbottom,i\gamma}\}_1 + G_{bottom,i}\{e^{-t,bottom,i/(Tbottom,i\gamma)}\}\{e^{-t,bottom,i/(Tbottom,i\gamma)}\}_2 + G_{bottom,i}\{e^{-t,bottom,i/(Tbottom,i\gamma)}\}\{e^{-t,bottom,i/(Tbottom,i\gamma)}\}\{e^{-t,bottom,i/(Tbottom,i\gamma)}\}_3 + \ldots + G_{bottom,i}\{e^{-t,bottom,i/(Tbottom,i\gamma)}\}\{e^{-t,bottom,i/(Tbottom,i\gamma)}\}\ldots\{e^{-t,bottom,i/(Tbottom,i\gamma)}\}_{m,i}\}\}/dt_{ship}>$$

$$= <d\{\sum(i = 1;\ i = M)\ (E_i)\ \{G_{bottom,i}\{(1/2)\ e^{-t,bottom,i/(\gamma t,bottom,i1/2)}\}_1 + G_{bottom,i}\{(1/2)\ e^{-t,bottom,i/(\gamma t,bottom,i1/2)}\}\{(1/2)\ e^{-t,bottom,i/(\gamma t,bottom,i1/2)}\}_2 + G_{bottom,i}\{(1/2)\ e^{-t,bottom,i/(\gamma t,bottom,i1/2)}\}\{(1/2)\ e^{-t,bottom,i/(\gamma t,bottom,i1/2)}\}\{(1/2)\ e^{-t,bottom,i/(\gamma t,bottom,i1/2)}\}_3 + \ldots + G_{bottom,i}\{(1/2)\ e^{-t,bottom,i/(\gamma t,bottom,i1/2)}\}\{(1/2)\ e^{-t,bottom,i/(\gamma t,bottom,i1/2)}\}\ldots\{(1/2)\ e^{-t,bottom,i/(\gamma t,bottom,i1/2)}\}_{m,i}\}\}/dt_{ship}>$$

For cases where the average bottomlet of the ith species would not decay with a probability of $n_{,i}/100$ after traveling the mean free path, and the probability of baryon or nuclei conversion upon contact is $k_{,i}/100$, but where the bottomlets are always destroyed when contacting the nuclei or converted into a non-bottom-transmutative form with a probability of unity, the number of catalyzed baryons or nuclei is equal to in first order:

$$N_i = G_{bottom,i}\{n_{,i}/100\}\{k_i/100\}_1 + G_{bottom,i}\{n_{,i}/100\}\{k_{,i}/100\}_1\{n_{,i}/100\}\{k_{,i}/100\}_2 + G_{bottom,i}\{n_{,i}/100\}\{k_{,i}/100\}_1\{n_{,i}/100\}\{k_{,i}/100\}_2\{n_{,i}/100\}\{k_{,i}/100\}_3 + \ldots + G_{bottom,i}\{n_{,i}/100\}\{k_{,i}/100\}_1\{n_{,i}/100\}\{k_{,i}/100\}_2\ldots\{n_{,i}/100\}\{k_{,i}/100\}_{m,i},$$

Here, *m* is the number of time steps for the overall reaction extinction.

Where the average yield energy per nucleon or nucleus conversion caused by transformation by particles if the ith species is E_i.

The total yield energy for a given QCD bomb is in first order:

$$\sum(i = 1; i = M) [(N_i)(E_i)] = \sum(i = 1; i = M) (E_i) \{G_{bottom,i}\{n_{,i}/100\}\{k_{,i}/100\}_1 + G_{bottom,i}\{n_{,i}/100\}\{k_{,i}/100\}_1\{n_{,i}/100\}\{k_{,i}/100\}_2 + G_{bottom,i}\{n_{,i}/100\}\{k_{,i}/100\}_1 \{n_{,i}/100\}\{k_{,i}/100\}_2\{n_{,i}/100\}\{k_{,i}/100\}_3 + \ldots + G_{bottom,i}\{n_{,i}/100\}\{k_{,i}/100\}_1\{n_{,i}/100\}\{k_{,i}/100\}_2 \ldots\{n_{,i}/100\}\{k_{,i}/100\}_{m,i}\},$$

The time averaged explosive power of a given QCD bomb in the ship frame is in first order:

$$<d\{\sum(i = 1; i = M) [(N_i)(E_i)]\}/dt_{ship}> = <d\{\sum(i = 1; i = M) (E_i) \{G_{bottom,i}\{n_{,i}/100\}\{k_{,i}/100\}_1 + G_{bottom,i}\{n_{,i}/100\}\{k_{,i}/100\}_1\{n_{,i}/100\}\{k_{,i}/100\}_2 + G_{bottom,i}\{n_{,i}/100\}\{k_{,i}/100\}_1 \{n_{,i}/100\}\{k_{,i}/100\}_2\{n_{,i}/100\}\{k_{,i}/100\}_3 + \ldots + G_{bottom,i}\{n_{,i}/100\}\{k_{,i}/100\}_1\{n_{,i}/100\}\{k_{,i}/100\}_2 \ldots\{n_{,i}/100\}\{k_{,i}/100\}_m\}\}/dt_{ship}>$$

Considering the length of time, $t_{m,i}$, traveled per mean free path where the average non-decay probability is n/100 for each mean path traveled, the velocity of the ith species of bottomlet is $L_i/t_{m,i}$, where the velocity of the bottomlet is negligibly relativistic.

For cases where the ith species of bottomlets produced are relativistic, the distance traveled by a given bottomlet before it decays is given by:

$$[L_i/t_{mi}](\gamma) = [L_i/t_{m,ii}]\{1/\{1 - [(v_{,i}/C)^2]\}^{1/2}\}$$

Once again, $N_i\ (t_i) = N_{0i}\ e^{-t,bottom,i/Tbottom,i} = N_{0i}\ (1/2)e^{-t,bottom,i/t,bottom,i1/2}$, where T_i is the mean lifetime of a particle such as an atom or subatomic particle, $t_{1/2}$ is the half-life of the particle, and N_{0i} is the starting population. $N_i\ (t_i)$ is the quantity that still remains and has not yet decayed after a time t and half-life $t_{1/2} = \ln(2)/\lambda_i = T_i \ln(2)$, where λ_i is a positive number called the decay constant of the decaying quantity.

In the case of a bottomlet species having a probability of nondecay along a mean path length of $P_i = n_i/100$, the number of bottomlets out of a starting population of $N_{0i} = G_{bottom,i}$ that contact a baryon or atomic nuclei is $N_{0i}\{n_i/100\}$ where $\{n_{,i}/100\} = e^{-t,bottom,i/Tbottom,i} = (1/2)\ e^{-t,bottom,i/t,bottom,i1/2}$. Therefore, the number of catalyzed baryons becomes in first order:

$$N_i = G_{bottomi}\{e^{-t,bottom,i/Tbottom,i}\}\{k_{,i}/100\}_1 + G_{bottom,i}\{e^{-t,bottom,i/Tbottom,i}\}\{k_{,i}/100\}_1\{e^{-t,bottom,i/Tbottom,i}\}\{k_{,i}/100\}_2 + G_{bottom,i}\{e^{-t,bottom,i/Tbottom,i}\}\{k_{,i}/100\}_1 \{e^{-t,bottom,i/Tbottom,i}\}\{k_{,i}/100\}_2 \{e^{-}$$

$$^{-t,bottom,i/Tbottom,i}\}\{k_{,i}/100\}_3 + \ldots + G_{bottom,i}\{e^{-t,bottom,i/Tbottom,i}\}\{k_{,i}/100\}_1\{e^{-t,bottom,i/Tbottom,i}\}\{k_{,i}/100\}_2 \ldots\{e^{-t,bottom,i/Tbottom,i}\}\{k_{,i}/100\}_{m,i}$$

$$= G_{bottom,i}\{(1/2)\, e^{-t,bottom,i/t,bottom,i1/2}\}\{k_{,i}/100\}_1 + G_{bottom,i}\{(1/2)\, e^{-t,bottom,i/t,bottom,i1/2}\}\{k_{,i}/100\}_1\{(1/2)\, e^{-t,bottom,i/t,bottom,i1/2}\}\{k_{,i}/100\}_2 + G_{bottom,i}\{(1/2)\, e^{-t,bottom,i/t,bottom,i1/2}\}\{k_{,i}/100\}_1\{(1/2)\, e^{-t,bottom,i/t,bottom,i1/2}\}\{k_{,i}/100\}_2\ \{(1/2)\, e^{-t,bottom,i/t,bottom,i1/2}\}\{k_{,i}/100\}_3 + \ldots + G_{bottom,i}\{(1/2)\, e^{-t,bottom,i/t,bottom,i1/2}\}\{k_{,i}/100\}_1\{(1/2)\, e^{-t,bottom,i/t,bottom,i1/2}\}\{k_{,i}/100\}_2 \ldots\{(1/2)\, e^{-t,bottom,i/t,bottom,i1/2}\}\{k_{,i}/100\}_{m,i}$$

Where the average yield energy per nucleon or nucleus conversion caused by transformation by particles if the ith species is E_i.

The total yield energy for a given QCD bomb is in first order:

$$\sum(i = 1;\ i = M)\ [(N_i)(E_i)] = \sum(i = 1;\ i = M)\ (E_i)\ \{G_{bottom,i}\{e^{-t,bottom,i/Tbottom,i}\}\{k_{,i}/100\}_1 + G_{bottom,i}\{e^{-t,bottom,i/Tbottom,i}\}\{k_{,i}/100\}_1\{e^{-t,bottom,i/Tbottom,i}\}\{k_{,i}/100\}_2 + G_{bottom,i}\ \{e^{-t,bottom,i/Tbottom,i}\}\{k_{,i}/100\}_1\ \{e^{-t,bottom,i/Tbottom,i}\}\{k_{,i}/100\}_2\ \{e^{-t,bottom,i/Tbottom,i}\}\{k_{,i}/100\}_3 + \ldots + G_{bottom,i}\{e^{-t,bottom,i/Tbottom,i}\}\{k_{,i}/100\}_1\{e^{-t,bottom,i/Tbottom,i}\}\{k_{,i}/100\}_2 \ldots\{e^{-t,bottom,i/Tbottom,i}\}\{k_{,i}/100\}_{m,i}\}$$

$$= \sum(i = 1;\ i = M)\ (E_i)\ \{G_{bottom,i}\{(1/2)\, e^{-t,bottom,i/t,bottom,i1/2}\}\{k_{,i}/100\}_1 + G_{bottom,i}\{(1/2)\, e^{-t,bottom,i/t,bottom,i1/2}\}\{k_{,i}/100\}_1\{(1/2)\, e^{-t,bottom,i/t,bottom,i1/2}\}\{k_{,i}/100\}_2 + G_{bottom,i}\{(1/2)\, e^{-t,bottom,i/t,bottom,i1/2}\}\{k_{,i}/100\}_1\{(1/2)\, e^{-t,bottom,i/t,bottom,i1/2}\}\{k_{,i}/100\}_2\ \{(1/2)\, e^{-t,bottom,i/t,bottom,i1/2}\}\{k_{,i}/100\}_3 + \ldots + G_{bottom,i}\{(1/2)\, e^{-t,bottom,i/t,bottom,i1/2}\}\{k_{,i}/100\}_1\{(1/2)\, e^{-t,bottom,i/t,bottom,i1/2}\}\{k_{,i}/100\}_2 \ldots\{(1/2)\, e^{-t,bottom,i/t,bottom,i1/2}\}\{k_{,i}/100\}_{m,i}\}$$

The time averaged explosive power of a given QCD bomb in the ship frame is in first order:

$$<d\{\sum(i = 1;\ i = M)\ [(N_i)(E_i)]\}/dt_{ship}> = <d\{\sum(i = 1;\ i = M)\ (E_i)\ \{G_{bottom,i}\{e^{-t,bottom,i/Tbottom,i}\}\{k_{,i}/100\}_1 + G_{bottom,i}\{e^{-t,bottom,i/Tbottom,i}\}\{k_{,i}/100\}_1\{e^{-t,bottom,i/Tbottom,i}\}\{k_{,i}/100\}_2 + G_{bottom,i}\ \{e^{-t,bottom,i/Tbottom,i}\}\{k_{,i}/100\}_1\ \{e^{-t,bottom,i/Tbottom,i}\}\{k_{,i}/100\}_2\ \{e^{-t,bottom,i/Tbottom,i}\}\{k_{,i}/100\}_3 + \ldots + G_{bottom,i}\{e^{-t,bottom,i/Tbottom,i}\}\{k_{,i}/100\}_1\{e^{-t,bottom,i/Tbottom,i}\}\{k_{,i}/100\}_2 \ldots\{e^{-t,bottom,i/Tbottom,i}\}\{k_{,i}/100\}_{m,i}\}\}/dt_{ship}>$$

$$= <d\{\sum(i = 1;\ i = M)\ (E_i)\ \{G_{bottom,i}\{(1/2)\, e^{-t,bottom,i/t,bottom,i1/2}\}\{k_{,i}/100\}_1 + G_{bottom,i}\{(1/2)\, e^{-t,bottom,i/t,bottom,i1/2}\}\{k_{,i}/100\}_1\{(1/2)\, e^{-}$$

$_{t,bottom,i/t,bottom,i1/2}\}\{k_{,i}/100\}_2 + G_{bottom,i}\{(1/2)\ e^{-t,bottom,i/t,bottom,i1/2}\}\{k_{,i}/100\}_1\{(1/2)\ e^{-t,bottom,i/t,bottom,i1/2}\}\{k_{,i}/100\}_2\ \{(1/2)\ e^{-t,bottom,i/t,bottom,i1/2}\}\{k_{,i}/100\}_3 + \ldots + G_{bottom,i}\{(1/2)\ e^{-t,bottom,i/t,bottom,i1/2}\}\{k_{,i}/100\}_1\{(1/2)\ e^{-t,bottom,i/t,bottom,i1/2}\}\{k_{,i}/100\}_2 \ldots\{(1/2)\ e^{-t,bottom,i/t,bottom,i1/2}\}\{k_{,i}/100\}_{m,i}\}\}/dt_{ship}>.$

If $t_{i1/2}$ and T_i are background stationary frames and the bottomlets are relativistic, the number of catalyzed baryons becomes in first order:

$$N_i = G_{bottom,i}\{e^{-t,bottom,i/(Tbottom,i\gamma}\}\{k_{,i}/100\}_1 + G_{bottom,i}\{e^{-t,bottom,i/(Tbottom,i\gamma)}\}\{k_{,i}/100\}_1\{e^{-t,bottom,i/(Tbottom,i\gamma)}\}\{k_{,i}/100\}_2 + G_{bottom,i}\{e^{-t,bottom,i/(Tbottom,i\gamma)}\}\{k/100\}_1\{e^{-t,bottom,i/(Tbottom,i\ \gamma)}\}\{k_{,i}/100\}_2\{e^{-t,bottom,i/(Tbottom,i\gamma)}\}\{k_{,i}/100\}_3 + \ldots + G_{bottom,i}\{e^{-t,bottom,i/(Tbottom,i\gamma)}\}\{k_{,i}/100\}_1\{e^{-t,bottom,i/(Tbottom,i\gamma)}\}\{k_{,i}/100\}_2 \ldots\{e^{-t,bottom,i/(Tbottom,i\gamma)}\}\{k_{,i}/100\}_{m,i}$$

$$= G_{bottom,i}\{(1/2)\ e^{-t,bottom,i/(\gamma t,bottom,i1/2)}\}\{k_{,i}/100\}_1 + G_{bottom,i}\{(1/2)\ e^{-t,bottom,i/(\gamma t,bottom,i1/2)}\}\{k_{,i}/100\}_1\{(1/2)\ e^{-t,bottom,i/(\gamma t,bottom,i1/2)}\}\{k_{,i}/100\}_2 + G_{bottom,i}\{(1/2)\ e^{-t,bottom,i/(\gamma t,bottom,i1/2)}\}\{k_{,i}/100\}_1\{(1/2)\ e^{-t,bottom,i/(\gamma t,bottom,i1/2)}\}\{k_{,i}/100\}_2\{(1/2)\ e^{-t,bottom,i/(\gamma t,bottom,i1/2)}\}\{k_{,i}/100\}_3 + \ldots + G_{bottom,i}\{(1/2)\ e^{-t,bottom,i/(\gamma t,bottom,i1/2)}\}\{k_{,i}/100\}_1\{(1/2)\ e^{-t,bottom,i/(\gamma t,bottom,i1/2)}\}\{k_{,i}/100\}_2 \ldots\{(1/2)\ e^{-t,bottom,i/(\gamma t,bottom,i1/2)}\}\{k_{,i}/100\}_{m,i}$$

Where the average yield energy per nucleon or nucleus conversion caused by transformation by particles if the ith species is E_i.

The total yield energy for a given QCD bomb is in first order:

$$\sum(i = 1; i = M)\ [(N_i)(E_i)] = \sum(i = 1; i = M)\ (E_i)\ \{G_{bottom,i}\{e^{-t,bottom,i/(Tbottom,i\gamma}\}\{k_{,i}/100\}_1 + G_{bottom,i}\{e^{-t,bottom,i/(Tbottom,i\gamma)}\}\{k_{,i}/100\}_1\{e^{-t,bottom,i/(Tbottom,i\gamma)}\}\{k_{,i}/100\}_2 + G_{bottom,i}\{e^{-t,bottom,i\ /(Tbottom,i\ \gamma)}\}\{k_i/100\}_1\{e^{-t,bottom,i\ /(Tbottom,i\ \gamma)}\}\{k_{,i}/100\}_2\{e^{-t,bottom,i/(Tbottom,i\gamma)}\}\{k_{,i}/100\}_3 + \ldots + G_{bottom,i}\{e^{-t,bottom,i/(Tbottom,i\gamma)}\}\{k_{,i}/100\}_1\{e^{-t,bottom,i/(Tbottom,i\gamma)}\}\{k_{,i}/100\}_2 \ldots\{e^{-t,bottom,i/(Tbottom,i\gamma)}\}\{k_{,i}/100\}_{m,i}\}$$

$$= \{\sum(i = 1; i = M)\ (E_i)\ \{G_{bottom,i}\{(1/2)\ e^{-t,bottom,i/(\gamma t,bottom,i1/2)}\}\{k_{,i}/100\}_1 + G_{bottom,i}\{(1/2)\ e^{-t,bottom,i/(\gamma t,bottom,i1/2)}\}\{k_{,i}/100\}_1\{(1/2)\ e^{-t,bottom,i/(\gamma t,bottom,i1/2)}\}\{k_{,i}/100\}_2 + G_{bottom,i}\{(1/2)\ e^{-t,bottom,i/(\gamma t,bottom,i1/2)}\}\{k_{,i}/100\}_1\{(1/2)\ e^{-t,bottom,i/(\gamma t,bottom,i1/2)}\}\{k_{,i}/100\}_2\{(1/2)\ e^{-t,bottom,i/(\gamma t,bottom,i1/2)}\}\{k_{,i}/100\}_3 + \ldots + G_{bottom,i}\{(1/2)\ e^{-t,bottom,i/(\gamma t,bottom,i1/2)}\}\{k_{,i}/100\}_1\{(1/2)\ e^{-t,bottom,i/(\gamma t,bottom,i1/2)}\}\{k_{,i}/100\}_2 \ldots\{(1/2)\ e^{-t,bottom,i/(\gamma t,bottom,i1/2)}\}\{k_{,i}/100\}_{m,i}\}$$

The time averaged explosive power of a given QCD bomb in the ship frame is in first order:

$$<d\{\sum(i = 1;\ i = M)\ [(N_i)(E_i)]\}\ /dt_{ship} > = <d\{\sum(i = 1;\ i = M)\ (E_i)\ \{G_{bottom,i}\{e^{-t,bottom,i/(Tbottom,i\gamma}\}\{k_{,i}/100\}_1 + G_{bottom,i}\{e^{-t,bottom,i/(Tbottom,i\gamma)}\}\{k_{,i}/100\}_1\{e^{-t,bottom,i/(Tbottom,i\gamma)}\}\{k_{,i}/100\}_2 + G_{bottom,i}\{e^{-t,bottom,i\ /(Tbottom,i\ \gamma)}\}\{k_i/100\}_1\{e^{-t,bottom,i\ /(Tbottom,i\ \gamma)}\}\{k_{,i}/100\}_2\{e^{-t,bottom,i/(Tbottom,i\gamma)}\}\{k_{,i}/100\}_3 + \ldots + G_{bottom,i}\{e^{-t,bottom,i/(Tbottom,i\gamma)}\}\{k_{,i}/100\}_1\{e^{-t,bottom,i/(Tbottom,i\gamma)}\}\{k_{,i}/100\}_2 \ldots\{e^{-t,bottom,i/(Tbottom,i\gamma)}\}\{k_{,i}/100\}_{m,i}\}/dt_{ship} >$$

$$= <d\{\sum(i = 1;\ i = M)\ (E_i)\ \{G_{bottom,i}\{(1/2)\ e^{-t,bottom,i/(\gamma t,bottom,i1/2)}\}\{k_{,i}/100\}_1 + G_{bottom,i}\{(1/2)\ e^{-t,bottom,i/(\gamma t,bottom,i1/2)}\}\{k_{,i}/100\}_1\{(1/2)\ e^{-t,bottom,i/(\gamma t,bottom,i1/2)}\}\{k_{,i}/100\}_2 + G_{bottom,i}\{(1/2)\ e^{-t,bottom,i/(\gamma t,bottom,i1/2)}\}\{k_{,i}/100\}_1\{(1/2)\ e^{-t,bottom,i/(\gamma t,bottom,i1/2)}\}\{k_{,i}/100\}_2\{(1/2)\ e^{-t,bottom,i/(\gamma t,bottom,i1/2)}\}\{k_{,i}/100\}_3 + \ldots + G_{bottom,i}\{(1/2)\ e^{-t,bottom,i/(\gamma t,bottom,i1/2)}\}\{k_{,i}/100\}_1\{(1/2)\ e^{-t,bottom,i/(\gamma t,bottom,i1/2)}\}\{k_{,i}/100\}_2 \ldots\{(1/2)\ e^{-t,bottom,i/(\gamma t,bottom,i1/2)}\}\{k_{,i}/100\}_{m,i}\}/dt_{ship} >$$

We would definitely not want stable bottomlets that are transmutative into-bottom-matter, or ones that are sufficiently long lived so as to produce a reaction that could consume the entire planet or propagate interstellar distances to reach other stars and planets where they could wreak havoc. Such bottomlets may even convert interstellar gas and dust in a gradual propagating wavefront and literally gradually eat away at the universe or multiverse completely in a progressive reaction wavefront. However, to produce such bottomlets or a black hole capable of doing this would require a galactic-sized accelerator—in other words, it is highly unlikely at best.

Regardless, bottomlet producing nuclear explosives might be used for the following:

1. Interstellar ramjet propulsion
2. Intergalactic ramjet propulsion
3. Hyperspatial rocket propulsion
4. Nuclear bottom matter bomblet pulse drive
5. Nuclear bottom matter bomblet pellet runway propulsion
6. Mass fuel beam nuclear bottom matter bomblet propulsion

7. Einsteinian 4-D photon, electron, muon, tauon, positron, anti-muon, anti-tauon, proton, antiproton, ion, anti-ion, charged exotic meson, charged exotic baryon, electron neutrino, anti-electron neutrino, muon neutrino, antimuon neutrino, tauon neutrino, anti-tauon neutrino, gravity wave, and quantum scale gravity wave or graviton rockets
8. Hyperspatial photon, electron, muon, tauon, positron, anti-muon, anti-tauon, proton, antiproton, ion, anti-ion, charged exotic meson, charged exotic baryon, electron neutrino, anti-electron neutrino, muon neutrino, antimuon neutrino, tauon neutrino, anti-tauon neutrino, gravity wave, and quantum scale gravity wave or graviton rockets. For rocket vehicle applications, the exothermic energy produced in bottom-matter reactors can be used to energize thrust-stream generators

Note that gravitons are hypothetical at this point. Their existence would require the discovery of the spin-2 boson, which is yet to be accomplished.

So for stable bottomlet technology, only assured-to-die-out chain reactions must be produced; otherwise, we risk universal destruction. The ramifications of the technology as a power source, however, are too profound not to be mentioned. This is true not only from a philosophical perspective of enabling technology that is far superior in power than pure matter-antimatter conversion, but more to the point, a technology that can revolutionize commercial energy production for manned starship propulsion. Consider the case where the matter within any hyperspatial dimensions that are coupled to our universe would not have periodic table elements or baryons having the exact properties as those found within our ordinary 4-D Einsteinian space-time. For such hyperspatial matters that have quantum properties that are similar to the baryonic mass within our ordinary 4-D universe, such exothermic conversion may still be possible, thereby enabling the production of hyperspatial ramjets and hyperspatial photon, electron, muon, tauon, positron, anti-muon, anti-tauon, proton, antiproton, ion, anti-ion, charged exotic meson, charged exotic baryon, electron neutrino, anti-electron neutrino, muon neutrino, antimuon neutrino, tauon neutrino, anti-tauon neutrino, gravity wave, and quantum-scale gravity wave or graviton rockets.

There are several plausible candidates for mechanisms that can be used to perform nuclear fusion in order to produce the required temperatures and particle kinetic energies so as to enable the bulk production of metastable bottomlets to provide for substantial but die-out forms of bottom-chain reactions. The devices are shaped charge nuclear fusion and/or fission devices. To the best of our knowledge, none of the associated nuclear-explosive types have yet been developed.

What can be accomplished with bottom reactions might also work for charmonium catalysis and strangeium catalysis. Toponium catalysis may prove possible in the long run; however, the production of toplets as yet have been unverified and may remain so for a long time. This is because toplets, which are bound states comprised of top quarks, would require the strong nuclear force to travel the distances to other quarks cocreated in particle accelerator collisions. All top quark-producing particle collisions to date have produced top quarks that decay too quickly for the light-speed-limited strong force to travel the distance between any cocreated top quarks, thereby preventing any stabilizing bound states.

Now combining strange, charmed, and bottom-particle-based transformations, we obtain:

$<d\{\sum(i = 1;\ i = M_{strange})\ [(N_i)(E_i)]\}\ /dt_{ship} > + <d\{\sum(i = 1;\ i = M_{charmed})\ [(N_i)(E_i)]\}/dt> + <d\{\sum(i = 1;\ i = M_{bottom})\ [(N_i)(E_i)]\}\ /dt_{ship} >$

$= <d\{\sum(i = 1;\ i = M)\ (E_i)\ \{G_{strange,i}\{e^{-t,strange,i/(Tstrange,i\gamma}\}\{k_{,i}/100\}_1 + G_{strange,i}\{e^{-t,strange,i/(Tstrange,i\gamma)}\}\{k_{,i}/100\}_1\{e^{-t,strange,i/(Tstrange,i\gamma)}\}\{k_{,i}/100\}_2 + G_{strange,i}\{e^{-t,strange,i\ /(Tstrange,i\ \gamma)}\}\{k/100\}_1\{e^{-t,strange,i\ /(Tstrange,i\ \gamma)}\}\{k_{,i}/100\}_2\{e^{-t,strange,i/(Tstrange,i\gamma)}\}\{k_{,i}/100\}_3 + \ldots + G_{strange,i}\{e^{-t,strange,i/(Tstrange,i\gamma)}\}\{k_{,i}/100\}_1\{e^{-t,strange,i/(Tstrange,i\gamma)}\}\{k_{,i}/100\}_2 \ldots\{e^{-t,strange,i/(Tstrange,i\gamma)}\}\{k_{,i}/100\}_{m,i}\}/dt_{ship} >$

$+ <d\{\sum(i = 1;\ i = M)\ (E_i)\ \{G_{charm,i}\{e^{-t,charm,i/(Tcharm,i\gamma}\}\{k_{,i}/100\}_1 + G_{charm,i}\{e^{-t,charm,i/(Tcharm,i\gamma)}\}\{k_{,i}/100\}_1\{e^{-t,charm,i/(Tcharm,i\gamma)}\}\{k_{,i}/100\}_2 + G_{charm,i}\{e^{-t,charm,i\ /(Tcharm,i\ \gamma)}\}\{k_i/100\}_1\{e^{-t,charm,i\ /(Tcharm,i\ \gamma)}\}\{k_{,i}/100\}_2\{e^{-t,charm,i/(Tcharm,i\gamma)}\}\{k_{,i}/100\}_3 + \ldots + G_{charm,i}\{e^{-t,charm,i/(Tcharm,i\gamma)}\}\{k_{,i}/100\}_1\{e^{-t,charm,i/(Tcharm,i\gamma)}\}\{k_{,i}/100\}_2 \ldots\{e^{-t,charm,i/(Tcharm,i\gamma)}\}\{k_{,i}/100\}_{m,i}\}/dt_{ship} >$

$+ <d\{\sum(i = 1;\ i = M_{bottom})\ (E_i)\ \{G_{bottom,i}\{e^{-t,bottom,i/(Tbottom,i\gamma}\}\{k_{,i}/100\}_1 + G_{bottom,i}\{e^{-t,bottom,i/(Tbottom,i\gamma)}\}\{k_{,i}/100\}_1\{e^{-t,bottom,i/(Tbottom,i\gamma)}\}\{k_{,i}/100\}_2 + G_{bottom,i}\{e^{-t,bottom,i\ /(Tbottom,i\ \gamma)}\}\{k_i/100\}_1\{e^{-t,bottom,i\ /(Tbottom,i\ \gamma)}\}\{k_{,i}/100\}_2\{e^{-t,bottom,i/(Tbottom,i\gamma)}\}\{k_{,i}/100\}_3 + \ldots + G_{bottom,i}\{e^{-t,bottom,i/(Tbottom,i\gamma)}\}\{k_{,i}/100\}_1\{e^{-t,bottom,i/(Tbottom,i\gamma)}\}\{k_{,i}/100\}_2 \ldots\{e^{-t,bottom,i/(Tbottom,i\gamma)}\}\{k_{,i}/100\}_{m,i}\}/dt_{ship} >$

$= <d\{\sum(i = 1;\ i = M)\ (E_i)\ \{G_{strange,i}\{(1/2)\ e^{-t,strange,i/(\gamma t,strange,i1/2)}\}\{k_{,i}/100\}_1 + G_{strange,i}\{(1/2)\ e^{-t,strange,i/(\gamma t,strange,i1/2)}\}\{k_{,i}/100\}_1\{(1/2)\ e^{-t,strange,i/(\gamma t,strange,i1/2)}\}\{k_{,i}/100\}_2 + G_{strange,i}\{(1/2)\ e^{-t,strange,i/(\gamma t,strange,i1/2)}\}\{k_{,i}/100\}_1\{(1/2)\ e^{-}$

$$^{t,strange,i/(\gamma t,strange,i1/2)}\} \{k_{,i}/100\}_2\{(1/2)\ e^{-t,strange,i/(\gamma t,strange,i1/2)}\} \{k_{,i}/100\}_3 + \ldots + G_{strange,i}\{(1/2)\ e^{-t,strange,i/(\gamma t,strange,i1/2)}\} \{k_{,i}/100\}_1\{(1/2)\ e^{-t,strange,i/(\gamma t,strange,i1/2)}\} \{k_{,i}/100\}_2 \ldots\{(1/2)\ e^{-t,strange,i/(\gamma t,strange,i1/2)}\} \{k_{,i}/100\}_{m,i}\}/dt_{ship} >$$

$$+ <d\{\sum(i = 1; i = M)\ (E_i)\ \{G_{charm,i}\{(1/2)\ e^{-t,charm,i/(\gamma t,charm,i1/2)}\} \{k_{,i}/100\}_1 + G_{charm,i}\{(1/2)\ e^{-t,charm,i/(\gamma t,charm,i1/2)}\} \{k_{,i}/100\}_1\{(1/2)\ e^{-t,charm,i/(\gamma t,charm,i1/2)}\} \{k_{,i}/100\}_2 + G_{charm,i}\{(1/2)\ e^{-t,charm,i/(\gamma t,charm,i1/2)}\} \{k_{,i}/100\}_1\{(1/2)\ e^{-t,charm,i/(\gamma t,charm,i1/2)}\} \{k_{,i}/100\}_2\{(1/2)\ e^{-t,charm,i/(\gamma t,charm,i1/2)}\} \{k_{,i}/100\}_3 + \ldots + G_{charm,i}\{(1/2)\ e^{-t,charm,i/(\gamma t,charm,i1/2)}\} \{k_{,i}/100\}_1\{(1/2)\ e^{-t,charm,i/(\gamma t,charm,i1/2)}\} \{k_{,i}/100\}_2 \ldots\{(1/2)\ e^{-t,charm,i/(\gamma t,charm,i1/2)}\} \{k_{,i}/100\}_{m,i}\}/dt_{ship} >$$

$$+ <d\{\sum(i = 1; i = M)\ (E_i)\ \{G_{bottom,i}\{(1/2)\ e^{-t,bottom,i/(\gamma t,bottom,i1/2)}\} \{k_{,i}/100\}_1 + G_{bottom,i}\{(1/2)\ e^{-t,bottom,i/(\gamma t,bottom,i1/2)}\} \{k_{,i}/100\}_1\{(1/2)\ e^{-t,bottom,i/(\gamma t,bottom,i1/2)}\} \{k_{,i}/100\}_2 + G_{bottom,i}\{(1/2)\ e^{-t,bottom,i/(\gamma t,bottom,i1/2)}\} \{k_{,i}/100\}_1\{(1/2)\ e^{-t,bottom,i/(\gamma t,bottom,i1/2)}\} \{k_{,i}/100\}_2\{(1/2)\ e^{-t,bottom,i/(\gamma t,bottom,i1/2)}\} \{k_{,i}/100\}_3 + \ldots + G_{bottom,i}\{(1/2)\ e^{-t,bottom,i/(\gamma t,bottom,i1/2)}\} \{k_{,i}/100\}_1\{(1/2)\ e^{-t,bottom,i/(\gamma t,bottom,i1/2)}\} \{k_{,i}/100\}_2 \ldots\{(1/2)\ e^{-t,bottom,i/(\gamma t,bottom,i1/2)}\} \{k_{,i}/100\}_{m,i}\}/dt_{ship} >$$

Chapter 14

Top Matter Reactors and Bombs and Other Things Nice

This chapter includes a description of highly conjectural forms of QCD/QED reactors and bombs. Such reactors and bombs would be used to power highly relativistic starships by converting a large fuel mass at least partially into energy and/or exothermic conversion of interstellar ordinary baryonic matter into strange, charm, and/or bottom-quark-based matter, and perhaps even top matter, if top quarks can be baryonized or mesonized.

Now that the nuclear and the eventual exotic QCD energy applications genie is permanently out of the bottle and exotic QCD energy applications loom on the horizon,, how might such applications be used to reach the stars?

Below is a brief summary of thoughts on how exotic applications of nuclear and subnuclear reactions can open up the cosmos for colonization and exploration by humanity. The technologies considered have likely already been pondered and studied to some degree; but if not, they definitely will be in the future.

Imagine a technology that could convert the elements in the periodic table atomic to other exotic forms of baryonic matter, releasing prodigious quantities of energy in a mass-specific reaction energy that is higher than that released by nuclear fusion. It has been proposed that certain species of some stable toplets that might be produced in particle accelerators may be able to convert periodic table atoms into top matter. An unfortunate consequence of this may be a runaway reaction that could cause the planet to be exothermically converted into top matter or matter comprised mostly of top quarks with some additional species of quarks mixed in.

Stable catalytic toplets may be ideal for powering future interstellar ramjet starships (ISRs). Imagine a large mass of stable top matter for which interstellar and intergalactic matter would be funneled into the intake of a reaction chamber where the intake mass would be converted to quickly decaying toplets, which would convert even more of the intake mass into energy through decay processes. The drag energy in the form of heat could be recycled to power photon, ion, electron, positron, muon, antimuon, tauon, antitauon, proton, antiproton, exotic charged meson, and/or exotic charged baryon rockets. The stable or metastable toplets produced could be used as an accelerated thrust stream. A starship using this form of propulsion might accelerate and achieve high Lorentz factors and/or near light speeds in any hyperspatial dimension where periodic table atomic matter or other convertible baryonic matter is present.

Let us consider a scenario where a species of toplet would not decay on average n percent of the time before making contact with an ordinary baryon or atomic nucleus, and where the probability of the conversion of the baryon or atomic nucleus is 100 percent. The average probability of the initial toplet converting an ordinary nucleon or atomic nucleus to another catalytic toplet is n/100. After m time steps where, for which each time step the probability of the average catalytic toplet converting a baryon or nucleus it first contacts is n/100, the number of catalyzed baryons or nuclei is equal to in first order:

$$N = G_{top}\{n/100\}_1 + G_{top}\{n/100\}_1\ \{n/100\}_2 + G_{top}\{n/100\}_2\ \{n/100\}_2\{n/100\}_3 + \ldots + G_{top}\{n/100\}_1\ \{n/100\}_2 \ldots \{\ n/100\}_m$$

Where m is the number of time steps for the overall reaction extinction and G_{top} is the initial population of toplets.

Another way of presenting reaction-wave propagation is to consider the average lifetime of the toplets and the average distance traveled by the toplets relative to the mean free path for toplet and nucleon or atomic nuclei contact. For cases where the average toplet would not decay with a probability of n/100 before completing the travel of its mean free path, the number of catalyzed baryons or nuclei is also equal to in first order:

$$N = G_{top}\{n/100\}_1 + G_{top}\{n/100\}_2\ \{n/100\}_2 + G_{top}\{n/100\}_2\ \{n/100\}_2\{n/100\}_3 + \ldots + G_{top}\{n/100\}_1\ \{n/100\}_2 \ldots \{\ n/100\}_m$$

Where m is the number of time steps for the overall reaction extinction.

Considering the length of time, t_m, traveled per mean free path, where the average nondecay probability is n/100 for each mean path traveled, the velocity of the toplet is L/t_m, where the velocity of the toplet is negligibly relativistic.

For cases where the toplets produced are relativistic, the average distance traveled by such toplets before it decays is given by:

$$D = [L/t_m](\gamma)\ (n)/(100) = [L/t_m]\{1/\{1 - [(v/C)^2]\}^{1/2}\}(n)/(100)$$

Now, applying the exponetial decay formula:

$$N(t) = N_0\, e^{-t,top/Ttop} = N_0\, (1/2) e^{-t,top/t,top1/2}$$

Where *T* is the mean lifetime of a particle (such as an atom or subatomic particle), $t_{1/2}$ is the half-life of the particle, and N_0 is the starting population. *N(t)* is the quantity that still remains and has not yet decayed after a time *t*.

Therefore the half-life is expressed as:

$$t_{1/2} = \ln(2)/\lambda = T \ln(2)$$

Here, λ is a positive number called the decay constant of the decaying quantity.

For a toplet having a probability of nondecay along a mean path length of P = n/100, the number of toplets out of a starting population of $N_0 = G_{top}$ that contact a baryon or atomic nuclei is $N_0\{n/100\}$ where $\{n/100\} = e^{-t,top/Ttop} = (1/2)\, e^{-t,top/t,top1/2}$. Therefore, the number of catalyzed baryons becomes in first order:

$$N = G_{top}\{e^{-t,top/Ttop}\}_1 + G_{top}\{e^{-t,top/Ttop}\}\{e^{-t,top/Ttop}\}_2 + G_{top}\, e^{-t,top/Ttop}\}\{e^{-t,top/Ttop}\}\{e^{-t,top/Ttop}\}_3 + \ldots + G_{top}\{e^{-t,top/Ttop}\}\{e^{-t,top/Ttop}\}\ldots\{e^{-t,top/Ttop}\}_m$$

$$= G_{top}\{(1/2)\, e^{-t,top/t,top1/2}\}_1 + G_{top}\{(1/2)\, e^{-t,top/t,top1/2}\}\{(1/2)\, e^{-t,top/t,top1/2}\}_2 + G_{top}\{(1/2)\, e^{-t,top/t,top1/2}\}\{(1/2)\, e^{-t,top/t,top1/2}\}\{(1/2)\, e^{-t,top/t,top1/2}\}_3 + \ldots + G_{top}\{(1/2)\, e^{-t,top/t,top1/2}\}\{(1/2)\, e^{-t,top/t,top1/2}\}\ldots\{(1/2)\, e^{-t,top/t,top1/2}\}_m,$$

Where *t* is the travel time of the toplets for the mean free path for toplet to ordinary baryon contact.

Now, if $t_{1/2}$ and T are background stationary frames and the toplets are relativistic, the number of catalyzed baryons becomes in first order:

$$N = G_{top}\{e^{-t,top/(Ttop\gamma}\}_1 + G_{top}\{e^{-t,top/(Ttop\gamma)}\}\{e^{-t,top/(Ttop\gamma)}\}_2 + G_{top}\{e^{-t,top/(Ttop\gamma)}\}\{e^{-t,top/(Ttop\gamma)}\}\{e^{-t,top/(Ttop\gamma)}\}_3 + \ldots + G_{top}\{e^{-t,top/(Ttop\gamma)}\}\{e^{-t,top/(Ttop\gamma)}\}\ldots\{e^{-t,top/(Ttop\gamma)}\}_m$$

$$= G_{top}\{(1/2)\, e^{-t,top/(\gamma t,top1/2)}\}_1 + G_{top}\{(1/2)\, e^{-t,top/(\gamma t,top1/2)}\}\{(1/2)\, e^{-t,top/(\gamma t,top1/2)}\}_2 + G_{top}\{(1/2)\, e^{-t,top/(\gamma t,top1/2)}\}\{(1/2)\, e^{-t,top/(\gamma t,top1/2)}\}\{(1/2)\, e^{-t,top/(\gamma t,top1/2)}\}_3 + \ldots + G_{top}\{(1/2)\, e^{-t,top/(\gamma t,top1/2)}\}\{(1/2)\, e^{-t,top/(\gamma t,top1/2)}\}\ldots\{(1/2)\, e^{-t,top/(\gamma t,top1/2)}\}_m$$

For cases where the average toplet would not decay with a probability of n/100 after traveling the mean free path, and the probability of baryon or nuclei conversion upon contact is k/100, but where the toplets are always destroyed when contacting the nuclei or converted into a non-top-transmutative form with a probability of unity, the number of catalyzed baryons or nuclei is equal to in first order:

$$N = G_{top}\{n/100\}\{k/100\}_1 + G_{top}\{n/100\}\{k/100\}_1\{n/100\}\{k/100\}_2 + G_{top}\{n/100\}\{k/100\}_1 \{n/100\}\{k/100\}_2\{n/100\}\{k/100\}_3 + \ldots + G_{top}\{n/100\}\{k/100\}_1\{n/100\}\{k/100\}_2 \ldots \{n/100\}\{k/100\}_m,$$

Here, *m* is the number of time steps for the overall reaction extinction.

Considering the length of time, t_m, traveled per mean free path, where the average nondecay probability is n/100 for each mean path traveled, the velocity of the toplet is L/t_m, where the velocity of the toplet is negligibly relativistic.

For cases where the toplets produced are relativistic, the distance traveled by a given toplet before it decays is given by:

$$[L/t_m](\gamma) = [L/t_m]\{1/\{1 - [(v/C)^2]\}^{1/2}\}$$

Once again, $N(t) = N_0\, e^{-t,top/Ttop} = N_0\, (1/2)e^{-t,top/t,top1/2}$, where T is the mean lifetime of a particle (such as an atom or subatomic particle), $t_{1/2}$ is the half-life of the particle, and N_0 is the starting population. $N(t)$ is the quantity that still remains and has not yet decayed after a time t, and half-life $t_{1/2} = \ln(2)/\lambda = T \ln(2)$, where λ is a positive number called the decay constant of the decaying quantity.

In the case of a toplet species having a probability of nondecay along a mean path length of $P = n/100$, the number of toplets out of a starting population of $N_0 = G_{top}$ that contact a baryon or atomic nuclei is $N_0\{n/100\}$ where $\{n/100\} = e^{-t,top/Ttop} = (1/2)\, e^{-t,top/t,top1/2}$. Therefore, the number of catalyzed baryons becomes in first order:

$$N = G_{top}\{e^{-t,top/Ttop}\}\{k/100\}_1 + G_{top}\{e^{-t,top/Ttop}\}\{k/100\}_1\{e^{-t,top/Ttop}\}\{k/100\}_2 + G_{top}\{e^{-t,top/Ttop}\}\{k/100\}_1\{e^{-t,top/Ttop}\}\{k/100\}_2\{e^{-t,top/Ttop}\}\{k/100\}_3 + \ldots + G_{top}\{e^{-t,top/Ttop}\}\{k/100\}_1\{e^{-t,top/Ttop}\}\{k/100\}_2 \ldots \{e^{-t,top/Ttop}\}\{k/100\}_m$$

$$= G_{top}\{(1/2)\, e^{-t,top/t,top1/2}\}\{k/100\}_1 + G_{top}\{(1/2)\, e^{-t,top/t,top1/2}\}\{k/100\}_1\{(1/2)\, e^{-t,top/t,top1/2}\}\{k/100\}_2 + G_{top}\{(1/2)\, e^{-t,top/t,top1/2}\}\{k/100\}_1\{(1/2)\, e^{-t,top/t,top1/2}$$

$$\}\{k/100\}_2 \{(1/2)\ e^{-t,top/t,top1/2}\}\{k/100\}_3 + \ldots + G_{top}\{(1/2)\ e^{-t,top/t,top1/2}\}\{k/100\}_1\{(1/2)\ e^{-t,top/t,top1/2}\}\{k/100\}_2 \ldots\{(1/2)\ e^{-t,top/t,top1/2}\}\{k/100\}_m$$

If $t_{1/2}$ and T are background stationary frames and the toplets are relativistic, the number of catalyzed baryons becomes in first order:

$$N = G_{top}\{e^{-t,top/(Ttop\gamma}\}\{k/100\}_1 + G_{top}\{e^{-t,top/(Ttop\gamma)}\}\{k/100\}_1\{e^{-t,top/(Ttop\gamma)}\}\{k/100\}_2 + G_{top}\{e^{-t,top/(Ttop\gamma)}\}\{k/100\}_1\{e^{-t,top/(Ttop\gamma)}\}\{k/100\}_2\{e^{-t,top/(Ttop\gamma)}\}\{k/100\}_3 + \ldots + G_{top}\{e^{-t,top/(Ttop\gamma)}\}\{k/100\}_1\{e^{-t,top/(Ttop\gamma)}\}\{k/100\}_2 \ldots\{e^{-t,top/(Ttop\gamma)}\}\{k/100\}_m$$

$$= G_{top}\{(1/2)\ e^{-t,top/(\gamma t,top1/2)}\}\{k/100\}_1 + G_{top}\{(1/2)\ e^{-t,top/(\gamma t,top1/2)}\}\{k/100\}_1\{(1/2)\ e^{-t,top/(\gamma t,top1/2)}\}\{k/100\}_2 + G_{top}\{(1/2)\ e^{-t,top/(\gamma t,top1/2)}\}\{k/100\}_1\{(1/2)\ e^{-t,top/(\gamma t,top1/2)}\}\{k/100\}_2\{(1/2)\ e^{-t,top/(\gamma t,top1/2)}\}\{k/100\}_3 + \ldots + G_{top}\{(1/2)\ e^{-t,top/(\gamma t,top1/2)}\}\{k/100\}_1\{(1/2)\ e^{-t,top/(\gamma t,top1/2)}\}\{k/100\}_2 \ldots\{(1/2)\ e^{-t,top/(\gamma t,top1/2)}\}\{k/100\}_m$$

Let us consider a scenario where a species, i, of toplet would not decay on average n percent of the time before making contact with an ordinary baryon or atomic nucleus and where the probability of the conversion of the baryon or atomic nucleus is 100 percent. The average probability of the initial toplet converting an ordinary nucleon or atomic nucleus to another catalytic toplet is $n_{,i}/100$. After m time steps where, for which each time step, the probability of the average catalytic toplet converting a baryon or nucleus it first contacts is $n_{,i}/100$, the number of catalyzed baryons or nuclei is equal to in first order:

$$N_{,i} = G_{top,i}\{n_{,i}/100\}_1 + G_{top,i}\{n_i/100\}_1\ \{n_{,i}/100\}_2 + G_{top,\ i}\{n_{,i}/100\}_2\ \{n_{,i}/100\}_2\{n_{,i}/100\}_3 + \ldots + G_{top,i}\{n_{,i}/100\}_1\ \{n_{,i}/100\}_2 \ldots\{\ n_{,i}/100\}_{m,i}$$

Where $m_{,i}$ is the number of time steps for the overall reaction extinction, and $G_{top,i}$ is the initial population of toplets.

So for M species of toplets, where each species has its unique mean free-path, initial population, and number of time steps for overall reaction extinctions, the number of catalyzed baryons or nuclei is equal to in first order:

$$\sum(i = 1;\ i = M)\ N_i = \sum(i = 1;\ i = M)\ \{G_{top,i}\{n_{,i}/100\}_1 + G_{top,i}\{n_{,i}/100\}_1\ \{n_{,i}/100\}_2 + G_{top,\ i}\{n_{,i}/100\}_2\ \{n_{,i}/100\}_2\{n_{,i}/100\}_3 + \ldots + G_{top,i}\{n_{,i}/100\}_1\ \{n_{,i}/100\}_2 \ldots\{\ n_{,i}/100\}_{m,i}\}$$

Where the average yield energy per nucleon or nucleus conversion caused by transformation by particles if the ith species is E_i.

The total yield energy for a given QCD bomb is in first order:

$$\sum(i = 1;\ i = M)\ [(N_i)(E_i)] = \sum(i = 1;\ i = M)\ (E_i)\ \{\{G_{top,i}\{n_{,i}/100\}_1 + G_{top,i}\{n_{,i}/100\}_1\ \{n_{,i}/100\}_2 + G_{top,\ i}\{n_{,i}/100\}_2\ \{n_{,i}/100\}_2\{n_{,i}/100\}_3 + \ldots + G_{top,i}\{n_{,i}/100\}_1\ \{n_{,i}/100\}_2 \ldots \{\ n_{,i}/100\}_{m,i}\}\}$$

The time averaged explosive power of a given QCD bomb in the ship frame is in first order:

$$<d\{\sum(i = 1;\ i = M)\ [(N_i)(E_i)]\}\ /dt_{ship} > = <d\{\sum(i = 1;\ i = M)\ (E_i)\{\{G_{top,i}\{n_{,i}/100\}_1 + G_{top,i}\{n_{,i}/100\}_1\ \{n_{,i}/100\}_2 + G_{top,\ i}\{n_i/100\}_2\ \{n_{,i}/100\}_2\{n_{,i}/100\}_3 + \ldots + G_{top,i}\{n_{,i}/100\}_1\ \{n_{,i}/100\}_2 \ldots \{\ n_{,i}/100\}_{m,i}\}\}/dt_{ship} >$$

Another way of presenting reaction-wave propagation is to consider the average lifetime of the toplets and the average distance traveled by the toplets relative to the mean free path for toplet and nucleon or atomic nuclei contact. For cases where the toplet of the ith species of toplets would nondecay with a probability of $n_{,i}/100$ before completing the travel of its mean free path, the number of catalyzed baryons or nuclei is also equal to in first order:

$$N_i = G_{top,i}\{n_{,i}/100\}_1 + G_{top,i}\{n_{,i}/100\}_1\ \{n_{,i}/100\}_2 + G_{top,\ i}\{n_{,i}/100\}_2\ \{n_{,i}/100\}_2\{n_{,i}/100\}_3 + \ldots + G_{top,i}\{n_{,i}/100\}_1\ \{n_{,i}/100\}_2 \ldots \{\ n_{,i}/100\}_{m,i}$$

Where $m_{,i}$ is the number of time steps for the overall reaction extinction for the ith species of toplet.

So, for M species of toplets, where each species has its unique mean free path, initial population, and number of time steps for overall reaction extinctions, the number of catalyzed baryons or nuclei is equal to in first order:

$$\sum(i = 1;\ i = M)\ N_i = \sum(i = 1;\ i = M)\ \{G_{top,i}\{n_{,i}/100\}_1 + G_{top,i}\{n_{,i}/100\}_1\ \{n_{,i}/100\}_2 + G_{top,\ i}\{n_{,i}/100\}_2\ \{n_{,i}/100\}_2\{n_{,i}/100\}_3 + \ldots + G_{top,i}\{n_{,i}/100\}_1\ \{n_{,i}/100\}_2 \ldots \{\ n_{,i}/100\}_{m,i}\}$$

Where the average yield energy per nucleon or nucleus conversion caused by transformation by particles if the ith species is E_i.

The total yield energy for a given QCD bomb is in first order:

$$\sum(i = 1;\ i = M)\ [(N_i)(E_i)] = \sum(i = 1;\ i = M)\ (E_i)\{\{G_{top,i}\{n_{,i}/100\}_1 + G_{top,i}\{n_{,i}/100\}_1\ \{n_{,i}/100\}_2 + G_{top,\ i}\{n_{,i}/100\}_2\ \{n_{,i}/100\}_2\{n_{,i}/100\}_3 + \ldots + G_{top,i}\{n_{,i}/100\}_1\ \{n_{,i}/100\}_2 \ldots\{\ n_{,i}/100\}_{m,i}\}\}$$

The time-averaged explosive power of a given QCD bomb in the ship frame is in first order:

$$<d\{\sum(i = 1;\ i = M)\ [(N_i)(E_i)]\}\ /dt_{ship} > = <d\{\sum(i = 1;\ i = M)\ (E_i)\{\{G_{top,i}\{n_{,i}/100\}_1 + G_{top,i}\{n_{,i}/100\}_1\ \{n_{,i}/100\}_2 + G_{top,\ i}\{n_{,i}/100\}_2\ \{n_{,i}/100\}_2\{n_{,i}/100\}_3 + \ldots + G_{top,i}\{n_{,i}/100\}_1\ \{n_{,i}/100\}_2 \ldots\{\ n_{,i}/100\}_{m,i}\}\}\}/dt_{ship} >$$

Considering the length of time, $t_{m,i}$, traveled per mean free path where the average nondecay probability is $n_{,i}/100$ for each mean path traveled, the velocity of the toplet is $L_i/t_{m.i}$, where the velocity of the toplet is negligibly relativistic for the ith species of toplets.

For cases where the ith species of toplets produced are relativistic, the average distance traveled by such toplets before it decays is given by:

$$D_i = [L_i/t_{m,i}](\gamma_{,i})\ (n_{,i})/(100) = [L_i/t_{m,i}]\{1/\{1 - [(v_{,i}/C)^2]\}^{1/2}\}(n_{,i})/(100)$$

Now, applying the exponetial decay formula:

$$N_i(t_i) = N_{0i}\ e^{-t,top.i/Ttop,i} = N_{0i}\ (1/2)e^{-t,top,i/t,top,i1/2}$$

Where T_i is the mean lifetime of a particle (such as an atom or subatomic particle), $t_{i1/2}$ is the half-life of the particle, and N_{0i} is the starting population. $N_i(t_i)$ is the quantity that still remains and has not yet decayed after a time t_i.

Therefore the half-life is expressed as:

$$t_{i1/2} = \ln(2)/\lambda_i = T_i \ln(2)$$

Here, λ_i is a positive number called the decay constant of the decaying quantity.

For a toplet of the ith toplet species having a probability of nondecay along a mean path length of $P_i = n_{,i}/100$, the number of toplets out of a starting population of $N_{0i} = G_{top,i}$ that contact a baryon or atomic nuclei is $N_{0i}\{n_{,i}/100\}$ where $\{n_{,i}/100\} = e^{-}$

$^{t,top,i/Ttop,i} = (1/2)\; e^{-t,top,i/t,top,i1/2}$. Therefore, the number of catalyzed baryons becomes in first order:

$$N_i = G_{top,i}\{e^{-t,top,i/Ttop,i}\}_1 + G_{top,i}\{e^{-t,top,i/Ttop,i}\}\{e^{-t,top,i/Ttop,i}\}_2 + G_{top,i}\, e^{-t,top,i/Ttop,i}\}\{e^{-t,top,i/Ttop,i}\}\{e^{-t,top,i/Ttop,i}\}_3 + \ldots + G_{top,i}\{e^{-t,top,i/Ttop,i}\}\{e^{-t,top,i/Ttop,i}\}\ldots\{e^{-t,top,i/Ttop,i}\}_{m,i}$$

$$= G_{top,i}\{(1/2)\, e^{-t,top,i/t,top,i1/2}\}_1 + G_{top,i}\{(1/2)\, e^{-t,top,i/t,top,i1/2}\}\{(1/2)\, e^{-t,top,i/t,top,i1/2}\}_2 + G_{top,i}\{(1/2)\, e^{-t,top,i/t,top,i1/2}\}\{(1/2)\, e^{-t,top,i/t,top,i1/2}\}\{(1/2)\, e^{-t,top,i/t,top,i1/2}\}_3 + \ldots + G_{top,i}\{(1/2)\, e^{-t,top,i/t,top,i1/2}\}\{(1/2)\, e^{-t,top,i/t,top,i1/2}\}\ldots\{(1/2)\, e^{-t,top,i/t,top,i1/2}\}_{m,i},$$

Where t_i is the travel time of the ith species of toplets for the mean free path for toplet to ordinary baryon contact.

Where the average yield energy per nucleon or nucleus conversion caused by transformation by particles if the ith species is E_i.

The total yield energy for a given QCD bomb is in first order:

$$\sum(i = 1;\, i = M)\, [(N_i)(E_i)] = \sum(i = 1;\, i = M)\, (E_i)\, \{G_{top,i}\{e^{-t,top,i/Ttop,i}\}_1 + G_{top,i}\{e^{-t,top,i/Ttop,i}\}\{e^{-t,top,i/Ttop,i}\}_2 + G_{top,i}\, e^{-t,top,i/Ttop,i}\}\{e^{-t,top,i/Ttop,i}\}\{e^{-t,top,i/Ttop,i}\}_3 + \ldots + G_{top,i}\{e^{-t,top,i/Ttop,i}\}\{e^{-t,top,i/Ttop,i}\}\ldots\{e^{-t,top,i/Ttop,i}\}_{m,i}\}$$

$$= \sum(i = 1;\, i = M)\, (E_i)\{G_{top,i}\{(1/2)\, e^{-t,top,i/t,top,i1/2}\}_1 + G_{top,i}\{(1/2)\, e^{-t,top,i/t,top,i1/2}\}\{(1/2)\, e^{-t,top,i/t,top,i1/2}\}_2 + G_{top,i}\{(1/2)\, e^{-t,top,i/t,top,i1/2}\}\{(1/2)\, e^{-t,top,i/t,top,i1/2}\}\{(1/2)\, e^{-t,top,i/t,top,i1/2}\}_3 + \ldots + G_{top,i}\{(1/2)\, e^{-t,top,i/t,top,i1/2}\}\{(1/2)\, e^{-t,top,i/t,top,i1/2}\}\ldots\{(1/2)\, e^{-t,top,i/t,top,i1/2}\}_{m,i}\},$$

The time averaged explosive power of a given QCD bomb in the ship frame is in first order:

$$<d\{\sum(i = 1;\, i = M)\, [(N_i)(E_i)]\}/dt_{ship}> = <d\{\sum(i = 1;\, i = M)\, (E_i)\, \{G_{top,i}\{e^{-t,top,i/Ttop,i}\}_1 + G_{top,i}\{e^{-t,top,i/Ttop,i}\}\{e^{-t,top,i/Ttop,i}\}_2 + G_{top,i}\, e^{-t,top,i/Ttop,i}\}\{e^{-t,top,i/Ttop,i}\}\{e^{-t,top,i/Ttop,i}\}_3 + \ldots + G_{top,i}\{e^{-t,top,i/Ttop,i}\}\{e^{-t,top,i/Ttop,i}\}\ldots\{e^{-t,top,i/Ttop,i}\}_{m,i}\}\}/dt_{ship}>$$

$$= <d\{\sum(i = 1;\, i = M)\, (E_i)\{G_{top,i}\{(1/2)\, e^{-t,top,i/t,top,i1/2}\}_1 + G_{top,i}\{(1/2)\, e^{-t,top,i/t,top,i1/2}\}\{(1/2)\, e^{-t,top,i/t,top,i1/2}\}_2 + G_{top,i}\{(1/2)\, e^{-t,top,i/t,top,i1/2}\}$$

$\{(1/2)\ e^{-t,top,i/t,top,i1/2}\}\ \{(1/2)\ e^{-t,top,i/t,top,i1/2}\}_3 + \ldots + G_{top,i}\{(1/2)\ e^{-t,top,i/t,top,i1/2}\}\ \{(1/2)\ e^{-t,top,i/t,top,i1/2}\}\ \ldots\{(1/2)\ e^{-t,top,i/t,top,i1/2}\}_{m,i}\}\}/dt_{ship}>$

Now, if $t_{i1/2}$ and T_i are background stationary frames and the ith species of toplets are relativistic, the number of catalyzed baryons becomes in first order:

$$N = G_{top,i}\{e^{-t,top,i/(Ttop,i\gamma}\}_1 + G_{top,i}\{e^{-t,top,i/(Ttop,i\gamma)}\}\{\ e^{-t,top,i/(Ttop,i\gamma)}\}_2 + G_{top,i}\{\ e^{-t,top,i/(Ttop,i\gamma)}\}\{\ e^{-t,top,i/(Ttop,i\gamma)}\}\{\ e^{-t,top,i/(Ttop,i\gamma)}\}_3 + \ldots + G_{top,i}\{\ e^{-t,top,i/(Ttop,i\gamma)}\}\{\ e^{-t,top,i/(Ttop,i\gamma)}\}\ \ldots\{\ e^{-t,top,i/(Ttop,i\gamma)}\}_{m,i}$$

$$= G_{top,i}\{(1/2)\ e^{-t,top,i/(\gamma t,top,i1/2)}\}_1 + G_{top,i}\{(1/2)\ e^{-t,top,i/(\gamma t,top,i1/2)}\}\{(1/2)\ e^{-t,top,i/(\gamma t,top,i1/2)}\}_2 + G_{top,i}\{(1/2)\ e^{-t,top,i/(\gamma t,top,i1/2)}\}\{(1/2)\ e^{-t,top,i/(\gamma t,top,i1/2)}\}\{(1/2)\ e^{-t,top,i/(\gamma t,top,i1/2)}\}_3 + \ldots + G_{top,i}\{(1/2)\ e^{-t,top,i/(\gamma t,top,i1/2)}\}\{(1/2)\ e^{-t,top,i/(\gamma t,top,i1/2)}\}\ \ldots\{(1/2)\ e^{-t,top,i/(\gamma t,top,i1/2)}\}_{m,i}$$

Where the average yield energy per nucleon or nucleus conversion caused by transformation by particles if the ith species is E_i.

The total yield energy for a given QCD bomb is in first order:

$$\sum(i = 1;\ i = M)\ [(N_i)(E_i)] = \sum(i = 1;\ i = M)\ (E_i)\ \{G_{top,i}\{e^{-t,top,i/(Ttop,i\gamma}\}_1 + G_{top,i}\{e^{-t,top,i/(Ttop,i\gamma)}\}\{\ e^{-t,top,i/(Ttop,i\gamma)}\}_2 + G_{top,i}\{\ e^{-t,top,i/(Ttop,i\gamma)}\}\{\ e^{-t,top,i/(Ttop,i\gamma)}\}\{\ e^{-t,top,i/(Ttop,i\gamma)}\}_3 + \ldots + G_{top,i}\{\ e^{-t,top,i/(Ttop,i\gamma)}\}\{\ e^{-t,top,i/(Ttop,i\gamma)}\}\ \ldots\{\ e^{-t,top,i/(Ttop,i\gamma)}\}_{m,i}\}$$

$$= \sum(i = 1;\ i = M)\ (E_{,i})\ \{G_{top,i}\{(1/2)\ e^{-t,top,i/(\gamma t,top,i1/2)}\}_1 + G_{top,i}\{(1/2)\ e^{-t,top,i/(\gamma t,top,i1/2)}\}\{(1/2)\ e^{-t,top,i/(\gamma t,top,i1/2)}\}_2 + G_{top,i}\{(1/2)\ e^{-t,top,i/(\gamma t,top,i1/2)}\}\{(1/2)\ e^{-t,top,i/(\gamma t,top,i1/2)}\}\{(1/2)\ e^{-t,top,i/(\gamma t,top,i1/2)}\}_3 + \ldots + G_{top,i}\{(1/2)\ e^{-t,top,i/(\gamma t,top,i1/2)}\}\{(1/2)\ e^{-t,top,i/(\gamma t,top,i1/2)}\}\ \ldots\{(1/2)\ e^{-t,top,i/(\gamma t,top,i1/2)}\}_{m,i}\}$$

The time-averaged explosive power of a given QCD bomb in the ship frame is in first order:

$$<d\{\sum(i = 1;\ i = M)\ [(N_i)(E_i)]\}\ /dt_{ship}> = <d\{\sum(i = 1;\ i = M)\ (E_i)\ \{G_{top,i}\{e^{-t,top,i/(Ttop,i\gamma}\}_1 + G_{top,i}\{e^{-t,top,i/(Ttop,i\gamma)}\}\{\ e^{-t,top,i/(Ttop,i\gamma)}\}_2 + G_{top,i}\{\ e^{-t,top,i/(Ttop,i\gamma)}\}\{\ e^{-t,top,i/(Ttop,i\gamma)}\}\{\ e^{-t,top,i/(Ttop,i\gamma)}\}_3 + \ldots + G_{top,i}\{\ e^{-t,top,i/(Ttop,i\gamma)}\}\{\ e^{-t,top,i/(Ttop,i\gamma)}\}\ \ldots\{\ e^{-t,top,i/(Ttop,i\gamma)}\}_{m,i}\}\}/dt_{ship}>$$

$$= <d\{\sum(i = 1;\ i = M)\ (E_i)\ \{G_{top,i}\{(1/2)\ e^{-t,top,i/(\gamma t,top,i1/2)}\}_1 + G_{top,i}\{(1/2)\ e^{-t,top,i/(\gamma t,top,i1/2)}\}\{(1/2)\ e^{-t,top,i/(\gamma t,top,i1/2)}\}_2 + G_{top,i}\{(1/2)\ e^{-}$$

$$^{t,top,i/(\gamma t,top,i1/2)}\}\{(1/2)\ e^{-t,top,i/(\gamma t,top,i1/2)}\}\{(1/2)\ e^{-t,top,i/(\gamma t,top,i1/2)}\}_3 + \ldots + G_{top,i}\{(1/2)\ e^{-t,top,i/(\gamma t,top,i1/2)}\}\{(1/2)\ e^{-t,top,i/(\gamma t,top,i1/2)}\} \ldots\{(1/2)\ e^{-t,top,i/(\gamma t,top,i1/2)}\}_{m,i}\}\}/dt_{ship} >$$

For cases where the average toplet of the ith species would not decay with a probability of $n_{,i}/100$ after traveling the mean free path, and the probability of baryon or nuclei conversion upon contact is $k_{,i}/100$, but where the toplets are always destroyed when contacting the nuclei or converted into a non-top-transmutative form with a probability of unity, the number of catalyzed baryons or nuclei is equal to in first order:

$$N_i = G_{top,i}\{n_{,i}/100\}\{k_i/100\}_1 + G_{top,i}\{n_{,i}/100\}\{k_{,i}/100\}_1\{n_{,i}/100\}\{k_{,i}/100\}_2 + G_{top,i}\{n_{,i}/100\}\{k_{,i}/100\}_1\ \{n_{,i}/100\}\{k_{,i}/100\}_2\{n_{,i}/100\}\{k_{,i}/100\}_3 + \ldots + G_{top,i}\{n_{,i}/100\}\{k_{,i}/100\}_1\{n_{,i}/100\}\{k_{,i}/100\}_2 \ldots\{n_{,i}/100\}\{k_{,i}/100\}_{m,i},$$

Here m is the number of time steps for the overall reaction extinction.

Where the average yield energy per nucleon or nucleus conversion caused by transformation by particles if the ith species is E_i.

The total yield energy for a given QCD bomb is in first order:

$$\sum(i = 1; i = M)\ [(N_i)(E_i)] = \sum(i = 1; i = M)\ (E_i)\ \{G_{top,i}\{n_{,i}/100\}\{k_{,i}/100\}_1 + G_{top,i}\{n_{,i}/100\}\{k_{,i}/100\}_1\{n_{,i}/100\}\{k_{,i}/100\}_2 + G_{top,i}\{n_{,i}/100\}\{k_{,i}/100\}_1 \{n_{,i}/100\}\{k_{,i}/100\}_2\{n_{,i}/100\}\{k_{,i}/100\}_3 + \ldots + G_{top,i}\{n_{,i}/100\}\{k_{,i}/100\}_1\{n_{,i}/100\}\{k_{,i}/100\}_2 \ldots\{n_{,i}/100\}\{k_{,i}/100\}_{m,i}\},$$

The time averaged explosive power of a given QCD bomb in the ship frame is in first order:

$$<d\{\sum(i = 1; i = M)\ [(N_i)(E_i)]\}\ /dt_{ship} > = <d\{\sum(i = 1; i = M)\ (E_i)\ \{G_{top,i}\{n_{,i}/100\}\{k_{,i}/100\}_1 + G_{top,i}\{n_{,i}/100\}\{k_{,i}/100\}_1\{n_{,i}/100\}\{k_{,i}/100\}_2 + G_{top,i}\{n_{,i}/100\}\{k_{,i}/100\}_1 \{n_{,i}/100\}\{k_{,i}/100\}_2\{n_{,i}/100\}\{k_{,i}/100\}_3 + \ldots + G_{top,i}\{n_{,i}/100\}\{k_{,i}/100\}_1\{n_{,i}/100\}\{k_{,i}/100\}_2 \ldots\{n_{,i}/100\}\{k_{,i}/100\}_m\}\}/dt_{ship} >$$

Considering the length of time, $t_{m,i}$, traveled per mean free path where the average nondecay probability is $n/100$ for each mean path traveled, the velocity of the ith species of toplet is $L_i/t_{m,i}$, where the velocity of the toplet is negligibly relativistic.

For cases where the ith species of toplets produced are relativistic, the distance traveled by a given toplet before it decays is given by:

$$[L_i/t_{mi}](\gamma) = [L_i/t_{m,ii}]\{1/\{1 - [(v_{,i}/C)^2]\}^{1/2}\}$$

Once again, $N_i\ (t_i) = N_{0i}\ e^{-t,top,i/Ttop,i} = N_{0i}\ (1/2)e^{-t,top,i/t,top,i1/2}$, where T_i is the mean lifetime of a particle (such as an atom or subatomic particle), $t_{1/2}$ is the half-life of the particle, and N_{0i} is the starting population. $N_i\ (t_i)$ is the quantity that still remains and has not yet decayed after a time t, and half-life $t_{1/2} = \ln(2)/\lambda_i = T_i \ln(2)$, where λ_i is a positive number called the decay constant of the decaying quantity.

In the case of a toplet species having a probability of nondecay along a mean path length of $P_i = n_i/100$, the number of toplets out of a starting population of $N_{0i} = G_{top,i}$ that contact a baryon or atomic nuclei is $N_{0i}\{n_i/100\}$ where $\{n_{,i}/100\} = e^{-t,top,i/Ttop,i} = (1/2)\ e^{-t,top,i/t,top,i1/2}$. Therefore, the number of catalyzed baryons becomes in first order:

$$N_i = G_{topi}\{e^{-t,top,i/Ttop,i}\}\{k_{,i}/100\}_1 + G_{top,i}\{e^{-t,top,i/Ttop,i}\}\{k_{,i}/100\}_1\{e^{-t,top,i/Ttop,i}\}\{k_{,i}/100\}_2 + G_{top,i}\{e^{-t,top,i/Ttop,i}\}\{k_{,i}/100\}_1\{e^{-t,top,i/Ttop,i}\}\{k_{,i}/100\}_2\{e^{-t,top,i/Ttop,i}\}\{k_{,i}/100\}_3 + \ldots + G_{top,i}\{e^{-t,top,i/Ttop,i}\}\{k_{,i}/100\}_1\{e^{-t,top,i/Ttop,i}\}\{k_{,i}/100\}_2 \ldots\{e^{-t,top,i/Ttop,i}\}\{k_{,i}/100\}_{m,i}$$

$$= G_{top,i}\{(1/2)\ e^{-t,top,i/t,top,i1/2}\}\{k_{,i}/100\}_1 + G_{top,i}\{(1/2)\ e^{-t,top,i/t,top,i1/2}\}\{k_{,i}/100\}_1\{(1/2)\ e^{-t,top,i/t,top,i1/2}\}\{k_{,i}/100\}_2 + G_{top,i}\{(1/2)\ e^{-t,top,i/t,top,i1/2}\}\{k_{,i}/100\}_1\{(1/2)\ e^{-t,top,i/t,top,i1/2}\}\{k_{,i}/100\}_2\{(1/2)\ e^{-t,top,i/t,top,i1/2}\}\{k_{,i}/100\}_3 + \ldots + G_{top,i}\{(1/2)\ e^{-t,top,i/t,top,i1/2}\}\{k_{,i}/100\}_1\{(1/2)\ e^{-t,top,i/t,top,i1/2}\}\{k_{,i}/100\}_2 \ldots\{(1/2)\ e^{-t,top,i/t,top,i1/2}\}\{k_{,i}/100\}_{m,i}$$

Where the average yield energy per nucleon or nucleus conversion caused by transformation by particles if the ith species is E_i.

The total yield energy for a given QCD bomb is in first order:

$$\sum(i = 1;\ i = M)\ [(N_i)(E_i)] = \sum(i = 1;\ i = M)\ (E_i)\ \{G_{top,i}\{e^{-t,top,i/Ttop,i}\}\{k_{,i}/100\}_1 + G_{top,i}\{e^{-t,top,i/Ttop,i}\}\{k_{,i}/100\}_1\{e^{-t,top,i/Ttop,i}\}\{k_{,i}/100\}_2 + G_{top,i}\{e^{-t,top,i/Ttop,i}\}\{k_{,i}/100\}_1\{e^{-t,top,i/Ttop,i}\}\{k_{,i}/100\}_2\{e^{-t,top,i/Ttop,i}\}\{k_{,i}/100\}_3 + \ldots + G_{top,i}\{e^{-t,top,i/Ttop,i}\}\{k_{,i}/100\}_1\{e^{-t,top,i/Ttop,i}\}\{k_{,i}/100\}_2 \ldots\{e^{-t,top,i/Ttop,i}\}\{k_{,i}/100\}_{m,i}\}$$

$= \sum(i = 1; i = M) (E_i) \{G_{top,i}\{(1/2) e^{-t,top,i/t,top,i1/2}\}\{k_{,i}/100\}_1 + G_{top,i}\{(1/2) e^{-t,top,i/t,top,i1/2}\}\{k_{,i}/100\}_1\{(1/2) e^{-t,top,i/t,top,i1/2}\}\{k_{,i}/100\}_2 + G_{top,i}\{(1/2) e^{-t,top,i/t,top,i1/2}\}\{k_{,i}/100\}_1\{(1/2) e^{-t,top,i/t,top,i1/2}\}\{k_{,i}/100\}_2 \{(1/2) e^{-t,top,i/t,top,i1/2}\}\{k_{,i}/100\}_3 + \ldots + G_{top,i}\{(1/2) e^{-t,top,i/t,top,i1/2}\}\{k_{,i}/100\}_1\{(1/2) e^{-t,top,i/t,top,i1/2}\}\{k_{,i}/100\}_2 \ldots\{(1/2) e^{-t,top,i/t,top,i1/2}\}\{k_{,i}/100\}_{m,i}\}$

The time averaged explosive power of a given QCD bomb in the ship frame is in first order:

$<d\{\sum(i = 1; i = M) [(N_i)(E_i)]\} /dt_{ship} > = <d\{\sum(i = 1; i = M) (E_i) \{G_{top,i}\{e^{-t,top,i/Ttop,i}\}\{k_{,i}/100\}_1 + G_{top,i}\{e^{-t,top,i/Ttop,i}\}\{k_{,i}/100\}_1\{e^{-t,top,i/Ttop,i}\}\{k_{,i}/100\}_2 + G_{top,i} \{e^{-t,top,i/Ttop,i}\}\{k_{,i}/100\}_1 \{e^{-t,top,i/Ttop,i}\}\{k_{,i}/100\}_2 \{ e^{-t,top,i/Ttop,i}\}\{k_{,i}/100\}_3 + \ldots + G_{top,i}\{ e^{-t,top,i/Ttop,i}\}\{k_{,i}/100\}_1\{ e^{-t,top,i/Ttop,i}\}\{k_{,i}/100\}_2 \ldots\{ e^{-t,top,i/Ttop,i}\}\{k_{,i}/100\}_{m,i}\}\}/dt_{ship} >$

$= <d\{\sum(i = 1; i = M) (E_i) \{G_{top,i}\{(1/2) e^{-t,top,i/t,top,i1/2}\}\{k_{,i}/100\}_1 + G_{top,i}\{(1/2) e^{-t,top,i/t,top,i1/2}\}\{k_{,i}/100\}_1\{(1/2) e^{-t,top,i/t,top,i1/2}\}\{k_{,i}/100\}_2 + G_{top,i}\{(1/2) e^{-t,top,i/t,top,i1/2}\}\{k_{,i}/100\}_1\{(1/2) e^{-t,top,i/t,top,i1/2}\}\{k_{,i}/100\}_2 \{(1/2) e^{-t,top,i/t,top,i1/2}\}\{k_{,i}/100\}_3 + \ldots + G_{top,i}\{(1/2) e^{-t,top,i/t,top,i1/2}\}\{k_{,i}/100\}_1\{(1/2) e^{-t,top,i/t,top,i1/2}\}\{k_{,i}/100\}_2 \ldots\{(1/2) e^{-t,top,i/t,top,i1/2}\}\{k_{,i}/100\}_{m,i}\}\}/dt_{ship} >$

If $t_{i1/2}$ and T_i are background stationary frames and the toplets are relativistic, the number of catalyzed baryons becomes in first order:

$N_i = G_{top,i}\{e^{-t,top,i/(Ttop,i\gamma}\}\{k_{,i}/100\}_1 + G_{top,i}\{e^{-t,top,i/(Ttop,i\gamma)}\}\{k_{,i}/100\}_1\{ e^{-t,top,i/(Ttop,i\gamma)}\}\{k_{,i}/100\}_2 + G_{top,i}\{ e^{-t,top,i/(Ttop,i\gamma)}\}\{k/100\}_1\{ e^{-t,top,i\ /(Ttop,i\ \gamma)}\}\{k_{,i}/100\}_2\{ e^{-t,top,i/(Ttop,i\gamma)}\}\{k_{,i}/100\}_3 + \ldots + G_{top,i}\{ e^{-t,top,i/(Ttop,i\gamma)}\}\{k_{,i}/100\}_1\{ e^{-t,top,i/(Ttop,i\gamma)}\}\{k_{,i}/100\}_2 \ldots\{ e^{-t,top,i/(Ttop,i\gamma)}\}\{k_{,i}/100\}_{m,i}$

$= G_{top,i}\{(1/2) e^{-t,top,i/(\gamma t,top,i1/2)}\}\{k_{,i}/100\}_1 + G_{top,i}\{(1/2) e^{-t,top,i/(\gamma t,top,i1/2)}\}\{k_{,i}/100\}_1\{(1/2) e^{-t,top,i/(\gamma t,top,i1/2)}\}\{k_{,i}/100\}_2 + G_{top,i}\{(1/2) e^{-t,top,i/(\gamma t,top,i1/2)}\}\{k_{,i}/100\}_1\{(1/2) e^{-t,top,i/(\gamma t,top,i1/2)}\}\{k_{,i}/100\}_2\{(1/2) e^{-t,top,i/(\gamma t,top,i1/2)}\}\{k_{,i}/100\}_3 + \ldots + G_{top,i}\{(1/2) e^{-t,top,i/(\gamma t,top,i1/2)}\}\{k_{,i}/100\}_1\{(1/2) e^{-t,top,i/(\gamma t,top,i1/2)}\}\{k_{,i}/100\}_2 \ldots\{(1/2) e^{-t,top,i/(\gamma t,top,i1/2)}\}\{k_{,i}/100\}_{m,i}$

Where the average yield energy per nucleon or nucleus conversion caused by transformation by particles if the ith species is E_i.

The total yield energy for a given QCD bomb is in first order:

$$\sum(i = 1;\ i = M)\ [(N_i)(E_i)] = \sum(i = 1;\ i = M)\ (E_i)\ \{G_{top,i}\{e^{-t,top,i/(Ttop,i\gamma)}\}\{k_{,i}/100\}_1 + G_{top,i}\{e^{-t,top,i/(Ttop,i\gamma)}\}\{k_{,i}/100\}_1\{e^{-t,top,i/(Ttop,i\gamma)}\}\{k_{,i}/100\}_2 + G_{top,i}\{e^{-t,top,i\ /(Ttop,i\ \gamma)}\}\{k_i/100\}_1\{e^{-t,top,i\ /(Ttop,i\ \gamma)}\}\{k_{,i}/100\}_2\{e^{-t,top,i/(Ttop,i\gamma)}\}\{k_{,i}/100\}_3 + \ldots + G_{top,i}\{e^{-t,top,i/(Ttop,i\gamma)}\}\{k_{,i}/100\}_1\{e^{-t,top,i/(Ttop,i\gamma)}\}\{k_{,i}/100\}_2 \ldots \{e^{-t,top,i/(Ttop,i\gamma)}\}\{k_{,i}/100\}_{m,i}\}$$

$$= \{\sum(i = 1;\ i = M)\ (E_i)\ \{G_{top,i}\{(1/2)\ e^{-t,top,i/(\gamma t,top,i1/2)}\}\ \{k_{,i}/100\}_1 + G_{top,i}\{(1/2)\ e^{-t,top,i/(\gamma t,top,i1/2)}\}\{k_{,i}/100\}_1\{(1/2)\ e^{-t,top,i/(\gamma t,top,i1/2)}\}\ \{k_{,i}/100\}_2 + G_{top,i}\{(1/2)\ e^{-t,top,i/(\gamma t,top,i1/2)}\}\{k_{,i}/100\}_1\{(1/2)\ e^{-t,top,i/(\gamma t,top,i1/2)}\}\{k_{,i}/100\}_2\{(1/2)\ e^{-t,top,i/(\gamma t,top,i1/2)}\}\{k_{,i}/100\}_3 + \ldots + G_{top,i}\{(1/2)\ e^{-t,top,i/(\gamma t,top,i1/2)}\}\{k_{,i}/100\}_1\{(1/2)\ e^{-t,top,i/(\gamma t,top,i1/2)}\}\{k_{,i}/100\}_2 \ldots \{(1/2)\ e^{-t,top,i/(\gamma t,top,i1/2)}\}\ \{k_{,i}/100\}_{m,i}\}$$

The time averaged explosive power of a given QCD bomb in the ship frame is in first order:

$$< d\{\sum(i = 1;\ i = M)\ [(N_i)(E_i)]\}\ /dt_{ship} > = < d\{\sum(i = 1;\ i = M)\ (E_i)\ \{G_{top,i}\{e^{-t,top,i/(Ttop,i\gamma}\}\{k_{,i}/100\}_1 + G_{top,i}\{e^{-t,top,i/(Ttop,i\gamma)}\}\{k_{,i}/100\}_1\{e^{-t,top,i/(Ttop,i\gamma)}\}\{k_{,i}/100\}_2 + G_{top,i}\{e^{-t,top,i\ /(Ttop,i\ \gamma)}\}\{k_i/100\}_1\{e^{-t,top,i\ /(Ttop,i\ \gamma)}\}\{k_{,i}/100\}_2\{e^{-t,top,i/(Ttop,i\gamma)}\}\{k_{,i}/100\}_3 + \ldots + G_{top,i}\{e^{-t,top,i/(Ttop,i\gamma)}\}\{k_{,i}/100\}_1\{e^{-t,top,i/(Ttop,i\gamma)}\}\{k_{,i}/100\}_2 \ldots \{e^{-t,top,i/(Ttop,i\gamma)}\}\{k_{,i}/100\}_{m,i}\}/dt_{ship} >$$

$$= < d\{\sum(i = 1;\ i = M)\ (E_i)\ \{G_{top,i}\{(1/2)\ e^{-t,top,i/(\gamma t,top,i1/2)}\}\ \{k_{,i}/100\}_1 + G_{top,i}\{(1/2)\ e^{-t,top,i/(\gamma t,top,i1/2)}\}\{k_{,i}/100\}_1\{(1/2)\ e^{-t,top,i/(\gamma t,top,i1/2)}\}\ \{k_{,i}/100\}_2 + G_{top,i}\{(1/2)\ e^{-t,top,i/(\gamma t,top,i1/2)}\}\{k_{,i}/100\}_1\{(1/2)\ e^{-t,top,i/(\gamma t,top,i1/2)}\}\{k_{,i}/100\}_2\{(1/2)\ e^{-t,top,i/(\gamma t,top,i1/2)}\}\{k_{,i}/100\}_3 + \ldots + G_{top,i}\{(1/2)\ e^{-t,top,i/(\gamma t,top,i1/2)}\}\{k_{,i}/100\}_1\{(1/2)\ e^{-t,top,i/(\gamma t,top,i1/2)}\}\{k_{,i}/100\}_2 \ldots \{(1/2)\ e^{-t,top,i/(\gamma t,top,i1/2)}\}\ \{k_{,i}/100\}_{m,i}\}/dt_{ship} >$$

We would definitely not want stable toplets that are transmutative into-top-matter or ones that are sufficiently long lived so as to produce a reaction that could consume the entire planet or propagate interstellar distances to reach other stars and planets where they could wreak havoc. Such toplets may even convert interstellar gas and dust in a gradual propagating wavefront and literally gradually eat away at the universe or multiverse completely in a progressive reaction wavefront. However, to produce such toplets or a black hole capable of doing this would require a galactic-sized accelerator—in other words, it is highly unlikely at best.

Regardless, toplet-producing nuclear explosives might be used for the following:

1. Interstellar ramjet propulsion
2. Intergalactic ramjet propulsion
3. Hyperspatial rocket propulsion
4. Nuclear top matter bomblet pulse drive
5. Nuclear top matter bomblet pellet runway propulsion
6. Mass fuel beam nuclear top matter bomblet propulsion
7. Einsteinian 4-D photon, electron, muon, tauon, positron, anti-muon, anti-tauon, proton, antiproton, ion, anti-ion, charged exotic meson, charged exotic baryon, electron neutrino, anti-electron neutrino, muon neutrino, antimuon neutrino, tauon neutrino, anti-tauon neutrino, gravity wave, and quantum-scale gravity wave or graviton rockets
8. Hyperspatial photon, electron, muon, tauon, positron, anti-muon, anti-tauon, proton, antiproton, ion, anti-ion, charged exotic meson, charged exotic baryon, electron neutrino, anti-electron neutrino, muon neutrino, antimuon neutrino, tauon neutrino, anti-tauon neutrino, gravity wave, and quantum-scale gravity wave or graviton rockets. For rocket vehicle applications, the exothermic energy produced in top matter reactors can be used to energize thrust-stream generators.

Note that gravitons are hypothetical at this point. Their existence would require the discovery of the spin-2 boson, which is yet to be accomplished.

So for stable toplet technology, only assured-to-die-out chain reactions must be produced; otherwise, we risk universal destruction. The ramifications of the technology as a power source, however, are too profound not to be mentioned. This is true not only from a philosophical perspective of enabling technology that is far superior in power than pure matter-antimatter conversion, but more to the point, a technology that can revolutionize commercial energy production for manned starship propulsion. Consider the case where the matter within any hyperspatial dimensions that are coupled to our universe would not have periodic table elements or baryons having the exact properties as those found within our ordinary 4-D Einsteinian space-time. For such hyperspatial matters that have quantum properties that are similar to the baryonic mass within our ordinary 4-D universe, such exothermic conversion may still be possible, thereby enabling the production of hyperspatial ramjets and hyperspatial photon, electron, muon, tauon, positron, anti-muon, anti-tauon, proton, antiproton, ion, anti-ion, charged exotic meson, charged exotic baryon, electron neutrino, anti-electron neutrino, muon neutrino, antimuon neutrino, tauon neutrino, anti-tauon neutrino, gravity wave, and quantum-scale gravity wave or graviton rockets.

There are several plausible candidates for mechanisms that can be used to perform nuclear fusion in order to produce the required temperatures and particle kinetic energies so as to enable the bulk production of meta-stable toplets to provide for substantial, but die out forms of top chain reactions. The devices are shaped charge nuclear fusion and/or fission devices. To the best of our knowledge, none of the associated nuclear explosive types have yet been developed.

What can be accomplished with top reactions might also work for toponoium catalysis and bottomonium catalysis. Toponium catalysis may prove possible in the long run; however, the production of toplets as yet have been unverified and may remain so for a long time. This is because toplets, which are bound states comprised of top quarks, would require the strong nuclear force to travel the distances to other quarks cocreated in particle accelerator collisions. All top-quark-producing particle collisions, to date, have produced top quarks that decay too quickly for the light-speed-limited strong force to travel the distance between any cocreated top quarks, thereby preventing any stabilizing bound states.

Nuclear energy comes in many forms.

Nuclear fission involves the splitting of atomic nuclei motivated mainly by neutron flux. Uranium-235 is a good nuclear-fission fuel. The complete nuclear fission of a portion of pure nuclear-fission fuel converts about 0.1 percent of its mass to energy.

Nuclear fusion involves fusing atomic nuclei together, which can start with simple hydrogen or protons and provides exothermic yields until the lowest energy isotope of iron is produced. Completed fusion processes convert about 1 percent of the mass of the reactants to energy.

Radioactive decay by beta or alpha processes, respectively, involves the emission of electrons or helium nuclei from various unstable isotopes. The fraction of mass converted to energy is usually roughly in the range of 0.001 percent.

Nuclear isomers provide about the same mass-specific energy yield as radioactive decay and involve stored energy in excited atomic nuclei of various, but not all, isotopes.

And there is a new kid on the block that is actively being studied. The energy forms involve hypernuclei which are atomic nuclei of atoms that have lumps of neutrons and/or protons in one or more groups separated but proximate to the primary interior portion of the nuclei.

Another less considered form of nuclear energy is more properly referred to as quantum-chromo-dynamics (QCD) energy. Accordingly, QCD energy may enable a greater portion of the reactants being converted to energy.

QCD energy, as anticipated above, would involve the production of stable strangelets, charmlets, bottomlets, and/or toplets, which could catalyze ordinary baryonic matter into strange, charmed, bottomed, and/or top matter. In some cases, such a reaction might result in the production of stable exotic quarkonium and may liberate most of the mass of the reactants as energy including heat, gamma radiation, neutrinos, beta, and alpha rays.

Other fascinating prospects for nuclear energy may involve the production of superheavy isotopes of the transuranic elements, as well as the actinides, rare earth elements, and lighter elements.

Accordingly, these exotic species may be produced in dedicated accelerators similar to the Facility for Rare Isotope Beams (FRIB). Some of these isotopes—in fact, most, have not yet been observed. However, if and when eventually produced, these conjectural isotopes may make excellent nuclear-fission fuels. Some of these fuels may even have more available latent energy than the best nuclear-fusion fuels.

Currently, there are about a couple of thousand known isotopes. However, nuclear models suggest that there are several thousand additional possible isotopes, many of which may be meta-stable and thus serve as outstanding nuclear-fission fuels.

When considering modifications to the currently best formulas for nuclear energy, there may be thousands of additional isotopes not yet predicted.

So the number of fissile species may be very large.

Other prospects of stable superheavies involves the production of extremely hard, high-strength, and refractory materials.

For example, isotopically optimized diamond has much greater optical transmissivity than the best naturally formed diamond. As such, isotopically optimized diamond may have excellent applications for extremely high-powered optical devices. By high-powered, I mean having the ability to transmit very high light-flux densities.

Yet another fascinating opportunity for nuclear energy includes metastable superheavy isotopes of ordinary elements that are fissionable but also have hypernuclei. As mentioned previously, hypernuclei are atomic nuclei that have at least some of their protons and neutrons off of but bonded to the main portions of the nuclei. Such materials may have still greater potential energy relative to ordinary nuclear fuels.

Yet still another possibility is metastable superheavy isotopes of ordinary elements that are both fissionable and also nuclear isomers.

And perhaps best of all are metastable superheavies that are fissionable, hypernuclei, and nuclear isomers at the same time. These prospects seem our best options for materials with greater mass-specific potential nuclear energy than the best nuclear-fusion fuels.

Now the specific impulse of the latter metastable superheavies, where one species of such fuel is considered, is defined by the following:

{{[(2)[(Mass specific fractional fission yield) + (Mass specific fractional hyper-nuclei yield) + (Mass specific fractional nuclear isomer yield)]] + [[(Mass specific fractional fission yield) + (Mass specific fractional hyper-nuclei yield) + (Mass specific fractional nuclear isomer yield)] EXP 2]} EXP (1/2)} c.

The accrued relativistic rocket velocity for the above fuel used is as follows:

Δv = c tanh [(Isp/c) ln (Mass-Ratio)] = c tanh {{{{[(2)[(Mass specific fractional fission yield) + (Mass specific fractional hyper-nuclei yield) + (Mass specific fractional nuclear isomer yield)]] + [[(Mass specific fractional fission yield) + (Mass specific fractional hyper-nuclei yield) + (Mass specific fractional nuclear isomer yield)] EXP 2]} EXP (1/2)} c

/c} ln (Mass-Ratio)}

The Lorentz factor of the above craft is determined in the usual way:

γ = {1/{1 – [(v/c) EXP 2]}} EXP (1/2)

In cases where more than one metastable superheavy is used, we simply take the total mass-specific yield energy of the combined fuels and substitute this into the equation of specific impulse.

To account for non-ideal efficiency, we use the following formula for specific impulse:

$$Isp = \{[(2)(e)(n)] + [[(e)(n)] \text{ EXP } 2]\} \text{ EXP } (1/2)$$

Here, *e* is the total efficiency of the fuel as a subunitary fraction, and *n* is the fractional mass-specific yield energy of the fuel.

It is rather ironic that starship propulsion may essentially be realized by the above simple computations.

Note that astrodynamic drag will reduce the spacecraft propulsion efficiency but non-zero-drag energy can be included in the computation of e. In cases where the specific impulse of the fuel varies with time, the above simple formula is not valid, but the accrued velocity of the spacecraft can simply be computed numerically by taking special relativistic velocity compositions, serially, for each leg of the journey. However, the actual specific impulse of each leg where the value differs must be included in the velocity term representing the specific impulse for the associated leg.

The basic math of relativistic rockets is so simple that I cannot see how rockets could not travel close to light-speed on nuclear fuels, even when all of the needed fuel is carried along from the start of a mission. I could teach young children how to compute these values using a good simple electronic calculator.

The simplicity of relativistic rocket velocity computations is a great hallmark of the advanced physics we humans now have to develop real starship missions.

Next we consider scenarios for which protons and neutrons in atomic nuclei can have excited energy states of at least a sizable fraction of their own invariant mass. Here we are considering scenarios for which the total rest energy of the excited states is at least equal to half an order of magnitude multiple greater than the ordinary nucleonic energy states.

Thus the usable energy content of the fuel is close to that provided by matter-antimatter fuels.

Additionally, the above enhanced nuclear fuels when made to undergo energy state decays and coincident nuclear fission or fusion may have somewhat too much greater effects on the background of higher-dimensional space than otherwise. Both the size of these effects and the varieties of these effects may be much greater than otherwise. Thus, if employed safely, perhaps by exotic effects shielding or filtering, these nuclear fuels fashioned into nuclear explosives of ordinary or shaped charge configurations may enable novel nuclear and electromagnetic effects that may be applied to novel transport phenomena and any associated space-time transportation.

Chapter 15

Catalytic Top-Matter Reactors for Top-Level Nuclear Propulsion Applications

This chapter covers speculative applications of catalytic top matter for maximizing mass-specific nuclear fuel yields and exotic higher-dimensional effects.

The top quark has a mass of 172.76±0.3 GeV/c^2, which is close to the rhenium atom mass. The mass of a rhenium nucleus is 186.207±0.001 AMU. The mass of a uranium 238 nucleus is about 238 AMU.

When transmuting U-238 nuclei to top quarks and no other particles, a toponium particle might conceivably convert a supply of U-238 to toponium or grow the mass of a lump of toponium with yield of [(238 – 186)/238] M c^2, or convert 21.8 percent of the mass of the U-238 to pure energy.

Say, an atomic nuclei of a stable superheavy isotope of a generic element has atomic mass of 300 AMU. Then a toponium reaction may have yield of [(300 – 186)/300] Mc^2, or convert 38 percent of the mass of the stable superheavy to pure energy.

Say, an atomic nuclei of a stable superheavy isotope of a generic element has an atomic mass of 300 AMU. Then a toponium reaction may have a yield of [(371 – 186)/371] Mc^2, or convert 49.8 .percent of the mass of the stable superheavy to pure energy. This is likely a first order limit to the yield of these reactions. Catalysis of a stable superheavy nuclei having a mass [[(2)(186)] + 1] AMU would likely result in the production of two top quarks and a yield energy of only 0.2681 percent of its mass to energy.

The above conjecture assumes only normal-matter top quarks are produced. If toponium reactions occur that convert only neutronium in neutron stars to top-antitop quarks in equal proportion, then the conversion of the matter into energy can increase to virtually 100 percent of the neutron star mass. Lumps of quarkonium made of lighter quarks, such as quark nuggets and other similar scale lumps of non-top-quark quarkonium, can likewise be almost completely converted to energy, as can extremely heavy atomic nuclei such as those in higher islands of stability. A good example would be an elemental isotope with a mass of 10,000 AMU. This efficiency is gained in cases where the top and anti-top quarks would self-annihilate in matter-antimatter reaction. In cases where the reactants would be carried along a relativistic rocket, the specific impulse of the fuel would approach 1 c. In cases where quark nuggets could be captured from the background of space and processes in a continuous manner, the specific impulse approaches ∞ c.

Toponium nuclear explosives may have enhanced higher-dimensional effects because of the boundary conditions manifested by toponium reactions and catalytic processes. This draws naturally from current Standard Model physics that asserts top quarks are the heaviest of the six known quark species.

Enhanced higher-dimensional effects would manifest in toponium reactions in a variety of gauge boson reactions as top quarks react by electromagnetic, weak, strong, and gravitational fields. So not only do we have the heaviest known quarks to consider, but then we also the wide variety of interactions that can come into play with top-quark thermodynamic processes.

Bottom quarks have a mass of 4.19 GeV. So atomic nuclei almost as heavy as (2)(4.19) GeV may be similarly processed for bottomium reactions that convert atomic nuclei to only bottom quarks. Here again, almost 50 percent of the reaction products would be free energy. For much larger periodic table nuclei, in cases where bottomiums would convert the nuclei to equal parts matter and antimatter bottom quarks, near-complete conversion of the atomic nuclei to energy is possible where the matter and antimatter bottom quarks would then be directed to pair-annihilate.

Charmed quarks have a mass of 1.270 GeV. So atomic nuclei almost as heavy as (2)(1.270) GeV may be similarly processed for charmonium reactions, which convert atomic nuclei to only charmed quarks. A large fraction of half of the mass of the atomic nuclei would be converted to free energy. There are only two species of atomic nuclei that would be most viable here. These are of deuterium, tritium, and helium-3. Deuterium- and tritium-heavy isotopes of hydrogen. For much larger periodic table nuclei, in cases where charmonium would convert the nuclei to equal parts matter and antimatter charm quarks, a near-complete conversion of the atomic nuclei to energy is possible, where the matter and antimatter charm quarks would then be directed to pair-annihilate.

Chapter 16

Living on Planets for the Cosmic Long Haul

This chapter includes concepts for maintaining stellar lifetimes indefinitely via waste extraction and refueling.

Most currently burning stars will be alive and well and burning a few trillion years from now. This is true even though stars like the sun will become red giants in about 5 billion years, periodically puff off gas as novas, and then settle down into cold white dwarfs.

Red dwarves are the most common star type in our universe, making up about 70 percent of all stars. Most red dwarves will live a few trillion years, some as long as 20 to 40 trillion years, and thus will still provide a means for supporting life.

Imagine a star-rise on a planet we humans may have colonized for which the star appears in the sky five to ten times the diameter of the sun. Now, imagine the deep-red sunlight shining warmly on your skin. This notion would be a good backdrop for a sci-fi space action flick with romantic themes.

Eventually, red dwarfs are going to burn out, but this will not be an end to our civilization.

We will then be able to colonize subsurface regions in icy-cold surface worlds for which the background neutrino flux and cold dark-matter flux will produce enough heat in these icy worlds to ordinarily support human habitation for upward of 10 EXP thirty years or more.

Now, what I am about to suggest is the slight distributed perturbation of our cosmic light-cone and more distant locations that will be inhabited by our species trillions of years hence in such manners that the residual hydrogen gas will be classically programmed to coalesce into proto-stellar disks and then form additional red dwarfs. The red dwarfs may be programmed to form at 30 trillion year intervals local times.

What is even cooler is that any cold dark non–Standard Model matter might be coaxed into gradual turbulent patterns of distribution so as to self-react to produce protons and electrons. The protons and electrons would be artificially separated from the antimatter versions. The resulting normal-matter hydrogen would be fashioned into red dwarf stars.

There is a remote chance that normal-matter gases of higher atomic number than hydrogen might be fashioned back into hydrogen by interaction with turbulent cold dark-matter eddies, currents, vortices, and the like.

Meanwhile, large cosmic microwave background photoelectric nets may capture CMBR having long been cosmically redshifted to huge wavelengths. The nets may also capture cosmically redshifted background starlight. Captured background electromagnetic radiation may then be converted to an electrical current to power accelerators to produce more protons and electrons and antimatter particle versions, which may in turn be used to restock the universe, ironically, even a universe that has continued to expand up to the then-present cosmic era.

Now, it may be possible to keep the sun and our solar system viable for habitation indefinitely.

Accordingly, large electrically and/or magnetically charged funnels may reach into the solar atmosphere to deposit hydrogen and helium while other funnels extract the fusion waste products. We will need to replace about 400 million tons of fusion fuel per second.

Fusion fuel may be collected from interstellar, and perhaps even intergalactic space, by funnels that undergo self-assembly–style growth to collect fusion fuel in a universe that becomes ever more rarified due to universal expansion.

Cosmic microwave background radiation may be converted via photoelectric cells to run accelerators that produce hydrogen and helium.

Large membranous or thin-walled fuel collection horns might be replaced by self-sustaining magnetic flux tubes that continue to grow and funnel gas to and from the sun.

So clear blue sunny skies may be here to stay, all the while human colonization teams and explorers travel ever farther out into the cosmos.

These are not my ideas of origin, but I do like contemplating them.

Chapter 17

Planning an Interplanetary Transit Systems

This chapter includes concepts for interplanetary fuel depots.

Manned Fueling Stations

Fuel pump operators may be included in space-based fuel depots. Maintenance and security personnel can ensure the safe operation of the depots.

Workspace, sleeping quarters, dining areas, hygiene rooms, exercise rooms, and recreations rooms can be built into the fuel depot stations.

Food can be regularly provided for smaller facilities. However, large fueling stations could include farming and waste recycling.

Security personnel likely will need hand-to-hand combat and restraint training, as well as firearms skills.

Firearms training in space-based environments may be quite different than on Earth's surface.

For example, in nonrotating stations, there is no significant gravitational-based dropping of fired rounds. So training for straight-line trajectories will be necessary.

Additionally, for rotating stations simulating full or partial Earth gravity, the rate of rotation and distance to targets need to be taken into account in different extents on Earth. This is especially true for long-range shots.

Depot Maintenance Personnel

Depots may require and would benefit from maintenance personnel. Such personnel can include any industrial positions needed for maintaining space habitats.

Industrial workers in space fuel depots can include equipment operator and repair technicians, medical staff, agricultural workers, food preparation workers, security personnel, and supervisory and management staff.

Chains of command are likely required for the orderly maintenance of the depots.

Security personnel may require training in close-quarter combat and small but effective fire arms usage. In cases of reduced or zero gravity, new close-quarter combat techniques will need development.

Nonlethal restraint tactics are also useful since use of deadly or injurious force may require unacceptable risks to life and health on the depots.

Some nonlethal restraint techniques can include the use of Tasers, adhesive sprays that solidify into rubberlike cables or ropes to restrain violent crew members or depot boarders, nets, and other similar nonlethal equipment

Automated Unmanned Fueling Stations

Fuel depots may be partially or fully automated.

For example, full supply docking may be similar to that of resupply missions to the International Space Station. The fueling process can also be automated. However, as nonlimiting examples, the depots can be manned at similar levels of sophistication and efficiency, as is the ISS.

Fuel Depots Resupply Missions

A number of current spacecraft may be used to resupply fuel to the depots. Some of these craft include SpaceX Falcon-9, SpaceX Falcon Heavy, NASA's Space Launch System, and the like.

Fuel Depots Maintenance Activities

Of greatest importance is the need to keep the fuel storage tanks full and well maintained.

For example, the tanks need to be periodically inspected for cracks and materials fatigue. Tank replacements, patching, re-lining, and other repair mechanism will likely be required at times.

Working quarters will need to be safely maintained.

For example, any cracks or holes due to pressure cycling, meteor dust, and the like will need repair. The entire crew quarters must be regularly inspected for such issues, and any repairs should be done on a timely basis.

Power generation for lighting, HVAC, sanitary, and cooking activities will be required. So electrical power systems will need periodic inspection for safety.

Fire suppression equipment and training will be required.

For large depots, power for agriculture and food processing may be designed into the depots.

Water Ice

Water present in ice on the moon, Mars, asteroids, and comets can be separated by electrolysis to produce hydrogen and oxygen.

Hydrogen and oxygen produced on planetary bodies and other smaller bodies can be concentrated and liquefied for use as rocket fuel.

Other rocket fuel blends may benefit from LOX production including LOX-methane, LOX-kerosene, LOX paraffin, and hydrazine.

Additionally, frozen water may be heated and used in liquid form for drinking, bathing, and industrial processes.

Engineering Design Issues for Propellant Storage

Propellant sloshing

Risk mitigation for tank rupture and other mission failure modes related to sloshing is needed. Sloshing on ascent and acceleration in space can result in hydrostatic shock and rupture or other damage to the tanks and conduits through which liquid propellants flow.

Sloshing can likewise occur in space-based deceleration.

A number of mechanisms can be used to mitigate propellant sloshing.

First, the reorientation of tanks with respect to acceleration vectors can be applied to reduce sloshing otherwise manifest in cases where fuel tanks have large unfilled vacancies. Additionally, the tanks may be reoriented so as to reduce maximum unit fuel column height in order to reduce hydrostatic loading and tank rupture potential.

A. Elastic fuel bladder squeezing of fuel

1) Rheo-elastic methods

Rheo-elastic members can be useful for squeezing fuel from liquid fuel bladders contained in propellant tanks. The volume of the bladders may be mission and vehicle specific.

Electrical properties of these fuel bladders may be bladder specific and/or adjusted to meet mission requirements. Some of these electrical properties include electrical resistivity, voltage capacities, electrical current handling capacities, electrical power handling capacities, electrical permittivity of the materials and so on.

Rheo-elastic bladders may in some cases need to remain very flexible under the cryogenic temperatures of the background of space and on the surface and atmospheres of the planets and planetary moons from which missions are launched. Note that missions may also be launched from space stations and artificial satellites, as well as from comets and asteroids.

However, in some cases, the bladders and fuels they contain may be preheated or simply exposed to ambient sunlight.

Methods of non-solar-thermal heating of the fuels and bladders can include electro-resistive heating, nuclear thermal heating, beamed energy heating, and the like. Beamed energy can include laser or maser energy, collimated sunlight, phased array-focused microwave energy, and the like.

Rheo-elastic bladders may have a multilayer or multifabric compositions.

For example, the bladders may contain a generic, hopefully durable, material of ordinary rheo-mechanical properties while being lined with or provided with interstitial rheo-elastic layers. The layers may be monolithic, porous, gridlike, or have other geometric properties. Some interesting options would include rheo-elastic nets, weaves, or knits. Note that these styles of fabrics are commonly included in membranous or sheetlike materials to increase the overall strength and tear resistance of the membranes.

2) Thermal-elastic methods

Thermal-elastic bladders can use thermal energy to induce bladder squeezing and shape changes.

The thermal conductivity, specific heats, materials of bladder composition, melting points, and other properties of thermal bladders can be chosen from a wide range of these parameters.

Thermal-elastic bladders may have a multilayer or multifabric compositions.

For example, the bladders may contain a generic, hopefully durable, material of ordinary thermal-mechanical properties while being lined with or provided with interstitial thermal-elastic layers. The layers may be monolithic, porous, gridlike, or have other geometric properties. Some interesting options would include thermal-elastic nets, weaves, or knits. Note that these styles of fabrics are commonly included in membranous or sheetlike materials to increase the overall strength and tear resistance of the membranes.

3) Reversible chemical elastic methods

Fuel bladders may also be made of reversible chemo-elastic materials.

For example, these flexing materials may be cycled in chemical composition by application of reagents, heat, light, and/or electrical energy to the respective chemical solutions of bladder composition.

As with the rheo-elastic and thermal-elastic methods, the chemo-elastic materials may form the entire bulk of the bladders or limited portions thereof to provide the rheo-elastic actuation.

Thermal-elastic bladders may have a multilayer or multifabric compositions.

For example, the bladders may contain a generic, hopefully durable, material of ordinary chemo-mechanical properties while being lined with or provided with interstitial chemo-elastic layers. The layers may be monolithic, porous, gridlike, or have other geometric properties. Some interesting options would include chemo-elastic nets, weaves, or knits. Note that these styles of fabrics are commonly included in membranous or sheetlike materials to increase the overall strength and tear resistance of the membranes.

Magnetoelastic methods

Fuel bladders may also be made of reversible magnetoelastic materials.

Magnetoelastic materials may have various and optionally adjustable magnetic properties. These properties can include magnetic permeability, degree(s) of magnetization, magnetic hysteresis, and so on.

Alternatively, magnetoelastic bladders may simply include permanent magnets and electromagnets of conventional and/or superconducting aspects, or a combination of permanent and electromagnet aspects. These magnetic features may include discreet elements distributed along the or within the surface of a non-magnetic bladder bulk material, or embedded within the bulk materials. Thus, magnetoelastic bladders may have a multilayer or multifabric compositions as nonlimiting options. So the bladders may contain a generic, hopefully durable, material of ordinary magneto-mechanical properties while being lined with or provided with interstitial magnetoelastic layers. The layers may be monolithic, porous, gridlike, or have other geometric properties. Some interesting options would include magnetoelastic nets, weaves, or knits. Note that these styles of fabrics are commonly included in membranous or sheetlike materials to increase the overall strength and tear resistance of the membranes.

Photoelastic methods

Fuel bladders actuated via photoelastic mechanisms are also possible.

Here, laser light, incandescent sources, photodiode, and other sources may irradiate the bulk of the bladders to induce bladder squeezing. Alternatively, limited portions of the bladders may be irradiated in ways that enable proper overall flexing of the bladders.

Both the photoelastic properties and the photon sources may be monochromatic or multispectral. The bladders may thus react differently in a frequency or spectrum specific manner.

The photoelastic bladders may have a multilayer or multifabric compositions as nonlimiting options. So the bladders may contain a generic, hopefully durable material of ordinary photomechanical properties while being lined with or provided with interstitial photoelastic layers. The layers may be monolithic, porous, gridlike, or have other geometric properties. Some interesting options would include photoelastic nets, weaves, or knits. Note that these styles of fabrics are commonly included in membranous or sheetlike materials to increase the overall strength and tear resistance of the membranes.

4) Need for cryogenically tolerant elastic or pliable materials for fuel bladders

Generally, the bladder materials need to be elastic and pliable while being cryogenically tolerant.

Material composites can be extremely useful.

5) Mechanical tensioning or squeezing of elastic fuel bladders

Fuel bladders may be squeezed by hydraulic, electromechanical, tensioned spring, and/or elastic cords, reo-elastic, thermoelastic, magnetoelastic, and/or photoelastic mechanisms. The caveats are that these systems may be actuated in the cryogenic temperatures of sun-shielded space and/or the high-temperature ambient-sunlight environments.

6) Mechanical tensioning or squeezing of nonelastic fuel bladders like tube of toothpaste.

Mechanically tensioned nonelastic fuel bladders may be squeezed in a manner analogous to a tube of common toothpaste.

Squeezing may be accomplished in a number of ways.

First, the bladder may be progressively squeezed, such that the unsqueezed volume is progressively reduced in volume and in extensity along at least one coordinate axis. This can be accomplished via rigid squeeze plates, elastic bands, rollers, and the like. Elastic bands may be naturally compressive or electrically, magnetically, thermally, optically, and/or chemically actuated.

Second, the bladder may be rolled up in a manner commonly used to extract the final portions of toothpaste within a tube. The required roller mechanism can include rigid, pliable, or liquid balloon-style rollers. The radius, temperature, and pressure exerted by the rollers on the bladders can be adjustable as nonlimiting options.

7) Corrosion and oxidation resistance for elastic bladders

Elastic bladders may have highly oxidized layers to resist corrosion and oxidation of oxidizer fuel components. Additionally, the exteriors of the bladders may also be highly oxidized to resist corrosion by interplanetary free radicals and naturally present oxidizers. All layers should be thin enough and sufficiently flexible so as to not crack or scale as the bladder is squeezed. Some highly oxidized compounds include oxides of silicon, aluminum, and the like.

Bladders also need to have external surfaces that are highly resistant to rocket exhaust, which may diffuse forward in some scenarios. Special care has to be made in choosing bladder materials and external surfaces that are resistant to hydrochloric and hydrofluoric acids in cases where the combustible fuel is liquid hydrogen and the oxidizers are liquid or solid forms of chlorine and fluorine, respectively. Protection from other combustion products may be necessary when other fuels are used.

8) Gas bladder negative-pressure mechanisms for squeezing out fuel

Negative-pressure mechanisms may also be employed for squeezing fuel.

For example, a fuel bladder with a high-elastic-modulus nonelastic design may be placed in a pressurized cavity for which the gas pressure, when gradually released, causes an overall unbalanced pressure, thus resulting in the compression of the volume within the bladder.

Elastic bladders may also activate to compress as a result of unbalanced external gas pressure. To facilitate the squeezing, the bladders may have more limited elasticity or substantially no elasticity along one or two of the three spatial coordinates of balloon extension.

Magnetic and electric fields may also supplement or complement gas-pressure mechanisms, or perhaps even used without pressure-activating gases. The bladders may themselves have magnetic and/or electrical charging. Alternatively, the bladders may be encased in piston-style squeezers that are activated by differential pressure gradients.

9) Filling of segmented tanks

Segmented tanks may be filled from intake valves.

The number of intake valves per segment may be one or greater.

Alternatively, only a limited portion of the set of segments may have filling valves.

It is even possible that segments would be symmetrically disposed in a planar or three-dimensional distribution that enables the filling of all segments by a single feed pipe or another conduit. The intake conduits may be rigid tubes, fiber-reinforced hoses, or elastic tubes.

Fuel tanks may be substantially planer or tubelike, in terms of the geometry of the segments. These tanks may rotate about an axis in manners such that the fuel is radially displaced to the outer portion of the rotating segments. As the segments fill, the fuel surfaces progress through the segment vacancies radially inward.

Of course, the fuel tanks need not necessarily rotate while being filled. So the segments may simply have cutoff valves that close or otherwise secure the fuel as the segments are filled.

Segmented tanks can have the same mechanical actuations as the elastic bladder considered previously.

10) Docking safely with rotating fuel depos

Rotating fuel depots would likely require that the ship docking would have rotation that matches the angular frequency of the depots. However, a nonrotating craft may have a docking coupling mounted on a magnetic bearing that rotates at the same angular frequency of the fuel depot.

11) Hardware for docking with rotating fuel depos

For spacecraft rotating at the same angular velocity as a rotating fuel depot, the docking mechanism may be similar to that employed on the International Space Station (ISS).

As for principally nonrotating spacecraft, a docking port affixed to the spacecraft via magnetic bearings may be the docking interface and may also include one or more hoses centered in the middle of the docking interface. The fuel may be uploaded into the docking interface in limited portions, which would have a temporary means for holding the fuel. Once the docking interface is filled with fuel, it may convey the fuel to the spacecraft via a leakproof interface that is mounted by bearings. This way, the fuel conduitions can be leakproof due to tight seals enabled, as typical of related bearing mountings. After each batch of fuel is downloaded to the spacecraft, any need for spin-up of the docking port can be achieved before reconnection to the fuel depot. Note that this system works best when the docking port detaches from the fuel depot between fuel downloads from the depot.

12) Numerous segmented tanks for fuel containment

Numerous segmented tanks for containment of fuels in pre- and post-fueled systems can reduce the risk of catastrophic fuel loss.

13) Spring style-limiting elastic fabric

This can be useful for holding fuel. However, the fabrics and tension springs should be able to operate in the environment of space. Moreover, the fabrics should be resistant to ultraviolet and X-ray radiation, as well as solar wind, if employed outside of a climate-controlled, enclosed environment.

Other elasticity-limiting elements can include one or more of the following: cables, chains, fiber mesh, and the like. Additionally, tensioning members may be elastic but less so than the overall bladder linings.

B) Rotating tanks

1. Nested circular trusswork may support rotating fuel tanks.

2. Support trusses should be made of metals or alloys that can handle the cold temperatures of shadowed locations, as well as the harsh temperatures experienced in broad sunlight. Additionally, the trusses should be resistant to erosion by ultraviolet, X-ray, and solar wind massive species. The solar wind includes electrons, and hydrogen and helium nuclei, or ions.

3. Nested circular trusswork with radially oriented overextending fuel-filled pistons serving as baffles may be included in the design of fuel depots. This way, the fuels can collect at the outer portions of the fuel tanks and be pushed out by the pistons for easier fueling of spacecraft.
4. Sphincter valves for fine-scale partitioning of fuel supply for fuel depots and rocket fuel tanks may be included in rotating tanks.
5. Plenum system for pressing squeezing fuel to reduce space for preventing sloshing is an important topic to address.

Accordingly, there is a need for low-friction contact or piston surfaces. Moreover, the plenums or pistons need to have fuel contact aspects that will not oxidize or become hydrogenically embrittled. Additionally, low-friction coatings on plenums would likely be needed.

Refilling fuel depots

Fuel depots will need to be refillable. The above hardware mechanisms can facilitate refilling of fuel depot.

Fuels can be transported to the depots by rocket methods as nonlimiting examples.

Such fuel transport methods can include chemical rockets, ion rockets, magneto-plasma rockets, Hall effect thrusters, and nuclear thermal rockets.

However, solar sails may be ideal transport propulsion systems since no reaction mass is needed for propulsion except for perhaps fine course adjustment.

Orbital planes and launch windows

Solar sail methods can be used for orbit change without the need for propellant.

Specific issues of cryogenic depots

Boil-off mitigation

There is a need to address cryogenic fuel boil-off.

Phase change material to mitigate boil-off, and cool fuel can include the following:

1. Solids, liquids, and gases

High-phase-change energy solids, liquids, and gases may be used to cool cryogenic fuels. Such coolants may be conditioned by refrigeration and freezing mechanisms.

The magneto-hysteresis phase-change solids may be activated by magnetic fields and otherwise controlled by magnetic fields. Magnetic hysteresis is an aspect of magnetic materials, whereby that maintains some magnetism in the absence of magnetic fields.

Chemical isomer exothermic and endothermic mechanisms can be employed for fuel refrigeration, as can spintronic materials for phase changes and thermal diodic materials.

Multiple-layer Dewars can be used for fuel containment. The following options are useful depending on the scale and solar radial coordinate of the fuel depots and nature of the fuels used.

1) External one-way highly IR reflective surfaces and one-way emissive surfaces
2) External one-way highly IR reflective but super-emissive surfaces and one-way emissive surfaces
3) Thermal-electric interspersed layers for passive cooling and sink for generated electrical power.
    a) Electrical power possibly running cryocoolers

4) Space shuttle–insulating material for reduced heat transfer from solar and geo-radiation sources.

Sun shields

One or more stages one-way IR, and visible light reflective/transmissive membranes to shield fuel depo tanks.

Conjectural materials with one-way super-emissive surfaces may be employed for extremely efficient shielding.

Other shields may include conventional rigid reflective materials.

The mitigation of oxidative processes on shielding membranes for LEO depots may be required depending on where the fuel depots are located

Other issues

Other issues are hydrogen embrittlement, a process by which some metals (including iron and titanium) become brittle and fracture following exposure to hydrogen. The resulting leaks makes storing cryogenic propellants in zero-gravity conditions difficult.

Evaluation of durability of other materials in the presence of hydrogen is required.

There is a risk of oxidation of carbonaceous supermaterials in the presence of LOX.

Oxidation risks include those of carbonaceous supermaterials in the presence of solid oxidizing agents and solid fuels.

Fuel may be used for VASIMR engine propellant storage

There are a wide variety of vehicles for LEO insertion, cis-lunar insertion, of fuel depot hardware. These include the SpaceX family of rockets, as well as rockets produced by Boeing and United Launch Alliance. Other new players in the field will likely include Blue Origin, Sierra Nevada Corporation, Virgin Galactic, Starfish Space, and others.

Solar sails and electrical propulsion methods may also be useful for moving and positioning fuel depots in cis-lunar space. Electrical propulsion methods include ion thrusters, Hall effect thrusters, magneto-plasma thrusters, and others.

The electronic stability of fuel depot control systems is also required. Control systems will need to be hardened for solar storms, microwave weapons, and electromagnetic pulses from nuclear explosives.

For LEO fuel depots, orbital-correction rockets are needed so as to prevent the atmospheric reentry of depots over the lifetime of their use.

Epilogue

Now, there are countable and uncountable infinities.

But before we go farther, consider the hyper-operator notation that was designed to express huge values not otherwise expressible.

For example, note that Hyper4(a, n) is equal to a tetrated *n* or *a* raised to the power of itself n-1 times. The latter value is symbolically written as n subscript a. For example, 3 EXP 4 = 81, but 4 subscript3 is approximately equal to 10 EXP (1,000,000,000,000).

Alternatively 4 subscript 2 = 2 EXP 2 EXP 2 EXP 2 = 2 EXP [2 EXP [2 EXP 2]] = 2 EXP (2 EXP 4) = 2 EXP 16 = 65,536.

For example, Hyper5(4, 4)is equal to 4 tetrated 4 tetrated 4 tetrated 4. This value is commonly referred to as 4 pentated 4.

Hyper 6, (4,4) is 4 pentated 4 pentated 4 pentated 4 and is also referred to as 4 hexated 4.

Hyper 7, (4,4) is 4 hexated 4 hexated 4 hexated 4 and so on.

Aleph 0 is the infinite number of integers.

Aleph 1 according to the perhaps-unprovable, and thus unfalsifiable, continuum hypotheses is the number of real numbers, which is greater than Aleph 0 by a multiplicative factor of infinity.

Aleph 2 is similarly greater than Aleph 1.

Aleph 3 is similarly greater than Aleph 2.

Aleph 4 is similarly greater than Aleph 3.

And so on

In general, Aleph n = 2 EXP [Aleph (n-1)]

Here, *n* is any counting number finite or infinite.

The number Ω is commonly stated as the least infinite positive integer or ordinal.

Now here is a real zinger.

So we can produce the abstraction of Hyper Aleph Ω (Aleph Ω, Aleph Ω).

So about expressions such as:

> Hyper [Hyper Aleph Ω (Aleph Ω, Aleph Ω)](Hyper Aleph Ω (Aleph Ω, Aleph Ω), Hyper Aleph Ω (Aleph Ω, Aleph Ω))

and

> Hyper [Hyper [Hyper Aleph Ω (Aleph Ω, Aleph Ω)](Hyper Aleph Ω (Aleph Ω, Aleph Ω), Hyper Aleph Ω (Aleph Ω, Aleph Ω))](Hyper [Hyper Aleph Ω (Aleph Ω, Aleph Ω)](Hyper Aleph Ω (Aleph Ω, Aleph Ω), Hyper Aleph Ω (Aleph Ω, Aleph Ω)), Hyper [Hyper Aleph Ω (Aleph Ω, Aleph Ω)](Hyper Aleph Ω (Aleph Ω, Aleph Ω), Hyper Aleph Ω (Aleph Ω, Aleph Ω))).

In fact, we could fill up all possible to produce computer electronic storage media with the resources available in our cosmic light cone to produce ever larger infinities.

So the reader can see that there is no limit to the size and number of infinities.

Now we can intuit spacecraft attaining the velocity of light in infinite Lorentz factors.

Accordingly, the spacecraft may run out of room to travel forward in time in ordinary 4-D space-time and may thus enter or "pop in" to another larger realm.

Eventually, the spacecraft may run out of room in said larger realm to enter yet another larger realm yet, and the process may continue perpetually from one realm to the next larger realm.

Now those familiar with my previous books will have likely noticed, if not read, my writings on the subject of advanced infinities, super-infinities, beyond super-infinities, outside of beyond super-infinities, and the like.

The readers may have noticed my writing on so-called hyperquantitative values in the sense that I refer to values that are not quantities but are distinct from qualities. As such, readers may recall I explained the nature of my conjectural infinities that are accordingly hyperquantitative.

Now, we move on to infinities of so-called "enhanced quantitative infinities," or enhanced-1-quantitative-infinities. These would be in a valid sense more quantitative than ordinary quantitative infinities and also commensurately greater to merit the classification of enhanced quantitativeness.

So we can imagine spacecraft having attained Lorentz factors of enhanced quantitative infinities.

We may continue further to consider so-called "enhanced, enhanced, quantitative infinities," or enhanced-2-quantitative-infinities. These would in a valid sense be more quantitative than enhanced quantitative infinities.

So we can imagine spacecraft having attained Lorentz factors of enhanced-2-quantitative-infinities.

We may continue to further consider so-called "enhanced, enhanced, enhanced, quantitative infinities," or enhanced-3-quantitative-infinities. These would in a valid sense be more quantitative than enhanced-2-quantitative-infinities.

So we can imagine spacecraft having attained Lorentz factors of enhanced-3-quantitative-infinities.

We may continue further to consider so-called "enhanced, enhanced, enhanced, enhanced, quantitative infinities," or enhanced-4-quantitative-infinities. These would in a valid sense be more quantitative than enhanced-3-quantitative-infinities.

We may continue this serial pattern forever as indicated in the following set-builder notation.

{enhanced-k-quantitative-infinities | k = 1, 2, 3, …}

Perhaps such extreme infinities are made available by Mother Nature to values of spacecraft Lorentz factors. Accordingly, as a first order consideration, such craft could cover numbers of light-years in one year ship-time, numbers of years into the future in one year ship-time, and velocity of translational travel effectively to numbers of light-speed multiples equal to any values within the set.

{enhanced-k-quantitative-infinities | k = 1, 2, 3, …}.

Having acquired such extreme Lorentz factors, perhaps the craft could travel into so-called enhanced materiality space-time-mass-energy realms or enhanced-1-materiality-space-time-mass-energy-realms.

Additionally, the increasingly infinite Lorentz factor–imbued spacecraft may travel into so-called enhanced, enhanced, materiality space-time-mass-energy realms or enhanced-2-materiality-space-time-mass-energy-realms.

Additionally, the increasingly infinite Lorentz factor–imbued spacecraft may travel into so-called enhanced, enhanced, enhanced, materiality space-time-mass-energy realms or enhanced-3-materiality-space-time-mass-energy-realms.

We may consider this digression forever as denoted in the following set-builder notation:

{enhanced-k-materiality-space-time-mass-energy-realms | k = 1, 2, 3, …}

Now, we of Western Abrahamic religious traditions, i.e., of Christian (and especially Catholic, Judaism, and Islam) tend to place GOD on the pedestal of extreme spiritual composition.

Perhaps a better approach, now that I have broken the ice is to consider GOD's nature as infinitely, and indeed, the ultimate in infinitely more extreme compositions outside of but loosely along the spectrum of the set of {enhanced-k-materiality-space-time-mass-energy-realms | k = 1, 2, 3, …}.

What's more is that god would even be described in concepts that are follow-ons but including those in the ultimate of infinities of more steps beyond the above digressions which we will assign as simply the first step. Such a GOD would be utterly extremely absolute to an ultimate infinite extent.

Science is indeed the new way of knowing and the nature of the cosmos, likely never to be fully understood, is intelligible by science and mathematics. GOD, however, has to be accepted by faith but is intelligible at primitive levels by science and math.

Perhaps explaining our souls more accurately as pure energy may supersede the notion of created spiritual entities.

Regarding the essence of GOD, perhaps the notion of the physical or the spiritual does not capture the nature of GOD.

For example, instead of only considering GOD's nature as infinitely, and indeed, the ultimate in infinitely more extreme compositions outside of but loosely along the spectrum of the set of {enhanced-k-materiality-space-time-mass-energy-realms | k = 1, 2, 3, …}, GOD may have a forever-undefinable additional categories of being aside from those implied by the latter set and aside from the notion of the spiritual where said additional categories include those that are as different from or more different from the physical and what we refer to as the spiritual, as the physical and spiritual are different from each other.

Even making the above conjectures would not even barely begin to adequately define GOD.

Regarding our visible portion of our universe, there are about 10 EXP 24 stars, an estimated 10 EXP 25 planets orbiting said stars, and, by symmetry to our solar system, a projectable roughly 10 EXP 26 planetary moons. These values are about equal to the number of fine grains of table sugar that would cover the entire United States, 100 meters, 1,000 meters, and 10,000 meters deep.

If we are simply able to attain a Lorentz factor of merely 5, which corresponds to about 98 percent of light-speed, this will place most of these worlds in our accessibility, even considered most probable models of continued universal expansion.

We may even find mineral deposits of stable or metastable superheavy elements and isotopes of known elements. These materials would likely have higher mass-specific fission energy than the best known pure cutting-edge fission fuels.

What's more, if the universal expansion begins to slow, we may be able to travel so far from the Milky Way galaxy so as to discover galaxies made of pure antimatter. This would be a wonderful discovery and likely cement our abilities to attain, at the very least, sub-canonical ensemble Lorentz factors, if not infinite Lorentz factors as background time would approach infinite numbers of years.

Most stars in the known universe are red dwarfs, and these will live if left unattended at least 3 trillion years and some as long as 30 trillion years. If we figure out how to dump stellar ash and refresh these stars with new hydrogen, then these stars might be made to shine forever.

As for the time being, we have a solar system to colonize. This will prove to be lots of fun and filled with amazing discoveries.

Appendix

Mathematical Details of Mass Driver Travel Tubes

Note again that the formula for non-varying centripetal force is as follows.

$F = \gamma m v^2 / r$

Here, γ is the Lorentz factor of the moving body, m is the relativistic mass of the moving body, v is the velocity of the body, and r is the radius of revolution of the body.

In the following text, we use a different method of computing relativistic centripetal force which gives close but only approximate results compared to the above relativistic formulation. The approximations are chosen for convenience based on the context of the argument but otherwise provide a good estimate of centripetal force acting on the body in circular motion.

For non-relativistic centripetal force, the above formula simply reduces to,

$F = m v^2 / r$

Here m is the invariant mass of the body.

As with the relativistic cases, the same approximate solutions are provided in the context of the presentation and the intuitive descriptions provided thereof. Note that the purpose of this book is to provide good understanding of the topics covered without being to mathematical. Afterall, the book is intended for a wider audience.

Now consider the formula:

$\gamma = \cosh(aT/c) = [1 + [(at/C)^2]]^{1/2} = [ad/(C^2)] + 1$

So, assuming a tangential acceleration of 1-G in the car frame, and a curvilinear travel path of 100 light-years or about 10^{18} meters, the accrued Lorentz factor will be $\gamma = \{[9.81\ m/s^2][10^{18}\ m]/\{[9 \times (10^{16})]\ m^2 s^{2}\}\} + 1 = 110$.

Assuming a tangential of 1-G in the car frame, and a curvilinear travel path of 10,000 light-years or about 10^{20} meters, the accrued Lorentz factor will be $\gamma = \{[9.81\ m/s^2][10^{20}\ m]/\{[9 \times (10^{16})]\ m^2 s^{2}\}\} + 1 = 10{,}900$.

Assuming a tangential of 1-G in the car frame, and a curvilinear travel path of 1,000,000 light-years or about 10^{22} meters, the accrued Lorentz factor will be $\gamma = \{[9.81 \text{ m/s}^2][10^{22} \text{ m}]/\{[9 \text{ x } (10^{16})] \text{ m}^2 \text{ s}^{-2}\}\} + 1 = 1{,}090{,}000$.

Assuming a tangential of 1-G in the car frame, and a curvilinear travel path of 100,000,000 light-years or about 10^{24} meters, the accrued Lorentz factor will be $\gamma = \{[9.81 \text{ m/s}^2][10^{24} \text{ m}]/\{[9 \text{ x } (10^{16})] \text{ m}^2 \text{ s}^{-2}\}\} + 1 = 109{,}000{,}000$.

Assuming a tangential acceleration of 10-Gs in the car frame, and a curvilinear travel path of 100 light-years or about 10^{18} meters, the accrued Lorentz factor will be $\gamma = \{[98.1 \text{ m/s}^2][10^{18} \text{ m}]/\{[9 \text{ x } (10^{16})] \text{ m}^2 \text{ s}^{-2}\}\} + 1 = 1{,}091$.

Assuming a tangential of 10-Gs in the car frame, and a curvilinear travel path of 10,000 light-years or about 10^{20} meters, the accrued Lorentz factor will be $\gamma = \{[98.1 \text{ m/s}^2][10^{20} \text{ m}]/\{[9 \text{ x } (10^{16})] \text{ m}^2 \text{ s}^{-2}\}\} + 1 = 109{,}000$.

Assuming a tangential of 10-Gs in the car frame, and a curvilinear travel path of 1,000,000 light-years or about 10^{22} meters, the accrued Lorentz factor will be $\gamma = \{[98.1 \text{ m/s}^2][10^{22} \text{ m}]/\{[9 \text{ x } (10^{16})] \text{ m}^2 \text{ s}^{-2}\}\} + 1 = 10{,}900{,}000$.

Assuming a tangential of 10-Gs in the car frame, and a curvilinear travel path of 100,000,000 light-years or about 10^{24} meters, the accrued Lorentz factor will be $\gamma = \{[98.1 \text{ m/s}^2][10^{24} \text{ m}]/\{[9 \text{ x } (10^{16})] \text{ m}^2 \text{ s}^{-2}\}\} + 1 = 1{,}090{,}000{,}000$.

Assuming a tangential acceleration of 10-Gs in the car frame, and a curvilinear travel path of 100 light-years or about 10^{18} meters, the accrued Lorentz factor will be $\gamma = \{[98.1 \text{ m/s}^2][10^{18} \text{ m}]/\{[9 \text{ x } (10^{16})] \text{ m}^2 \text{ s}^{-2}\}\} + 1 = 1{,}091$.

Assuming a tangential of 100-Gs in the car frame, and a curvilinear travel path of 10,000 light-years or about 10^{20} meters, the accrued Lorentz factor will be $\gamma = \{[981 \text{ m/s}^2][10^{20} \text{ m}]/\{[9 \text{ x } (10^{16})] \text{ m}^2 \text{ s}^{-2}\}\} + 1 = 1{,}090{,}000$.

Assuming a tangential acceleration of 100-Gs in the car frame, and a curvilinear travel path of 1,000,000 light-years or about 10^{22} meters, the accrued Lorentz factor will be $\gamma = \{[981 \text{ m/s}^2][10^{22} \text{ m}]/\{[9 \text{ x } (10^{16})] \text{ m}^2 \text{ s}^{-2}\}\} + 1 = 109{,}000{,}000$.

Assuming a tangential acceleration of 100-Gs in the car frame, and a curvilinear travel path of 100,000,000 light-years or about 10^{24} meters, the accrued Lorentz factor will be $\gamma = \{[981 \text{ m/s}^2][10^{24} \text{ m}]/\{[9 \text{ x } (10^{16})] \text{ m}^2 \text{ s}^{-2}\}\} + 1 = 10{,}900{,}000{,}000$.

Assume that an acceleration profile could be maintained such that the craft could tangentially accelerate at a constant 1 G, ship's reference frame over a path length equal to that of the circumference of the currently observable universe.

Substituting 10 m/(s^2) for g to simplify calculations, 3 x 10^8 m/s for C, and [76 x (10^9)] [3.1 x (10^7)] seconds for t, we obtain:

T_o = {[3 x (10^8)]/(10)} ln {[3 x (10^8)] {[[[3 x (10^8)]^2] + [[(10)[2.356 x 10^{18}]]^2]$^{1/2}$] + [(10)[2.356 x 10^{18}]]} /[[3 x (10^8)]^2]} seconds

= [3 x (10^7)] ln {[3 x (10^8)]{{{[9 x (10^{16})] + [[2.356 x (10^{19})]^2]}$^{1/2}$} + [2.356 x (10^{19})]}/[9 x (10^{16})]} seconds

= {[3 x (10^7)] ln [1.57 x (10^{11})]} seconds = 773 million seconds = 24.9 years

The ship would have aged only 24.9 years' ship time for an overall averaged Lorentz factor or time dilation of 3.05 billion in 76 x 10^9 years' background reference frame time. For slowing down within the tube, the spacecraft could use electrodynamic braking to cover 152 billion light-years' travel path-length in 49.8 years' ship time.

Now assume that a tangential acceleration profile could be maintained at a constant 0.1 G, ship's reference frame.

Substituting 1.0 m/(s^2) for g, 3 x 10^8 m/s for C, and (76 x 10^9) (3.1 x 10^7 seconds) for t, we obtain the following:

To = {[3 x (10^8)]/(1.0)} ln {[3 x (10^8)] {[[[3 x (10^8)]^2] + [[(1.0)[2.356 x 10^{18}]]^2]$^{1/2}$] + [(1.0)[2.356 x 10^{18}]]} /[[3 x (10^8)]^2] } seconds

= [3 x (10^8)] ln {[3 x (10^8)]{{{[9 x (10^{16})] + [[2.356 x (10^{18})]^2]}$^{1/2}$} + [2.356 x (10^{18})]}/[9 x (10^{16})]} seconds = [3 x (10^8)] ln [1.57 x (10^{10})] seconds

= 7.043 billion seconds = 227.2 years.

The ship would have aged only 227.2 years' ship time for an overall averaged Lorentz factor or time dilation of 334.5 million in 76 x 10^9 years' background reference frame time. For slowing down, the spacecraft could use electrodynamic braking to cover a 152 billion light-years' travel path-length in 454.4 years' ship time.

Assume that a tangential acceleration profile could be maintained at a constant 10 G, cars' reference frame.

We substitute 100 m/(second^2) for g, 3 x 10^8 m/s for C, and [76 x (10^9)] [3.1 x (10^7)] seconds for t, and obtain the following:

$$To = \{[3 \text{ x } (10^8)]/(100)\}\ ln\ \{[3 \text{ x } (10^8)]\ \{[[[3 \text{ x } (10^8)]^2] + [[(100)[2.356 \text{ x } 10^{18}]]^2]^{1/2}] + [(100)[2.356 \text{ x } 10^{18}]]\}\ /[[3 \text{ x } (10^8)]^2]\} \text{ seconds}$$

$$= [3 \text{ x } (10^6)]\ ln\ \{[3 \text{ x } (10^8)]\{\{\{[9 \text{ x } (10^{16})] + [[2.356 \text{ x } (10^{20})]^2]\}^{1/2}\} + [2.356 \text{ x } (10^{20})]\}/[9 \text{ x } (10^{16})]\}\text{seconds} = \{[3 \text{ x } (10^6)]\ ln\ [1.57 \text{ x } (10^{12})]\} \text{ seconds}$$

$$= 84.25 \text{ million seconds} = 2.7176 \text{ years}$$

In 76 x 10^9 years' background reference frame time, the car would have aged only 2.7176 years' car time for an overall averaged Lorentz factor or time dilation of 27.97 billion. For deceleration, the car could use electrodynamic braking to cover a 152 billion light-years' travel path-length in 5.435 years' ship time.

The travel velocity of the cars at high mass-driver Lorentz factors will induce very large centripetal forces within the contents of the cars. Thus, a means to reduce the effective G forces for the centripetal motions in circular mass-drivers is required.

Among these is the enclosure of crew members' bodies in hydrostatically sealed breathable oxygenated liquid-containing vessels. Alternatively, perhaps nanotechnology types of pressure suits could completely encase the crew members' bodies. The pressure suits might optionally pump high-pressure air into the lungs of the crew members wearing them and gradually relax the lung pressure as the rate of acceleration was reduced to more manageable levels. The crew members may be enclosed in smart fabric types of whole-body pressure suits such as might optionally be constructed from rheo-elastic or other electro-elastic materials.

Another possibility includes placing the crew and any passengers in either a cryogenic form of sleep. The bulk modulus of the sleeping human bodies can approach that of hard steel for freezing temperatures near absolute zero.

Perhaps some sort of G-force cancellation techniques can be used such as an instilled electrical charge within the car and where a reactive electrical field emitted within the car or externally would cancel out the forces due to centripetal acceleration. The charge created within human bodies can be deposited by nanotechnology mechanisms for precise control.

Alternatively, a magnetic field generation mechanism within the cars or external to the cars can induce a dipole moment within the passengers' bodies, thus magnetizing the bodies. The magnetic field can be adjusted in order to effect nearly complete G-force cancellation. Alternatively, a gravity field generator or antigravity field generator can be used to cancel out the centripetal acceleration experienced by car passengers. Furthermore, the cars may optionally utilize two or more of the above centripetal-force cancellation mechanisms simultaneously.

Now, how on earth is a 1-metric-ton car traveling a 20-light-year-radius loop going to be contained within a circular accelerator without rupturing the accelerator via centripetally induced tensile stress! The simple answer may involve none other than ten-kilometer-thick wall tubes made of solid carbonaceous supermaterials. Such tubes when square in cross section would have a tensile strength of $(10^7)(10^{12})(4)$ newtons = $[4 \times 10^{19}]$ newtons. A one-kilometer-long section would have a mass of 400 billion metric tons. A 120-light-year-long circular track would have a mass of $(120)(10^{13})[4 \times 10^{11}]$ metric tons = $[4.8 \times 10^{26}]$ metric tons.

Now consider the formula:

$$\gamma = \cosh(aT/c) = [1 + [(at/C)^2]]^{1/2} = [ad/(C^2)] + 1 \approx 30$$

Where we replace d with 3×10^{17} meters or 30 light-years. We assume an acceleration of one G = 9.81 m/s^2.

Thus,

$$[ad/(C^2)] + 1 = \{[9.81 \text{ m/s}^2][3 \times 10^{17} \text{ m}] / \{[9 \times 10^{16}] \text{ m}^2/\text{s}^2\}\} + 1 = 33.7 \approx 30.$$

Since one complete circuit involves the car going from a Lorentz factor of 33.7 in one direction to a Lorentz factor of 33.7 in an antiparallel direction and then back to a Lorentz factor of 33.7 in the former direction, we assume that the centripetal acceleration of the spacecraft would likewise impose a G force of 1 G for a spacecraft revolving in the tube at a constant Lorentz factor of 33.7. However, a spacecraft orbiting at a Lorentz factor of 33.7 will have a total relativistic mass with respect to the tubular stationary frame of 33.7 M_{rest}. Therefore, the tube will need to resist a centripetal force of 33.7 metric tons where the car has an invariant mass of one metric ton for which the revolutional Lorentz factor of the car will be 33.7.

Now consider the formula:

$$\gamma = \cosh(aT/c) = [1 + [(at/C)^2]]^{1/2} = [ad/(C^2)] + 1$$

Where we replace d with 3×10^{17} meters or 30 light-years and assume an acceleration of 10,900,000 Gs = [(10,900,000) (9.81)] m/s^2 = 1.06929×10^8 m/s^2.

Therefore:

$\gamma = \cosh(aT/c) = [1 + [(at/C)^2]]^{1/2} = [ad/(C^2)] + 1 = \{\{[1.06929 \times 10^8]m/s^2\} [3 \times 10^{17} m] / \{[9 \times 10^{16}] m^2/s^2\}\} + 1$

$= 3.5643 \times 10^8$

Consider one complete circuit involves the car going from a Lorentz factor of 3.5643 x 10^8 in one direction to a Lorentz factor of 3.5643 x 10^8 in an antiparallel direction and then back to a Lorentz factor of 3.5643 x 10^8 in the former direction, we assume that the centripetal acceleration of the spacecraft would likewise impose an acceleration force of 10,900,000 Gs on the tube for a car revolving in the tube at a constant Lorentz factor of 3.5643 x 10^8 under the condition where there would be no relativistic mass increase. However, a spacecraft orbiting at a Lorentz factor of 3.5643 x 10^8 will have a total relativistic mass with respect to the tubular stationary frame of 3.5643 x 10^8 M_{rest}. Therefore, the tube will need to resist a centripetal force of [3.5643 x 10^8] [10,900,000] metric tons, or [3.885 x 10^{15}] metric tons, where the car has an invariant mass of one metric ton for which the revolutional Lorentz factor of the car will be 3.5643 x 10^8. This is equal to 3.811 x 10^{19} newtons. Therefore, the loop should handle the load with a small safety margin.

Now consider again the formula:

$\gamma = \cosh(aT/c) = [1 + [(at/C)^2]]^{1/2} = [ad/(C^2)] + 1$

Where we replace d with 3 x 10^{17} meters or 30 light-years and assume an acceleration of 10,900,000,000 Gs = [(10,900,000,000) (9.81)] m/s^2 = 1.06929 x 10^{11}.

Therefore:

$\gamma = \cosh(aT/c) = [1 + [(at/C)^2]]^{1/2} = [ad/(C^2)] + 1 = \{\{[1.06929 \times 10^{11}]m/s^2\} [3 \times 10^{17} m] / \{[9 \times 10^{16}] m^2/s^2\}\} + 1$

Since one complete circuit involves the car going from a Lorentz factor of 3.5643 x 10^{11} in one direction to a Lorentz factor of 3.5643 x 10^{11} in an antiparallel direction and then back to a Lorentz factor of 3.5643 x 10^{11} in the former direction, we assume that the centripetal acceleration of the spacecraft would likewise impose an acceleration force of 10,900,000,000 Gs on the tube for a car revolving in the tube at a constant Lorentz factor of 3.5643 x 10^{11} under the condition where there would be no relativistic mass increase. However, a spacecraft orbiting at a Lorentz factor of 3.5643 x 10^{11} will have a total relativistic mass with respect to the tubular stationary frame of 3.5643 x 10^{11} M_{rest}. Therefore, the tube will need to resist a centripetal force of [3.5643 x 10^{11}] [10,900,000,000] metric tons or [3.885 x 10^{21}]

metric tons, where the car has an invariant mass of one metric ton for which the revolutional Lorentz factor of the car will be 33.7. This is not a trivial force at 3.8112×10^{25} newtons. So we need something stronger than carbonaceous supermaterials.

Neutronium to the rescue!

We now assume tube construction out of extreme materials such as neutroniums, quarkoniums, higgsiniums, and similar supernuclear-density materials.

Now the force of attraction between nucleons is about 10,000 newtons. Therefore, the tensile strength of neutronium would be approximately $\{(10{,}000 \text{ newtons})[(10^{15})^2] = 10^{34}$ newtons per square meter. Since carbonaceous supermaterials have a tensile strength that is roughly 10^{11} newtons per square meter, but which have a density that is only 10^{-15} times that of neutronium, neutronium tubes have a strength to invariant mass ratio that is 100 million times that of carbonaceous supermaterials. Consequently, neutronium subway tubes may have a mass that is 100 million times less than carbonaceous supermaterials having the same tensile strength.

Nature already has precedence for neutronium in the context of neutron stars. However, the neutronium in such stars is continuously being regenerated under the enormous gravitationally self-induced pressures within neutron stars.

Large assemblages of neutrons and protons have been proven to exist in the form of the atomic nuclei of heavy periodic table elements.

The yield strength of protons and neutrons is about 100 times greater than the force of attraction among these nucleons within the composition of the atomic nucleus. The latter force of attraction is about 10,000 newtons. Therefore, the yield strength of protons and neutrons is about 1,000,000 newtons. A differential cross-sectional area of a column of up-down quarkonium where the column is $1{,}000^{1/2}$ neutrons by $1{,}000^{1/2}$ neutrons wide is about $(3)[1{,}000^{1/2}]$ up and/or down quarks wide by $(3)[1{,}000^{1/2}]$ up and/or down quarks thick. The tensile strength of typical up-down quarkonium will thereby be about $\{\{[(3)^{1/2}]\}^2\}(100)$ times greater than that of neutronium or on the order of 1,000 times greater than that of neutronium. There are likely a large number of possible up-down quarkonium types considering the range of plausible crystalline patterns, quarkonium excited or isomer states, and the like. Thus, we may have some freedom in the designed strength of specific quarkoniums by up to 3, or perhaps even 4 orders of magnitude. Note the analogous cases of comparison for the element carbon that includes soft pencil graphite or activated charcoal or the much stronger forms of carbonaceous supermaterials. Here, the yield strength of the materials spans a range of several orders of magnitude.

Non-hadronically differentiated up-down quarkonium having the same density as neutronium may in theory have a strength to weight ratio that is 10 billion times that of carbonaceous supermaterials.

Quarkoniums in the form of quark nuggets left over as relics from the early stages of the big bang are plausible and have been the subject of serious theoretical studies. The quark nuggets would theoretically be on the order of one to two meters in size and have a mass of roughly that of Earth's moon.

Some theories of stellar evolution involving intermediate stages between neutron stars and black holes have posited the existence of quark stars. Either way, many theoretical considerations regarding the interior of neutron stars suggest the existence of a quarkonium core within at least some neutron stars.

The ratio $[3.8112 \times 10^{25}]/ [3.8112 \times 10^{19}]$ is 1,000,000. Therefore, about $\{(10^{-8})(10^{6}) [4.8 \times 10^{26}]/[2 \times 10^{27}]\}$ solar masses or 0.0024 solar masses of neutronium should suffice.

For non-hadronically differentiated up-downium having a mass-specific yield of that of the proton, only 0.000024 solar masses of neutronium would suffice.

Using 0.24 solar masses of neutronium to construct the ring should enable one metric ton car Lorentz factors of $(10)[3.5643 \times 10^{11}] = 3.5643 \times 10^{12}$. Using 0.24 solar masses of non-hadronically differentiated up-downium should enable Lorentz factors of 3.5643×10^{13}.

Now here comes the zinger.

Consider communication lines set up between the circular tube and thus the extreme Lorentz factor cars and a repeater located near or at the very axial center of the tube ring. In other words, no further than 10 light-years from the car at any time. Further, consider that an identical system is eventually assembled at a distance of 1 billion light-years from the first habitat, and indeed, yet farther on, where quintillions of identical systems are assembled in a spherical patterns having a radius of 1 billion light-years but where some such systems shadow in part systems located closely behind the former systems.

Further consider scenarios where the communications from each car of a first constructed ring system centered near the sun can be uploaded in microsecond-long pulses or shorter pulses, which may be digitally modulated, amplitude modulated, frequency modulated, polarity modulated, and/or super-chirality modulated, and then are broadcast to each of the spherically disposed subway worlds located a distance of 1 billion light-years from the sun.

Assume each car of the Sun-centered ring world is time dilated by a factor of 3.5643×10^{11} with respect to the ring and, consequently, with respect to the ring's central repeater. Further, consider that each of the 1 billion light-year-distant spherically disposed identical systems have revolving cars that are also traveling at a Lorentz factor of 3.5643×10^{11} with respect to its own host ring. The two-way communications time experienced by such car passengers sending out messages to the sun-centered system would be a mere 0.005611 years or about 2.04949 days in the car's frame.

In fact, the longest complete signal time between locations along the 1-billion-light-year-radius sphere, which are diametrically opposed to each other would be only 4.09898 days in the car's frame.

For a similar sphere with a central ring hab-centered around the sun, but which has a radius of 10 billion light-years, and which contains, on the order of 10^{21} subway hab-tube systems identical to the ones proposed previously, the two-way radial communication time will be a mere 20.4949 days car time. For diametrically opposed ring-worlds, the two-way communication will be a mere 40.9898 days, or 1.36 months' car time.

Now, assume each car of the sun-centered ring world is time dilated by a factor of 3.5643×10^{12} with respect to the ring, and consequently with respect to the ring's central repeater. Further, consider that each of the 1-billion-light-year-distant spherically disposed identical systems have revolving cars that are also traveling at a Lorentz factor of 3.5643×10^{12} with respect to its own host ring. The two-way communications time experienced by such car passengers sending out messages to the sun-centered system would be a mere 0.0005611 years, or about 0.204949 days in the car's frame car time.

In fact, the longest complete signal time between locations along the 1-billion-light-year-radius sphere, which are diametrically opposed to each other, would be only 0.409898 days' car time in the car's frame.

For a similar sphere with a central ring hab-centered around the sun, but which has a radius of 10 billion light-years, and which contains on the order of 10^{21} subway hab-tube systems identical to the ones proposed previously, the two-way radial communication time will be a mere 2.04949 days' car time. For diametrically opposed ring-worlds, the two-way communication will be a mere 4.09898 days' car time.

Assume 1-billion-light-year-radius rings so constructed that support 1-metric-ton cars revolving at a Lorentz factor of 3.5643×10^{13} with respect to the rings and its central transceiving devices. The two-way communication time experienced by such car passengers sending out messages to the sun-centered system would be a mere 0.00005611 years, or about 0.49187 hours in the car's frame.

In fact, the longest complete signal time between locations along the 1-billion-light-year-radius sphere, which are diametrically opposed to each other and would be only 0.983755 hours car time in the car's frame.

For a similar sphere with a central ring hab-centered around the sun, but which has a radius of 10 billion light-years, and which contains on the order of 10^{21} subway hab-tube systems identical to the ones proposed previously, the two-way radial communication time will be a mere 0.204949 days' car time. For diametrically opposed ring-worlds, the two-way communication will be a mere 0.409898 hours car time.

I have verified the correctness of the above reasoning in principle but kindly noted that the carbonaceous supermaterial rings capable of supporting cars having a Lorentz factor of 3.5643×10^{8} would have a mass beyond the Jeans limit and thereby collapse under its own weight into a disposition of sausage-like links, which would then collapse into a collection of spheres that would in turn collapse into a sphere.

What if we could set the ring rotating at a sufficient Keplerian velocity so that it does not self-collapse? This rotational motion would not be enough to significantly alter the Lorentz factor capacities of the cars or the communication upload and download time, but it could be employed to increase the effective Jeans limit of the ring.

Alternatively, repulsive laser-like magnetic fields or electric field flux patterns concentrated radially toward paired portions of the ring-worlds may provide the necessary force to overcome tendencies for ring collapse. The ring may be suitably magnetically and/or electrically charged for the associated self-repulsion.

The neutronium rings would likely be self-supporting even while nonrotating and the neutronium density, non-hadronically differentiated up-downium, would certainly be stable against collapse even for non-rotating systems.

Assuming a neutronium ring having a mass of 2,400 solar masses and the ring could support a one metric ton car revolving at a Lorentz factor of 3.5643 x 10^{14}. For non-hadronically differentiated nuclear density up-downium, the ring could support a car revolving at gamma = 3.5643 x 10^{15}!

Considering once again the above 1-billion-light-year-radius format, but this time for the neutronium 2,400 solar mass ring case except where the number of ring-word tube systems is reduced by a factor of 10,000 so as to avoid gravitational collapse, the two-way communication between the cars and the hab-ring-world centered near the sun would be a mere [2 x 10^9]/ [3.5643 x 10^{14}] years, or 2.9512 minutes' car time. The communication time between diametrically opposed locations on the sphere would be a mere 5.9025 minutes' car time. For the 10-billion-light-year-radius ring, the two respective communications times would be a mere 29.512 minutes and 59.025 minutes, car time, respectively.

Considering once again the above 1-billion-light-year-radius format, but this time for the up-downium 2,400 solar mass ring case, the two-way communication between the cars and the hab-ring-world centered near the sun would be a mere [2 x 10^9]/ [3.5643 x 10^{15}] years or 17.394 seconds. The communication time between diametrically opposed locations on the sphere would be a mere 34.788 seconds' car time. For the 10-billion-light-year-radius ring, the two respective communications times would be a mere 173.94 seconds and 347.88 seconds, car time, respectively.

Should the big bang's expansion rate slow to a relative distance specific crawl over the next trillion years of so, within say about 2 trillion years from now, perhaps a little longer, we should be able to construct 1-trillion-light-year-radius sphere analogues consisting of [10^{21}][10^{-4}][10^3] 2,400 solar mass rings. This is a whopping 10^{20} ring-worlds. For neutronium ring world tubular systems, the two-way communication time between a spherically disposed hab and the sun-centered hab will still be very short at a mere 4.8316 hours' car time. Two-way communications entirely across the sphere will take place in 9.6633 days' car time. For the up-downium cases, the two-way communication time between a spherically disposed hab and the sun-centered hab will still be very short at a mere 0.48316 hours' car time. Two-way communications entirely across the sphere will take place in only 0.96633 days' car time.

We cannot permit too great of mass concentration in the construction of the sphere assemblage, otherwise, we will reach the Jeans limit analogue to gravitating masses disposed over the associated cosmic-distance scales. Black-hole radius grows in proportion to black hole mass, and so we would be wise to limit the number of such 2,400 solar mass ring-worlds in constructed spheres to a number that scales with the diameter of the sphere.

Consider the prospects of extending a subway system to the scale of the scale of the Milky Way galaxy. The subway tubes may be distributed to coincide with the rotation of the spiral arms of the Milky Way. Additionally, one or more rotating rings of roughly the same radius of the Milky Way may be concentrically disposed. Other rings may be located at fractional values of the radius of the galaxy.

Since the average star is about 1/5 of the mass of the sun, the total available construction material ranges from (0.00002)(0.2)[4 x 10^{11}] to (0.0001)(0.2)[4 x 10^{11}] solar masses or 1,600,000 to 8,000,000 solar masses. This would be about [2 x (10^{27})](1,600,000) to [2 x (10^{27})](8,000,000) metric tons or [3.2 x (10^{33})] to [1.6 x (10^{34})] metric tons. Assuming that a single human person only requires 1,000 metric tons of mass for life support, [3.2 x (10^{30})] to [1.6 x (10^{31})] human persons may be simultaneously supported by the Milky Way. Assuming that most stars are red dwarfs with a lifetime on the order of $10^{12.5}$ years, in cases where the human life expectancy can be increased to 1,000 years, the Milky Way alone could nurture 1.0119 x 10^{43} to 5.0596 x 10^{43} persons. This can be considered an ensemble of souls. This huge number is roughly equal to the number of atoms in 50 cubic kilometers of liquid water at STP conditions.

Assume circulinear travel tube circumference equal to the 76,000 light-years, or about the radius of the observable universe. Assume that an acceleration profile could be maintained such that the craft could accelerate at a constant 1 G, ship's reference frame so that the craft can make 1 million revolutions around the ring.

Substituting 10 m/(s^2) for *g* for computational simplicity, 3 x 10^8 m/s for C, and [76 x (10^9)] [3.1 x (10^7)] seconds for *t*, we obtain:

T_o = {[3 x (10^8)]/(10)} *ln* {[3 x (10^8)] {[[[3 x (10^8)]^2] + [[(10)[2.356 x 10^{18}]]^2]$^{1/2}$] + [(10)[2.356 x 10^{18}]]} /[[3 x (10^8)]^2]} seconds

= [3 x (10^7)] *ln* {[3 x (10^8)]{{{[9 x (10^{16})] + [[2.356 x (10^{19})]^2]}$^{1/2}$} + [2.356 x (10^{19})]}/[9 x (10^{16})]} seconds

= {[3 x (10^7)] *ln* [1.57 x (10^{11})]} seconds = 773 million seconds = 24.9 years

The car would have aged only 24.9 years' ship time for an overall averaged Lorentz factor or time dilation of 3.05 billion in only 76 x 10^9 years' background reference frame time. For slowing down, the car could use electrodynamic braking such as reverse mass driver modes to travel a total path length of 152 billion light-years' travel distance in 49.8 years' ship time. Now assume that an acceleration profile could be maintained at a constant 0.1 G, ship's reference frame.

Substituting 1.0 $m/(s^2)$ for *g* for computational simplicity, 3 x 10^8 m/s for C, and (76 x 10^9) (3.1 x 10^7 seconds) for *t*, we obtain:

$$To = \{[3 \text{ x } (10^8)]/(1.0)\} \ ln \ \{[3 \text{ x } (10^8)] \ \{[[[3 \text{ x } (10^8)]^2] + [[(1.0)[2.356 \text{ x } 10^{18}]]^2]^{1/2}] + [(1.0)[2.356 \text{ x } 10^{18}]]\} /[[3 \text{ x } (10^8)]^2] \ \} \text{ seconds}$$

$$= [3 \text{ x } (10^8)] \ ln \ \{[3 \text{ x } (10^8)]\{\{\{[9 \text{ x } (10^{16})] + [[2.356 \text{ x } (10^{18})]^2]\}^{1/2}\} + [2.356 \text{ x } (10^{18})]\}/[9 \text{ x } (10^{16})]\} \text{ seconds} = [3 \text{ x } (10^8)] \ ln \ [1.57 \text{ x } (10^{10})] \text{ seconds}$$

$$= 7.043 \text{ billion seconds} = 227.2 \text{ years}$$

The car would have aged only 227.2 years' car time for an overall averaged Lorentz factor or time dilation of 334.5 million in 76 x 10^9 years' background reference frame time. For slowing down, the car could use electrodynamic braking such as reverse mass driver modes to cover 152 billion light-years travel distance in 454.4 years' ship time. Assume that an acceleration profile could be maintained at a constant 10 G, ship's reference frame.

We substitute 100 m/(second^2) for *g* for computational simplicity, 3 x 10^8 m/s for C, and [76 x (10^9)] [3.1 x (10^7)] seconds for *t*, and obtain:

$$To = \{[3 \text{ x } (10^8)]/(100)\} \ ln \ \{[3 \text{ x } (10^8)] \ \{[[[3 \text{ x } (10^8)]^2] + [[(100)[2.356 \text{ x } 10^{18}]]^2]^{1/2}] + [(100)[2.356 \text{ x } 10^{18}]]\} /[[3 \text{ x } (10^8)]^2]\} \text{ seconds}$$

$$= [3 \text{ x } (10^6)] \ ln \ \{[3 \text{ x } (10^8)]\{\{\{[9 \text{ x } (10^{16})] + [[2.356 \text{ x } (10^{20})]^2]\}^{1/2}\} + [2.356 \text{ x } (10^{20})]\}/[9 \text{ x } (10^{16})]\}\text{seconds} = \{[3 \text{ x } (10^6)] \ ln \ [1.57 \text{ x } (10^{12})]\} \text{ seconds}$$

$$= 84.25 \text{ million seconds} = 2.7176 \text{ years}$$

In 76 x 10^9 years' background reference frame time, the car would have aged only 2.7176 years' ship time for an overall averaged Lorentz factor or time dilation of 27.97 billion. For deceleration, the spacecraft could use electrodynamic braking such as reverse mass driver modes to cover 152 billion light-years' travel distance 5.435 years' ship time. The actual distance of travel from the Milky Way galaxy would be considerably greater yet by roughly 2 orders of magnitude. The recessional velocity of the spacecraft would be roughly 2 orders of magnitude greater than C due to the expansion of space-time, assuming that the expansion of the universe continues, on average, at its current rate over the next 152 billion years.

Now yet again consider the formula:

$\gamma = \cosh(aT/c) = [1 + [(at/C)^2]]^{1/2} = [ad/(C^2)] + 1$

Where we replace *d* with 1.9 x 10^{26} meters or 19,000,000,000 light-years and assume an tangential acceleration of 1G = 9.81 m/s^2.

Therefore:

$\gamma = \cosh(aT/c) = [1 + [(at/C)^2]]^{1/2} = [ad/(C^2)] + 1 = \{[9.81 \text{ m/s}^2][1.9 \times 10^{26} \text{ m}] / \{[9 \times 10^{16}] \text{ m}^2/\text{s}^2\}\} + 1$

$= 2.071 \times 10^{10}$

The actual radial centripetal force exerted by the 1-metric-ton car were it not for special relativistic mass increase would be (10^6) metric tons force. However, the car is traveling at a Lorentz factor of with respect to the travel tube reference frame and so the centripetal force acting on the ring due to the car will be (10^6)[2.071 x 10^{10}](9,810) newtons or 2.03165 x 10^{20} newtons.

A track made of 45.0738-kilometer-thick wall tubes constructed of solid carbonaceous supermaterials should provide bursting resistance. Such tubes when square in cross-section would have a tensile strength of (10^7)(10^{10})(45.0738^2) newtons = [2.03165 x 10^{20}] newtons. A one kilometer long section would have a mass of 2.03165 x 10^{12} metric tons. A 76,000-light-year-long circular track would have a mass of (76,000)(10^{13})[2.03165 x 10^{12}] metric tons = [1.54405 x 10^{30}] metric tons, or about 772 solar masses.

However we would prefer that perhaps a population of 100 million cars could revolve around the tube simultaneously so as to transport humans from our cosmic era well into the future. Such humans may remain alive and awake during the transport process. Therefore, travel tubes made of neutronium, but having the same mass would suffice. In order to support 10 billion cars, non-hadronically differentiated up-downium may suffice.

Alternatively, since one complete circuit involves the car going from a Lorentz factor of 2.071 x 10^{10} in one direction to a Lorentz factor of 2.071 x 10^{10} in an antiparallel direction and then back to a Lorentz factor of 2.071 x 10^{10} in the former direction, we assume that the centripetal acceleration of the spacecraft would likewise impose an acceleration force of 1,000,000 Gs on the tube for a car revolving in the tube at a constant Lorentz factor of 2.071 x 10^{10} under the condition where there would be no relativistic mass increase. However, a spacecraft orbiting at a Lorentz factor of 2.071×10^{10} will have a total relativistic mass with respect to the tubular stationary frame of 2.071×10^{10} M_{rest}. Therefore, the tube will need to resist a centripetal force of [2.071×10^{10}] [1,000,000] metric tons or [2.071×10^{16}] metric tons where the car has an invariant mass of one metric ton for which the revolutional Lorentz factor of the car will be 2.071×10^{10}. This is not a trivial force at 2.03165×10^{20} newtons. So we need something stronger than carbonaceous supermaterials for the case where the car's Lorentz factor is 100 million times greater.

Once again consider the formula:

$$\gamma = \cosh(aT/c) = [1 + [(at/C)^2]]^{1/2} = [ad/(C^2)] + 1$$

Where we replace *d* with 1.9×10^{26} meters or 19,000,000,000 light-years and assume an tangential acceleration of 10,000 Gs = 98,100 m/s^2.

Therefore:

$$\gamma = \cosh(aT/c) = [1 + [(at/C)^2]]^{1/2} = [ad/(C^2)] + 1 = \{[98{,}100\ m/s^2][1.9 \times 10^{26}\ m] / \{[9 \times 10^{16}]\ m^2/s^2\}\} + 1$$

$$= 2.071 \times 10^{14}$$

The radial centripetal force exerted by the 1 metric ton car were it not for special relativistic mass increase would be 10^{10} metric tons force. However, the car is traveling at a Lorentz factor of 2.071×10^{14} with respect to the travel tube reference frame and so the centripetal force acting on the ring due to the car will be (10^{10})[2.071×10^{14}](9,810) newtons or 2.03165×10^{28} newtons. This is 100 million times the centripetal force exerted on the previous carbonaceous supermaterial system which is logical given the 100-million-fold mass specific strength of neutronium over that of carbonaceous supermaterials. For the case of the up-downium, the maximum supportable Lorentz factor is 2.071×10^{15} or ten times greater than in the latter neutronium case.

Now consider neutronium tubes having a mass of 772,000,000 solar masses and a circumference of 76,000 light-years. These tubes would support a one metric ton car revolving at a Lorentz factor of 2.071×10^{17}. Tubes constructed of the up-downium having the same mass and radius would support a Lorentz factor of 2.071×10^{18}.

One-billion-light-year-radius spherical dispositions of 21.4 billion such neutronium galactic ring-worlds centered around an identical Milky Way ring-world, could send and receive two-way communications to the Milky Way ring-world in only

$\{[2 \times 10^9]/[2.071 \times 10^{17}]\}[31{,}000{,}000]$ seconds or 0.2993 seconds. Two-way communications diametrically spanning the sphere could be completed in only 0.5987 seconds. For an otherwise-identical up-downium sphere, the respective communication times are 0.02993 seconds and 0.05987 seconds.

One-trillion-light-year-radius spherical dispositions of 21.4 trillion such neutronium galactic ring-worlds centered around an identical Milky Way ring-world, could send and receive two-way communications to the Milky Way ring-world in only $\{[2 \times 10^{12}]/[2.071 \times 10^{17}]\}[31{,}000{,}000]$ seconds or 299.3 seconds. Two-way communications diametrically spanning the sphere could be completed in only 598.7 seconds. For an otherwise-identical up-downium sphere, the respective communication times are 29.93 seconds and 59.87 seconds.

One-quadrillion-light-year-radius spherical dispositions of 21.4 quadrillion such neutronium galactic ring-worlds centered around an identical Milky Way ring-world, could send and receive two-way communications to the Milky Way ring-world in only $\{[2 \times 10^{15}]/[2.071 \times 10^{17}]\}[31{,}000{,}000]$ seconds, or 3.464 days. Two-way communications diametrically spanning the sphere could be completed in only 6.929 days. For an otherwise-identical up-downium sphere, the respective communication times are 0.3464 days and 0.6929 days.

Such 76,000 light-year-radius ring-worlds could each support rotating habitats traveling at a Lorentz factor on the order of millions to billions while still housing many, many trillions of forward time-traveling citizens as ambassadors to our cosmic future descendants.

Still once again, consider the formula:

$$\gamma = \cosh(aT/c) = [1 + [(at/C)^2]]^{1/2} = [ad/(C^2)] + 1$$

Except this time, we replace *d* with 1.9×10^{24} meters or 190,000,000 light-years and assume a Lorentz factor of 2.071×10^{17}.

Therefore:

$$\gamma = \cosh(aT/c) = [1 + [(at/C)^2]]^{1/2} = [ad/(C^2)] + 1 = \{(a)[1.9 \times 10^{24}\, \mathrm{m}] / \{[9 \times 10^{16}]\, \mathrm{m}^2/\mathrm{s}^2\}\} + 1$$

$$= 2.071 \times 10^{17}$$

Therefore, by solving for *a*, we obtain an acceleration equal to 10^9 Gs.

The radial centripetal force exerted by the 1 metric ton car, were it not for special relativistic mass increase, would be 10^9 metric tons. However, the car is traveling at a Lorentz factor of 2.071×10^{17} with respect to the travel tube reference frame, and so the centripetal force acting on the ring due to the car will be $(10^9)[2.071 \times 10^{17}](9{,}810)$ newtons or 2.03165×10^{31} newtons. Now consider neutronium tubes having a mass of 772,000,000 solar masses and a circumference of 760,000,000 light-years. These tubes would support a 1-metric-ton car revolving at a Lorentz factor of 2.071×10^{17}. Tubes constructed of the up-downium having the same mass and radius would support a Lorentz factor of 2.071×10^{18}.

Now consider neutronium tubes having a mass of 772,000,000,000,000 solar masses and a circumference of 760,000,000 light-years. These tubes would support a 1-metric-ton car revolving at a Lorentz factor of 2.071×10^{20}. Tubes constructed of the up-downium having the same mass and radius would support a Lorentz factor of 2.071×10^{21}.

One-billion-light-year-radius spherical dispositions of 214 such neutronium galactic ring-worlds centered around an identical Milky Way ring-world, could send and receive two-way communications to the Milky Way ring-world in only $\{[2 \times 10^9]/[2.071 \times 10^{20}]\}[31{,}000{,}000]$ seconds or 0.0002993 seconds. Two-way communications diametrically spanning the sphere could be completed in only 0.0005987 seconds. For an otherwise-identical up-downium sphere, the respective communication times are 0.00002993 seconds and 0.00005987 seconds.

One-trillion-light-year-radius spherical dispositions of 21,400 such neutronium galactic ring-worlds centered around an identical Milky Way ring-world, could send and receive two-way communications to the Milky Way ring-world in only $\{[2 \times 10^{12}]/[2.071 \times 10^{20}]\}[31{,}000{,}000]$ seconds or 0.2993 seconds. Two-way communications diametrically spanning the sphere could be completed in only 0.5987 seconds. For an otherwise-identical up-downium sphere, the respective communication times are 0.02993 seconds and 0.05987 seconds.

One-quadrillion-light-year-radius spherical dispositions of 21,400,000 such neutronium galactic ring-worlds centered around an identical Milky Way ring-world, could send and receive two-way communications to the Milky Way ring-world in only $\{[2 \times 10^{15}]/[2.071 \times 10^{20}]\}[31{,}000{,}000]$ seconds, or 4.9883 minutes. Two-way communications diametrically spanning the sphere could be completed in only 9.9766 minutes. For an otherwise-identical up-downium sphere, the respective communication times are 0.49883 minutes and 0.99766 hours.

One-quintillion-light-year-radius spherical dispositions of 21,400,000,000 such neutronium galactic ring-worlds centered around an identical Milky Way ring-world, could send and receive two-way communications to the Milky Way ring-world in only $\{[2 \times 10^{18}]/[2.071 \times 10^{20}]\}[31{,}000{,}000]$ seconds, or 3.4641 days.

Two-way communications diametrically spanning the sphere could be completed in only 6.92819 days. For an otherwise-identical up-downium sphere, the respective communication times are 0.34641 weeks and 0.692819 weeks.

Still yet again, consider the formula:

$\gamma = \cosh (aT/c) = [1 + [(at/C)\ ^2]]\ ^{1/2} = [ad/(C^2)] + 1$

Except this time, we replace *d* with 1.9 x 10^{28} meters or 190,000,000,000,000 light-years and assume a Lorentz factor of 2.071 x 10^{23}.

Therefore:

$$\gamma = \cosh (aT/c) = [1 + [(at/C)\ ^2]]\ ^{1/2} = [ad/(C^2)] + 1 = \{(a)[1.9 \text{ x } 10^{28} \text{ m}] / \{[9 \text{ x } 10^{16}] \text{ m}^2/\text{s}^2\}\} + 1$$

$$= 2.071 \text{ x } 10^{23}.$$

Therefore, by solving for *a*, we obtain an acceleration equal to 10^{11} Gs.

The radial centripetal force exerted by the 1-metric-ton car were it not for special relativistic mass increase would be 10^{11} metric tons. However, the car is traveling at a Lorentz factor of 2.071 x 10^{23} with respect to the travel tube reference frame and so the centripetal force acting on the ring due to the car will be (10^{11})[2.071 x 10^{23}](9,810) newtons or 2.03165 x 10^{38} newtons. Now consider neutronium tubes having a mass of 772,000,000,000,000,000,000 solar masses or 7 x 10^{20} solar masses and a circumference of 760,000,000,000,000 light-years or 7.6 x 10^{14} light-years. These tubes would support a 1-metric-ton car revolving at a Lorentz factor of 2.071 x 10^{23}. Tubes constructed of the up-downium having the same mass and radius would support a Lorentz factor of 2.071 x 10^{24}.

Now consider neutronium tubes having a mass of 7 x 10^{26} solar masses solar masses and a circumference of 7.6 x 10^{14} light-years. These tubes would support a 1-metric-ton car revolving at a Lorentz factor of 2.071 x 10^{26}. Tubes constructed of the up-downium having the same mass and radius would support a Lorentz factor of 2.071 x 10^{27}.

With 10^{16} light-year-radius spherical dispositions of 21,472, such neutronium galactic ring-worlds centered around an identical Milky Way ring-world, could send and receive two-way communications to the Milky Way ring-world in only {[2 x 10^{16}]/[2.071 x 10^{23}]}[31,000,000] seconds, or 2.993 seconds. Two-way communications diametrically spanning the sphere could be completed in only 5.987 seconds. For an otherwise-identical up-downium sphere, the respective communication times are 0.2993 seconds and 0.5987 seconds.

With 10^{19} light-year-radius spherical dispositions of 21,472,000, such neutronium galactic ring-worlds centered around an identical Milky Way ring-world, could send and receive two-way communications to the Milky Way ring-world in only {[2 x 10^{19}]/[2.071 x 10^{23}]}[31,000,000] seconds or 49.895 minutes. Two-way communications diametrically spanning the sphere could be completed in only 99.79 minutes. For an otherwise-identical up-downium sphere, the respective communication times are 4.9895 minutes and 9.979 minutes.

With 10^{22} light-year-radius spherical dispositions of 21,472,000,000, such neutronium galactic ring-worlds centered around an identical Milky Way ring-world, could send and receive two-way communications to the Milky Way ring-world in only {[2 x 10^{22}]/[2.071 x 10^{23}]}[31,000,000] seconds, or 34.649 days. Two-way communications diametrically spanning the sphere could be completed in only 69.299 days. For an otherwise-identical up-downium sphere, the respective communication times are 3.4649 days and 6.9299 days.

With 10^{25} light-year-radius spherical dispositions of 21,472,000,000,000, such neutronium galactic ring-worlds centered around an identical Milky Way ring-world, could send and receive two-way communications to the Milky Way ring-world in only {[2 x 10^{25}]/[2.071 x 10^{23}]}[31,000,000] seconds, or 94.863 years. Two-way communications diametrically spanning the sphere could be completed in only 189.727 years. For an otherwise-identical up-downium sphere, the respective communication times are 9.4863 years and 18.9727 years.

We can consider still more extreme cases in such detail, however, by now the reader likely has an intuitive understanding of the mathematical functions describing systems covered herein.

Thus, I have provided a thought problem solution with first-order mathematical analysis indicating in simple terms the plausibility of effectively extraordinarily super-luminal communication over cosmic distance scales. Special relativity ironically makes this possible.

One major caveat for such extreme Lorentz factors includes virtually complete evacuation of thermal emissions from within the travel tubes. CMBR photons would be blueshifted to extremely hard gamma radiation and may annihilate a travel car body made of neutronium and quarkonium panels. A second caveat is actually being able to fabricate such extreme materials on such massive scales. A third caveat is the assembly such huge quantities of nuclear density materials into the order and controlled extremely extensive dispositions. A fourth caveat is the need for a suitable power source.

Now, the radius of a black hole is

$R_{bh} = 2GM/c^2 \approx 2.95\ (M/M_{Sun})$ km

A black hole with the radius of the observable universe would have a mass of

> [13.75 ± 0.11 billion light-years][10^{13} km/1 light-year](1 solar mass)/(2.95 km) = 4.66 x 10^{22} solar masses. The mass of a critically open universe is about [10^{53} kg]/[2 x 10^{30} kg] solar mass = 5 x 10^{22} solar masses.

A marginally open universe will have a mass of about 10^{53} kg. Therefore, any black-hole travel tubes having a radius equal to that of the observable universe will be limited to a mass of about 5 x 10^{22} solar masses.

In fact, any circulinear travel tube car will be limited in radius $R > 2GM_{tt}/c^2 = 2.95\ (M_{tt}/M_{Sun})$ km.

Recall the following simple points.

The radial centripetal force exerted by the 1-metric-ton car were it not for special relativistic mass increase would be 10^{11} metric tons. However, the car is traveling at a Lorentz factor of 2.071 x 10^{23} with respect to the travel tube reference frame and so the centripetal force acting on the ring due to the car will be (10^{11})[2.071 x 10^{23}](9,810) newtons or 2.03165 x 10^{38} newtons. Now consider neutronium tubes having a mass of 7,720,000,000,000,000,000,000 solar masses or 7 x 10^{21} solar masses and a circumference of 760,000,000,000,000 light-years or 7.6 x 10^{14} light-years. These tubes would support a one metric ton car revolving at a Lorentz factor of 2.071 x 10^{23}. Tubes constructed of the up-downium having the same mass and radius would support a Lorentz factor of 2.071 x 10^{24}.

Consider neutronium tubes having a mass of 7 x 10^{27} solar masses solar masses and a circumference of 7.6 x 10^{14} light-years. These tubes would support a one metric ton car revolving at a Lorentz factor of 2.071 x 10^{26}. Tubes constructed of the up-downium having the same mass and radius would support a Lorentz factor of 2.071 x 10^{27}.

Now, the radius of the observable universe is 1.375 x 10^{10} light-years and a black hole having the same radius has a mass of 4.66 x 10^{22} solar masses.

Thus, we would expect the following boundary condition to hold.

> [1.375 x 10^{10} light-year]/[4.66 x 10^{22} solar mass] = [2.419 x 10^{14} light-year]/M_{tt}

→ M_{tt} = [2.419 x 10^{14} light-year] [4.66 x 10^{22} solar mass]/[1.375 x 10^{10} light-year] = 8.1982 x 10^{26} solar masses

Thus, M_{tt} is a little less than the Jean's limit mass for the previously computed exemplar ring, and so it is plausible that the ring can be upheld against collapse into a useless black hole.

W can provide a means by which the ring's tensile strength can remain constant for progressively larger rings thereby permitting constant minor cross-section of the ring. Thus, for every factoral increase, *n*, in ring growth beyond a circumference of 7.6 x 10^{14} light-years, we can support a factoral increase in car Lorentz factor by $n^{1/2}$. Thus, for an up-downium linear mass density maximized ring having a circumference of 7.6 x 10^{18} light-years, we could support a Lorentz factor of [2.071 x 10^{29}]. For an up-downium linear mass density maximized ring having a circumference of 7.6 x 10^{22} light-years, we could support a Lorentz factor of [2.071 x 10^{31}]. For an up-downium linear mass density maximized ring having a circumference of 7.6 x 10^{26} light-years, we could support a Lorentz factor of [2.071 x 10^{33}]. For an up-downium linear mass density maximized ring having a circumference of 7.6 x 10^{30} light-years, we could support a Lorentz factor of [2.071 x 10^{35}], and so on.

Now the length specific mass of an 8.1982 x 10^{26} solar mass ring having a circumference of7.6 x 10^{14} light-years is {[8.1982 x 10^{26}][2 x 10^{27}]}/{[7.6 x 10^{14}][10^{16}]} kg/m = 2.1574 x 10^{23} kg/m. Assuming an up-downium density equal to neutronium at 10^{18} kg/m^3, the travel tube material cross-sectional area will be 215,740 square meters. This is the maximum possible linear mass density for the up-downium considered herein. The tensile strength of the tube as specified will be 2.1574 x 10^{39} newtons.

It is conceivable that no quantized Standard Model–based material(s), Minimally Supersymmetric Standard Model material(s), or low-order Extended Supersymmetric Models material(s) denser than quarkoniums will be practical for the formation of travel tubes such as those described in the more extreme specific examples provided above. However, materials denser than quarkoniums are conceivable with much higher areal tensile strengths.

For example, consider the materials below:

X-Y Material

The X and Y bosons are predicted by the Georgi–Glashow model, a grand unified theory. This model proposes another force. The X and Y bosons would be analogous to the W and Z bosons of the weak force.

An X boson would decay into two up quarks or an anti-down quark and a positron.

A Y boson would decay into a positron and an anti-up-quark, a down quark, and an up quark, or an anti-down-quark and an anti-electron-neutrino.

The mass of the X and Y bosons would be extreme at 10^{15} GeV/c^2 = {[1.602 x 10^{-19}](10^{24})Joules}/{{[3 x 10^8]m/s}^{2}} = 1.78 x 10^{-12} kg.

The width of a somehow stabilized valence arrangement of one X boson per three Y bosons would plausibly be about (0.000511)/(10^{15}) times that of a typical neutral periodic table atom or 5.11 x 10^{-19} times that of a typical atom. Thus, the diameter of such a valence arrangement would be about 5.11 x 10^{-29} meters.

We can imagine a string comprised of (0.0001){[5.11 x 10^{-29}]$^{-1}$ } = 1.9569 x 10^{26} X-Y boson atoms, which would be about 0.0001 meter long and have a mass of 1.3933 x 10^{15} kilograms and which would be 10 X-Y bosonic atoms wide and 10 X-Y bosonic atoms thick. Now a black hole having a mass of 1.3933 x 10^{15} kilograms will have a radius of about 2.37468 x 10^{-13} meters. We could therefore plausibly compact the 0.0001 meter long string into a conical configuration having a lateral height about equal to 10^{-11} meters and a basic width of 10^{-15} meters and comprised of a cross-woven grid of X-Y boson threads separated by about 2 x 10^{-22} meters. The lateral area of the cone will be 1.5708 x 10^{-26} square meters. The reasoning behind this assumption is that the 0.0001-meter-thread could be partitioned into 10^{-7} sub-threads where each sub-thread has a cross-sectional area of 10 X-Y bosonic atoms by 10 X-Y bosonic atoms and is 10^{-11} meters long. A 10^{-18} meter by 10^{-18} meter section of the fabric will have a mass of [1.3933 x 10^{15}](10^{-14})(10,000) kg = 139,330 kg. This is the equivalent of 139,330 kg per 10^{-36} square meters. Thus, corrected mass of the cone will be (139,330 kg) [1.5708 x 10^{-26}]/(10^{-36}) = 2.1885 x 10^{15} kg.

The bond strength between the X and Y bosons should be roughly equal to:

$$F = \{1/[4\pi\varepsilon_0]\}[q_1 q_2]/[r^2]$$
$$= \{[8.987551787 \text{ x } 10^9] \text{ N m}^2 \text{ C}^{-2}\}[-1.602176565(35)\times10^{-19} \text{ Coulombs}]$$

$[1.602176565(35)\times10^{-19}$ Coulombs$]/\{[5.11 \times 10^{-29}\text{m}]^2\} = 8.83528 \times 10^{28}$ newtons

Therefore, the yield strength of each thread should be about $(10^2)[8.83528 \times 10^{28}]$ newtons $= 8.83528 \times 10^{30}$ newtons.

Such an X-Y material cone could be used as the leading tip of an astrodynamic space train like starship and could therefore bear the forward brunt of the cosmic ray, dust particle, gas and plasma particle, and photonic onslaught from the vacuum of space. The X-Y material could be the proverbial tip of the spear for truly extreme Lorentz factor starships.

One example of a plausible crystalline material formed from X and Y particles has the following pattern:

$X Y_3$	Y	$X Y_3$	Y	$X Y_3$	Y	$X Y_3$	Y
Y	$X Y_3$	Y	$X Y_3$	Y	$X Y_3$	Y	$X Y_3$
$X Y_3$	Y	$X Y_3$	Y	$X Y_3$	Y	$X Y_3$	Y
Y	$X Y_3$	Y	$X Y_3$	Y	$X Y_3$	Y	$X Y_3$
$X Y_3$	Y	$X Y_3$	Y	$X Y_3$	Y	$X Y_3$	Y
Y	$X Y_3$	Y	$X Y_3$	Y	$X Y_3$	Y	$X Y_3$

Another example of a plausible crystalline material formed from X and Y particles has the following pattern:

$-X-Y_3$	$-Y$	$-X-Y_3$	$-Y$	$-X-Y_3$	$-Y$	$-X-Y_3$	$-Y$
$-Y$	$-X-Y_3$	$-Y$	$-X-Y_3$	$-Y$	$-X-Y_3$	$-Y$	$-X-Y_3$
$-X-Y_3$	$-Y$	$-X-Y_3$	$-Y$	$-X-Y_3$	$-Y$	$-X-Y_3$	$-Y$
$-Y$	$-X-Y_3$	$-Y$	$-X-Y_3$	$-Y$	$-X-Y_3$	$-Y$	$-X-Y_3$
$-X-Y_3$	$-Y$	$-X-Y_3$	$-Y$	$-X-Y_3$	$-Y$	$-X-Y_3$	$-Y$
$-Y$	$-X-Y_3$	$-Y$	$-X-Y_3$	$-Y$	$-X-Y_3$	$-Y$	$-X-Y_3$

Monopolium

I grew up while my father was serving in the nuclear navy as an MIT-minted PhD in nuclear engineering who graduated at the very top of his class. He worked as an inspector at the Nuclear Power Training Unit that used to be located in or near Windsor Locks, Connecticut, within the United States. He then went on to work for the Naval Reactors Division of Naval Sea Systems Command and worked on nuclear propulsion system research and development and maintenance for the US Navy, first as a career navy officer, and then as a civilian, Department of Energy/Department of Defense employee. He would virtually always receive the grade of Outstanding in his periodic evaluations in just about every category for his job performance. I thus became enamored with the awesome power of nuclear energy, which persists to this very day.

By now the reader understands my writing attitudes and temperaments and is aware of my high regard for applications of nuclear, sub-nuclear, and high-energy physics research and/or development. I am especially interested in its applications for manned star travel that are so expansive that one might instead almost refer to the conjectural techniques as astral travel or travel among the metaphorical astral planes.

The mass of the still-theoretical magnetic monopole is at most about 10^{17} GeV/c^2 = $\{[1.602 \times 10^{-19}](10^{26})\text{Joules}\}/\{\{[3 \times 10^{8}]\text{m/s}\}^{2}\} = 1.78 \times 10^{-10}$ kg.. The width of a stable valence relationship between a south monopole and a north monopole would seem to be about $(0.000511)/(10^{17})$ times that of a typical neutral period table atom, or 5.11×10^{-21} times that of a typical period table atom. Thus, the diameter of such a valence arrangement would be 5.11×10^{-31} meters, which is approaching that of the Planck Length at:

$$l_p = \{[h(2\pi)]G/[C^3]\}^{1/2} = 1.616199 \times 10^{-35} \text{ meters}$$

X and/or Y Monopolium

A unit cell of solid crystalline materials formed of equal parts by particle of high-end mass range monopoles (10^{17} GeV/c^2 per particle), and 10^{15} GeV/c^2 X particles would have a width of about [5.11×10^{-29}] + [5.11×10^{-31}] anstroms = 5.1611×10^{-29} meters. Here is an example of a planar crystalline pattern for the monohiggsinium:

X M X M- X M X M-

M- X- M X- M- X- M X-

X M X M- X M X M-

M- X- M X- M- X- M X-

X M X M- X M X M-

M- X- M X- M- X- M X-

Here is another example of a planar crystalline form for the mono-higgsinium:

Y M Y M- Y M Y M-

M- Y- M Y- M- Y- M Y-

Y M Y M- Y M Y M-

M- Y- M Y- M- Y- M Y-

Y M Y M- Y M Y M-

M- Y- M Y- M- Y- M Y-

Another material plausibly formed from high-end of mass-range monopoles and X and Y particles having a similar density and mechanical strength has the following crystalline pattern:

XY_3 M XY_3 M- XY_3 M XY_3 M-

M- Y M Y M- Y M Y

XY_3 M XY_3 M- XY_3 M XY_3 M-

M- Y M Y M- Y M Y

XY_3 M XY_3 M- XY_3 M XY_3 M-

M- Y M Y M- Y M Y

Still another material plausibly formed from high-end of mass range monopoles and X and Y particles having a similar density and mechanical strength has the following crystalline pattern:

$-X-Y_3$	M	XY_3	M-	$-X-Y_3$	M	XY_3	M-
M-	-Y	M	Y	M-	-Y	M	Y
$-X-Y_3$	M	XY_3	M-	$-X-Y_3$	M	XY_3	M-
M-	-Y	M	Y	M-	-Y	M	Y
$-X-Y_3$	M	XY_3	M-	$-X-Y_3$	M	XY_3	M-
M-	-Y	M	Y	M-	-Y	M	-Y

Still yet another material plausibly formed from high-end of mass range monopoles and X and Y particles having a similar density and mechanical strength has the following crystalline pattern:

Y	M	-X	M-	Y	M	-X	M-
M-	Y-	M	X	M-	Y-	M	X
Y	M	-X	M-	Y	M	-X	M-
M-	Y-	M	X	M-	Y-	M	X
Y	M	-X	M-	Y	M	-X	M-
M-	Y-	M	X-	M-	Y-	M	X-

Still yet another material plausibly formed from high-end of mass range monopoles and X and Y particles having a similar density and mechanical strength has the following crystalline pattern:

XY_3	M	Y	M-	XY_3	M	Y	M-
M-	XY_3	M	Y	M-	XY_3	M	Y
XY_3	M	Y	M-	XY_3	M	Y	M-
M-	XY_3	M	Y	M-	XY_3	M	Y
XY_3	M	Y	M-	XY_3	M	Y	M-
M-	XY_3	M	Y	M-	XY_3	M	Y

Still yet another material plausibly formed from high-end of mass range monopoles and X and Y particles having a similar density and mechanical strength has the following crystalline pattern:

-X-Y_3	M	-Y	M-	-X-Y_3	M	-Y	M-
M-	-X-Y_3	M	-Y	M-	-X-Y_3	M	-Y
-X-Y_3	M	-Y	M-	-X-Y_3	M	-Y	M-
M-	-X-Y_3	M	-Y	M-	-X-Y_3	M	-Y
-X-Y_3	M	-Y	M-	-X-Y_3	M	-Y	M-
M-	-X-Y_3	M	-Y	M-	-X-Y_3	M	-Y

The specific crystalline patterns depicted above would resemble ionic-bonds in a manner analogous to bi-atomic salts such as ordinary table salt, or sodium chloride.

Note that the negative signs included in the above crystalline depictions indicate antimatter versions of the particle species as denoted by the uppercase letters.

Now, reconsider the yield strength of a ten X-Y boson atom by ten X-Y boson atom cross-sectioned string of about $(10^2)[8.83528 \times 10^{28}]$ newtons = 8.83528×10^{30} newtons. Consider that the mass-specific length of the string is equal to 1.3933×10^{19} kg/m. Now consider that the number of strings required to yield a tensile strength of 2.1574×10^{39} newtons is equal to $[2.1574 \times 10^{39}$ newtons$]/[8.83528 \times 10^{30}$ newtons$]$ = 244,180,000. The mass of the resulting X-Y boson string will be $[1.3933 \times 10^{19}$ kg$]$ [244,180,000] = 3.40216×10^{27} kg/m. This is $[3.40216 \times 10^{27}$ kg/m$]/[2.1574 \times 10^{23}$ kg/m$]$. This is 15,769 times more massive than a conjectured up-downium travel tube having the same tensile strength.

In short, as far as Standard Model, MSSM, and low-order ESUSY matter particles are concerned, quarkonium is king.

The following limits will hold in a very cosmically distant future marginally open universe for which expansion as slowed to a crawl.

Lim γ = ensemble
R_{tt} --- > ensemble meters.

Lim γ = infinity scraper
R_{tt} --- > infinity scraper meters.

Lim $\gamma = \Omega$
R_{tt} --- > Ω meters.

Lim γ = ensemble
M_{tt} --- > ensemble kilograms.

Lim γ = infinity scraper
M_{tt} --- > infinity scraper kilograms.

Lim $\gamma = \Omega$
M_{tt} --- > Ω kilograms.

The following limits are based on personal conjecture.

Lim γ = Aleph 0
R_{tt} --- > (Aleph 0) meters.

Lim γ = Aleph 1
R_{tt} --- > (Aleph 1) meters.

Lim γ = Aleph 2
R_{tt} --- > (Aleph 2) meters.

…

Lim γ = Aleph g
R_{tt} --- > (Aleph g) meters.

Lim γ = Aleph 0
M_{tt} --- > (Aleph 0) kilograms.

$\text{Lim}\ \gamma = \text{Aleph 1}$
M_{tt} --- > (Aleph 1) kilograms.

$\text{Lim}\ \gamma = \text{Aleph 2}$
M_{tt} --- > (Aleph 2) kilograms.

…

$\text{Lim}\ \gamma = \text{Aleph g}$
M_{tt} --- > (Aleph g) kilograms.

Paul Gilster has posted on his Tau Zero Centauri Dreams website another one of his wonderful articles, on the subject of a Galactic Transit System.

Paul creatively discusses the notion of a Galactic Transit System and its would be parallels to the subway system in London, UK at the following URL:

http://www.centauri-dreams.org/?p=6637

One can imagine quite literally a galactic subway system composed of a series of accelerator tubes or mass drivers that is somehow set rotating along with the galaxy.

The main caveat is that the connections of the tubes and the tubes themselves would need sufficient mechanical strength so as to not pull itself apart under their own rotation-induced centripetal tension.

The really cool thing about such a system is that the energy used to accelerate cars within the system could be extracted by magnetic-induction braking and recycled over and over to accelerate cars or trains down the transit tubes.

Car acceleration velocities of 1 G should enable one to sport about anywhere within the Milky Way in one average human familial generation car time. Higher Gs are possible, but some sort of G force cancellation would then be required to make the trips comfortable and survivable over the long transit times.

If there is a hyperdimensional component to the Milky Way that has been overlooked by general relativity, perhaps such a system could be extended into or interfaced with higher dimensional space-time wherein the quantity of traffic could be increased. The travel tubes might not even be required to take the form of wormholes but, rather, perhaps simply only electrodynamic mass drivers.

I have always been enamored with such concepts well before I even knew the basic equations of Lorentz transformations or for that matter, even high-school algebra. I remember having some dream as a six- or seven-year-old at night, wherein I imagined traveling down a highway in some sort of dark environment in a car at light speed. Obviously, my childish images of what it would be like to travel down such a road at C, were woefully inaccurate, but the dream left so much of an impression on me that I still think of the dream at times when I am driving long distances at night.

Either way, the concept of a Milky Way subway is indeed very whimsical and might actually be built at some future time frame. Every time I ride on the Washington DC Metro subway, I seem to start thinking about such travel as well as wormhole travel. The small subway fare is well worth the mental imagery.

Now, the mechanical strength properties of such a material based galactic tubes transit system would need to be much superior to that of ring-worlds or Dyson spheres, however, perhaps the issues of such strength requirements might be mitigated if the tubes were of a spiraling form of layout much like the spiral arms of the Milky Way galaxy. The tubes could more or less rotate at naturally set velocities with the natural rotation rates that exist at each differential radial position from the galactic center.

In the case that the transit tubes would stretch as a result of differing angular rotation rates at differing radial distances from the center of the Galaxy, perhaps the tubes could have some sort of nanotech based self-assembly feature by which they would incorporate interstellar gas into their composition thus causing a gradual elongation of the tubes. A growth in tube length of one micrometer per meter per year would enable the tubes to double in length in only 1 million years. A growth in tube length of only one nanometer per meter per year would result in the doubling of the tube length in 1 billion years, which should be more than adequate for matching the elongation rate of the spiral arms of the Milky Way galaxy.

The tubes might also be composed of some form of high-strain-capable elastic material(s) so that they can stretch as the natural dynamics of relative rotation rates at differing radial distances from the center of the Milky Way would tend to stretch the tubes out. If the tubes could stretch like rubber bands, they might be able to double their length every 3 billion years and still not snap after 10 billion years. The gradual incorporation of material into the tubes from the interstellar medium would no doubt mitigate this problem.

Alternatively, the tubes might be constructed of exotic materials such as some form of femtometer or so, thick stabilized neutronium or quarkonium. One square meter of a femtometer thick sheet of neutronium would have a mass of only one metric ton and so a 100,000 light yearlong tube of neutronium with a linear mass density of 1,000 metric tons per meter would have a mass of only $(10^5)(10^{13})(10^3)(10^3)$ metric tons or 10^{24} metric tons or about 0.001 solar masses. An assemblage composed of a billion of such highway travel tubes would only have a mass of one million solar masses or about one 3 millionth of the mass of the Milky Way Galaxy.

Alternatively, some sort of mass collection mechanism that attracts intragalactic matter in relative close proximity to the tubes might collect interstellar matter and incorporate the matter into the tubes through some sort of macroscale robot construction machincry that would add section upon section to the tubes as required to keep the tubes from being stretched apart. The collection and construction methodology might include both atomic and molecular matter, as well as neutron or quark density-matter-based materials.

Another option is that tube-length servicing stations might be placed or distributed along the length of the tubes, wherein the material to lengthen the tubes is brought to the tube length construction stations from remote sources.

It is conceivable that perhaps femtotechnology self-assembly could be used to permit the lengths of solid neutronium tubes or quarkonium tubes to grow so as to mitigate mechanical strength limitation issues of the neutronium or quarkonium. The neutronium, if configured at the solid state neutronium physics cellular level in a proper manner, might exists as an elastic high strain capable form of neutronium and thus permit gradual growth of the neutronium tubes to reduce the stresses due to natural tube lengthening effects based on galactic evolution dynamics.

Note that such travel tubes might be made of superconducting materials, thus producing a region of space within that has similar properties to the space between the conducting plates of a Casimir plate apparatus, wherein the electromagnetic zero point waves within the plates are limited to frequencies that have nodes on the inner surfaces of the conducting plates in accordance to the rules of classical electrodynamics that permit standing waves within a conducting cavity to have only zero amplitude on the inner surfaces of the conducting plates. As a result of the elimination of the remainder of the otherwise possibly-zero-point virtual electromagnetic waves from the space between the plates, a force pushing the plates inward toward each other is produced due to the imbalance of the zero-point electromagnetic waves between the plates with respect to those outside of the plates.

The result of the elimination of some of the zero-point energy waves from between the plates is theoretically a slight change of increase in the inverse square root of the product of epsilon naught and mu naught where epsilon naught is the electrical permittivity of free space and mu naught is the magnetic permeability of free space. The value of the inverse square root of these two quantities is equal to the speed of light in a vacuum. Thus, the speed of light between the plates, in theory increases ever so slightly as the separation of the plates is decreased.

Note, however, the measured Casimir force between the plates rapidly decreases to undetectability as the distance between the plates grows beyond the micrometer range, and so the effect on the zero-point energy electromagnetic properties within the vacuum in a transit tube such as the ones conjectured about above that would be able to accommodate a human car or rail-line vehicle, would be very minuscule.

The point is that if there is nonetheless an increase in the speed of light within the Casimir tubes, minuscule as the increase may be, perhaps superluminal travel within the tubes with respect to the outside interstellar vacuum is possible for finite-rest mass cars or trains that have reached a sufficiently extreme Lorentz factor with respect to the speed of light within the transit tubes. The net effect might be a very minuscule backward time travel component wherein the cars would arrive at the other end of the tube before the time that they left, thus effectively leading to quicker-than-instantaneous space travel.

In the event that the cars would arrive within one, or perhaps even a few, Planck time units into the past with respect to the present starting time of the journey, perhaps the paradoxes of backward time travel would be avoided based on the uncertainly or the poor definition of time that exists as the time period under consideration drops in length to near the Planck time unit. Note that the Planck time Unit is equal to $[(h/(2\ \pi))\ G/(C^5)]^{1/2}$ which is about $[5.39124(27) \times (10^{-44})]$ seconds.

Another way in which the kinematics of extreme Lorentz factor travel down such tubes might not involve backward time travel at all but, rather, a shortened time travel into the future as the craft travels down a given tube is based on a truncated or muted tendency for the craft to otherwise experience such a forward time travel reduction, such that in the unmitigated case, the forward time travel of the craft would become negative in value thus leading to effectively backward time travel.

The shortened time travel of the craft into the future would not simply be the result of the relativistic Lorentz factor–based time dilation according to the formula $\Delta t_1 = (\Delta t_0) \{1/ [1 - [(v/c)^2]]^{1/2}\}$. Instead, the effect would be due to the reduction in forward time travel (as the craft traveled down the tube) due to a reduced value or effect of the kinematical function, factor of the function, or term of the function that expresses the degree of backward time travel that in theory would result from superluminal travel through space.

Note that if such mass driver tubes can be constructed throughout the Milky Way, perhaps they can also be constructed in intergalactic space, or even throughout the now currently observable universe and beyond. The same nanotech and/or femtotech self-assembly methods might be used to construct the transit tubes and the materials out of which the tubes are constructed might be of high strain, high strength, elastic properties in order to permit tube stretching and growth along with the expansion of space-time and the perturbative effects of localized gravitational-force-based pulling by proximate galaxies or galactic or intergalactic gas clouds, including those made of baryonic matter and cold dark matter.

Naturally, some methods would be required to protect the tubes from collision with other objects and even dust within the vacuum of intragalactic and intergalactic space, as well as from stellar explosions such as nova, supernova, hypernova, and gamma ray bursts. Perhaps the tubes could have some forms of rockets, electrodynamic-reaction mechanism(s) that react against the ambient space-based dust and gas, magnetic fields, and the like that are powered by nuclear fusion, nuclear fission, matter-antimatter reactions, or perhaps by stored photovoltaic or photo-electrodynamic energy collected gradually from starlight and cosmic microwave background radiation, respectively. These mechanisms would push and/or pull a section of tube out of the path of an incoming large object that could otherwise not be destroyed.

To destroy or neutralize the threats from large objects, perhaps nuclear bombs or matter-antimatter bombs could be used to destroy large objects on a collision course with a given tube, or perhaps fission, fusion, or matter-antimatter rockets could be attached to the large objects to nudge them out of the way of the transit tubes.

Now, what can be done with mass-driver travel tubes? Can they, in principle, be duplicated with relativistic rockets and the relativistic Lorentz turning force? The environments around large stars are an ideal starting point for such considerations. The next levels of such a rocket mechanism might include globular clusters, galaxies, galaxy clusters, then galaxy superclusters. Eventually, nucleations of galactic matter larger than superclusters may occur, which have more or less radially and circumferentially symmetrical and uniform poloidal magnetic fields.

In our last exemplar Lorentz factor of 2.071 x 10^{35}, we are only at a relativistic mass of about 15 orders of magnitude less than that of the currently observable universe. In theory, for 1-metric-ton-cars, a Lorentz factor of about 10^{50} should be possible for revolutional systems. For revolutional systems having a circumference of 7.6 x 10^{30} light-years, a Lorentz factor as high as about 10^{70} should be possible. From here on, the maximum possible Lorentz factor for revolutional systems should scale with the radius of revolution.

Now for closed universes, it is interesting to note that a spacecraft traveling in one direction long enough may eventually come back to its rough starting position, much as an aircraft flying around Earth along a meridian will eventually fly back to or over the location whence it began its flight. The relativistic mass limit of the universally circumnavigating craft may well be limited to that of the maximum non-black-hole mass concentration for a maximally expanded closed universe. In such a case, substantially all of the real mass content within the universal closed space time would have been converted into spacecraft kinetic energy.

Given the considerations in the previous paragraph, it may be possible to increase the rate of universal expansionary slowdown and collapse simply by powering the exemplar spacecraft up to super-general-relativistic masses via extraction of propulsion energy from the zero-point energy fields. Such a mechanism would result in the immediate formation of a black hole where the enclosing universe was not already one.

On a smaller scale, such a mechanism may enable the formation of black holes of cosmic radius scales within either a marginally open or open universe. For open universes, it would be of great help if the rate of universal expansion would non-trivially decrease over cosmic time scales. Such a slowing may be required so that the space-time revolution process of the spacecraft would not be diluted in its associated relativistic mass energy content as a result of space-time expansion.

We have an interesting result and point to ponder here. Assuming that black holes can be formed by such Lorentz turning-force effects from spacecraft accelerated to super-general-relativistic kinetic energies, perhaps the spacecraft can be interpreted as forming its own black hole. Should wormhole travel be possible in the external vicinity of the extremely stressed space-time near black-hole event horizons, or if not, within the black-hole event horizon itself, perhaps such spacecraft accelerated to super-relativistic kinetic energies may be interpreted as entering its own black hole. Since most of the energy of the spacecraft and the space-time enclosing the spacecraft's revolutional path would be contained within the spacecraft, perhaps due to special relativistic mass increase of the spacecraft, the spacecraft would enter a black hole formed immediately by its relativistic mass, thereby causing the spacecraft to disconnect from the space-time or external universe whence it accelerated up to speed. Such a self-imposed black-hole state may be interpretable as inter-universal travel, extra-dimensional travel, hyperspatial travel, wormhole travel, and the like.

Next time you are riding the tubes, especially an extensive metro-transit system, think about the above article. You, and most likely, here and now, very few people are aware of the above arguments in favor of cosmic-scale subways that do not involve wormholes.

Perhaps our cosmic infrastructure will keep getting bigger, even if we are limited to light-speed travel. However, it may be the case that light-speed travel, or nearly so, becomes the new cosmic norm for humanity, ETI, and UTI kind. Our local portion of the universe may gradually morph into a cosmic civilization of love traveling at light speed or near light speed with respect to the background.

To the extent that the future cosmic civilization of love would morph into an ensemble Lorentz factor dimensionality with respect to the background, systems such as those described above may enable Planck time communications over ensemble light-year distances.

To the extent that the future cosmic civilization of love would morph into an infinity-scraper Lorentz factor dimensionality with respect to the background, systems such as those described above may enable Planck time communications over infinity scraper light-years distances.

To the extent that the future cosmic civilization of love would morph into a light-speed dimensionality with respect to the background, systems such as those described above may enable Planck time communications over any low-order infinite distances, even those measured in small-end-range Aleph numbers of light-years.

It may be the case where several, many, an ensemble, an infinity scraper, an Ω, an Aleph 0, an Aleph 1, and Aleph 2 and so-on, number of "near light speed" and "light speed" dimensions can be realized, where there is a one-to-one correspondence between distinction promoting Lorentz factor ranges and the associated substantially light-speed dimensionality.

In a strong sense, many or all of these substantially light-speed infrastructural phases may each correspond to a kind of parallel universe, but in a different sense than the many-worlds interpretation of quantum theory types and the parallel history types.

Near-light-speed to substantially light-speed kinematics for future cosmic civilizations is going to become fun in the eons and aeons to follow.

Any of the ring-worlds presented in the previous digression may in principle rotate with similar tangential, and even radial, acceleration values, as would the cars in the above conjecture.

A number of mechanisms may be applied to enable such rapid ring rotation.

First, provided the rotation rate was sufficiently relativistic, Doppler blueshifted star-light, quasar-light, supernova light, cosmic-background radiation, and other electromagnetic sources could be applied in specialized meta-material optics to exert counterpressure on the rotating rings in order to prevent the rings from snapping under the centripetal pressure induced by the high levels of rotational inertia.

Second, a very large electrical charge may be installed at the center of the ring while the ring system is strongly oppositely charged so that the rotational inertial is balanced by the coulombic attraction.

Third, a very large black hole may be naturally present or artificially installed in the center of the ring so as to attract differential portions of the ring to counterbalance rotationally induced radial pressure.

Fourth, Casimir plate–like mechanisms that are able to produce unbalanced forces may also be of use, provided that unbalanced Casimir forces are possible or practical to produce.

Fifth, cosmic gravitational background radiation may be processed by gravitational wave-optics installed in discrete or continuous patterns along the rotating ring in such a manner so that differential portions of the ring are pushed and/or pulled inward so as to balance centripetal acceleration-induced pressures.

One can imagine quite literally a galactic subway system composed of a series of accelerator tubes or mass drivers that is somehow set rotating along with the galaxy.

The main caveat is that the connections of the tubes and the tubes themselves would need sufficient mechanical strength so as to not pull themselves apart under their own rotation-induced centripetal tension.

The really cool thing about such a system is that the energy used to accelerate cars within the system could be extracted by magnetic-induction braking and recycled over and over to accelerate cars or trains down the transit tubes.

Car acceleration velocities of 1 G should enable one to sport about anywhere within the Milky Way in one average human familial generation car time. Higher Gs are possible, but some sort of G force cancellation would then be required to make the trips comfortable and survivable over the long transit times.

If there is a hyperdimensional component to the Milky Way that has been overlooked by general relativity, perhaps such a system could be extended into or interfaced with higher-dimensional space-time wherein the quantity of traffic could be increased. The travel tubes might not even be required to take the form of wormholes but, rather, perhaps simply only electrodynamic mass drivers.

I have always been enamored with such concepts well before I even knew the basic equations of Lorentz transformations or, for that matter, even high-school algebra. I remember having some dream as a six- or seven-year-old at night, wherein I imagined traveling down a highway in some sort of dark environment in a car at light speed. Obviously, my childish images of what it would be like to travel down such a road at C, were woefully inaccurate, but the dream left so much of an impression on me that I still think of the dream at times when I am driving long distances at night.

Either way, the concept of a Milky Way subway is indeed very whimsical and might actually be built at some future time frame. Every time I ride on the Washington DC metro subway, I seem to start thinking about such travel as well as wormhole travel. The small subway fare is well worth the mental imagery.

Now, the mechanical strength properties of such a material-based galactic tubes transit system would need to be much superior to that of ring-worlds or Dyson spheres; however, perhaps the issues of such strength requirements might be mitigated if the tubes were of a spiraling form of layout much like the spiral arms of the Milky Way galaxy. The tubes could more or less rotate at naturally set velocities with the natural rotation rates that exist at each differential radial position from the galactic center.

In the case that the transit tubes would stretch as a result of differing angular rotation rates at differing radial distances from the center of the galaxy, perhaps the tubes could have some sort of nanotech-based self-assembly feature by which they would incorporate interstellar gas into their composition, thus causing a gradual elongation of the tubes. A growth in tube length of one micrometer per meter per year would enable the tubes to double in length in only 1 million years, a growth in tube length of only one nanometer per meter per year would result in the doubling of the tube length in 1 billion years, which should be more than adequate for matching the elongation rate of the spiral arms of the Milky Way galaxy.

The tubes might also be composed of some form of high-strain-capable elastic material(s) so that they can stretch as the natural dynamics of relative rotation rates at differing radial distances from the center of the Milky Way would tend to stretch the tubes out. If the tubes could stretch like rubber bands, they might be able to double their length every 3 billion years and still not snap after 10 billion years. The gradual incorporation of material into the tubes from the interstellar medium would no doubt mitigate this problem.

Alternatively, the tubes might be constructed of exotic materials such as some form of femtometer or so, thick stabilized neutronium or quarkonium. One square meter of a femtometer thick sheet of neutronium would have a mass of only 1 metric ton, and so a 100,000-light-year-long tube of neutronium with a linear mass density of 1,000 metric tons per meter would have a mass of only (10 EXP 5)(10 EXP 13)(10 EXP 3)(10 EXP 3) metric tons, or 10 EXP 24 metric tons, or about 0.001 solar masses. An assemblage composed of a billion of such highway travel tubes would only have a mass of 1 million solar masses, or about one-3 millionth of the mass of the Milky Way galaxy.

Alternatively, some sort of mass-collection mechanism that attracts intragalactic matter in relative close proximity to the tubes might collect interstellar matter and incorporate the matter into the tubes through some sort of macroscale robot construction machinery that would add section upon section to the tubes as required to keep the tubes from being stretched apart. The collection and construction methodology might include both atomic and molecular matter, as well as neutron or quark density-matter-based materials.

Another option is that tube-length servicing stations might be placed or distributed along the length of the tubes wherein the material to lengthen the tubes is brought to the tube length construction stations from remote sources.

It is conceivable that perhaps femtotechnology self-assembly could be used to permit the lengths of solid neutronium tubes or quarkonium tubes to grow so as to mitigate the mechanical-strength limitation issues of the neutronium or quarkonium. The neutronium, if configured at the solid state neutronium physics cellular level in a proper manner, might exist as an elastic high-strain-capable form of neutronium, and thus permit the gradual growth of the neutronium tubes to reduce the stresses due to the natural tube-lengthening effects based on galactic evolution dynamics.

Note that such travel tubes might be made of superconducting materials, thus producing a region of space within that has similar properties to the space between the conducting plates of a Casimir plate apparatus, wherein the electromagnetic zero point waves within the plates are limited to frequencies that have nodes on the inner surfaces of the conducting plates in accordance to the rules of classical electrodynamics that permit standing waves within a conducting cavity to have only zero amplitude on the inner surfaces of the conducting plates. As a result of the elimination of the remainder of the otherwise possible zero-point virtual electromagnetic waves from the space between the plates, a force pushing the plates inward toward each other is produced due to the imbalance of the zero-point electromagnetic waves between the plates with respect to those outside of the plates.

The result of the elimination of some of the zero-point energy waves from between the plates is theoretically a slight change of increase in the inverse square root of the product of epsilon naught and mu naught, where epsilon naught is the electrical permittivity of free space and mu naught is the magnetic permeability of free space. The value of the inverse square root of these two quantities is equal to the speed of light in a vacuum. Thus, the speed of light between the plates, in theory, increases ever so slightly as the separation of the plates is decreased.

Note, however, the measured Casimir force between the plates rapidly decreases to undetectability as the distance between the plates grows beyond the micrometer range, and so the effect on the zero-point energy electromagnetic properties within the vacuum in a transit tube such as the ones conjectured about above that would be able to accommodate a human car or rail-line vehicle, would be very minuscule.

The point is that if there is, nonetheless, an increase in the speed of light within the Casimir tubes, minuscule as the increase may be, perhaps superluminal travel within the tubes with respect to the outside interstellar vacuum is possible for finite rest mass cars or trains that have reached a sufficiently extreme Lorentz factor with respect to the speed of light within the transit tubes. The net effect might be a very minuscule backward time travel component wherein the cars would arrive at the other end of the tube before the time that they left, thus effectively leading to quicker-than-instantaneous space travel.

In the event that the cars would arrive within one, or perhaps even a few Planck time units into the past with respect to the present starting time of the journey, perhaps the paradoxes of backward time travel would be avoided based on the uncertainly or the poor definition of time that exists as the time period under consideration drops in length to near the Planck time unit. Note that the Planck time unit is equal to [(h/(2 pi)) G/(C EXP 5)] EXP (1/2), which is about [5.39124(27) x (10 EXP – 44)] seconds.

Another way in which the kinematics of extreme Lorentz factor travel down such tubes might not involve backward time travel at all but, rather, a shortened time travel into the future as the craft travels down a given tube is based on a truncated or muted tendency for the craft to otherwise experience such a forward time-travel reduction such that in the unmitigated case, the forward time travel of the craft would become negative in value, thus leading to effectively backward time travel.

The shortened time travel of the craft into the future would not simply be the result of the relativistic Lorentz factor–based time dilation according to the formula Delta t1 = (Delta t0) {1/[1 - [(v/c) EXP 2]] EXP (1/2)}. Instead, the effect would be due to the reduction in forward time travel (as the craft traveled down the tube) due to a reduced value or effect of the kinematical function, factor of the function, or term of the function that expresses the degree of backward time travel, which in theory would result from superluminal travel through space.

Note that if such mass driver tubes can be constructed throughout the Milky Way galaxy, perhaps they can also be constructed in intergalactic space, or even throughout the now-currently-observable universe and beyond. The same nanotech and/or femtotech self-assembly methods might be used to construct the transit tubes and the materials out of which the tubes are constructed might be of high-strain, high-strength elastic properties in order to permit tube stretching and growth along with the expansion of space time and the perturbative effects of localized gravitational force based pulling by proximate galaxies or galactic or intergalactic gas clouds including those made of baryonic matter and cold dark matter.

Naturally, some methods would be required to protect the tubes from collision with other objects, and even dust within the vacuum of intragalactic and intergalactic space, as well as from stellar explosions such as nova, supernova, hypernova, and gamma ray bursts. Perhaps the tubes could have some forms of rockets, electrodynamic reaction mechanism(s) that react against the ambient space-based dust and gas, magnetic fields, and the like that are powered by nuclear fusion, nuclear fission, matter-antimatter reactions, or perhaps by stored photovoltaic or photo-electrodynamic energy collected gradually from starlight and cosmic microwave background radiation, respectively. These mechanisms would push and/or pull a section of tube out of the path of an incoming large object that could otherwise not be destroyed.

To destroy or neutralize the threats from large objects, perhaps nuclear bombs or matter-antimatter bombs could be used to destroy large objects on a collision course with a given tube, or perhaps fission, fusion, or matter-antimatter rockets could be attached to the large objects to nudge them out of the way of the transit tubes.

Below, a mathematical presentation of relativistic magnetic mass drivers is provided.

Consider a wire or conductive coil that is axially oriented in a direction parallel to the spacecraft velocity vector. The linear induction coil generator will operate on more or less static background interstellar and intergalactic magnetic fields, as well as magnetic fields in solar systems. For relativistic spacecraft, the background magnetic field components will be as follows:

$$\mathbf{Bx} = \mathrm{Bx}$$

$$\mathbf{By} = \gamma\{B_y + \{[[v/[C^2]]E_z\}\} = \{1 - [(v/C)^2]\}^{-1/2}\{B_y + \{[[v/[C^2]]E_z\}\}$$

$$\mathbf{Bz} = \gamma\{B_z - \{[[v/[C^2]]E_y\}\} = \{1 - [(v/C)^2]\}^{-1/2}\{B_z - \{[[v/[C^2]]E_y\}\}$$

Here B_x is the component of the background magnetic field that is parallel to the spacecraft velocity vector.

Now, in the spacecraft reference frame, the voltage developed within a conductive coil is equal to $E_{mf} = -N\Delta\varphi/\Delta t = -N\Delta(BA)/\Delta t_{ship}$, where N is the number of cable turns, $\varphi = BA$ is the magnetic flux, and t_{ship} is the ship frame time.

Assuming N_c coils where the $N_{c,gth}$, coil has an plan form area $A_{c,gth,j}$, in the jth time step, and where the background magnetic field in the jth ship time step is

$$Bx_j + \gamma_j\{B_{y,j} + \{[[v_j/[C^2]]E_{z,j}\}\} + \gamma_j\{B_{z,j} - \{[[v_{,j}/[C^2]]E_{y,j}\}\} = Bx_j + \{1 - [(v_j/C)^2]\}^{-1/2} \{B_{y,j} + \{[[v_j/[C^2]]E_{z,j}\}\} + \{1 - [(v_j/C)^2]\}^{-1/2} \{B_{z,j} - \{[[v_j/[C^2]]E_{y,j}\}\}.$$

The electric power generated in the cable having a resistance R_j in the jth time step is equal to:

$$P_j = V_j^2/R_j = E_{mf,j}{}^2/R_j = [-N_{c,gth} \Delta\varphi j/\Delta t_{ship,j}]^2/R_j = \{-N_{c,gth} \Delta\{Bx_j + \gamma_j\{B_{y,j} + \{[[v_j/[C^2]]E_{z,j}\}\} + \gamma_j\{B_{z,j} - \{[[v_{,j}/[C^2]]E_{y,j}\}\} \} (A_{c,gth,j})/\Delta t_{ship,j}\}^2/R_j$$

In principle, the generated electrical power may be used to power super-relativistic specific-impulse chargon rockets, and/or as another nonlimiting option and electrodynamic-hydrodynamic-plasma drive system.

Mass drivers within Earth's orbit might be used for efficient travel between near-Earth space and Earth's surface.

For example, beanstalk-like mass drivers may be used to ferry persons, equipment, and supplies to interplanetary space-faring craft. Accordingly, the mass drivers would function as elevators.

Beanstalk mass drivers would most likely first be constructed of carbonaceous supermaterials. Such materials would include single-walled carbon nanotubes, multi-walled carbon nanotubes, graphene, graphene oxide paper, boron-nitride nanotubes, diamond fiber–based nano-tubes, beta-carbon-nitride–based nanotubes and the like.

These tubes could be lined or enmeshed with electromagnetic that are sequentially activated above the craft for ascent to near-Earth space. The magnets would be switched on ahead of the craft in a propagating manner so that the craft is continually pulled upward.

For controlled descent back to Earth's surface, a magnetic pulling mechanism may be activated to tug on a craft under the attractive influence of gravity so that the speed of descent may be safely controlled.

Such beanstalks may extend out beyond geosynchronous orbit and be stabilized by inertial mass anchors.

Beanstalks may have foodcourts, restaurants, lodging, observation posts and other amenities for travelers.

Additionally, military, police, and security personnel may use the beanstalks to ferry personnel and equipment in and out of near-Earth space.

Still further, general commerce and industrial support can be facilitated by the stalks.

Each beanstalk is perhaps best attached to Earth near the equator for greatest stability.

More than one stalk may be used.

For example, a network of regularly and/or irregularly equatorial disposed stalks may fan out into extra-geosynchronous space. Additionally, the stalks may have interconnections, or bridges, connecting one or more stalks to some or all of the other stalks.

Inter-stalk bridgework may also be fabricated from carbonaceous supermaterial and may provide azimuthal locomotion via mass drivers. Once again, any of the infrastructure generally included within, or alongside a stalk, may be placed within or alongside the bridges.

The stalks and bridges can be electrified by solar photovoltaic stations distributed along the length of these conduits in continuous or discreet manner. Solar panels can include rigid as well as membranous types. Additionally, solar panels that operate on concentrated sunlight may provide higher mass-specific power outputs for the panels themselves.

To facilitate efficient sunlight concentration, inflatable or light-pressure-deployed parabolic, spherical, catenary, trough-like, or other metalized reflective members may be used to concentrate sunlight onto the solar panels. Membranous reflectors may be tuned or detuned to include a variety of continuous or faceted reflective surfaces.

These metalized membranes may optionally be constructed of any of the carbonaceous supermaterials used for the structural aspects of the stalks and bridges. However, rather ordinary membranous materials may be used instead.

Examples of ordinary membranous structural materials include Mylar, Nylon, Kevlar, Kapton, and similar polymers.

The reflective membranes may range in thickness from a few mils (thousandths of an inch) to values on the order of microns for ordinary material compositions. Carbonaceous supermaterials may be used in the construction of reflectors as thin as one nanometer, and perhaps even less.

Alternatively, the electrical power supplied to the stalks and bridges and affixed infrastructure may originate from or below Earth's surface.

Some terrestrial and subterranean power supplies can include nuclear-fission reactor sources, yet to be developed fusion reactor sources, tidal generators, wind turbines, hydroelectric sources, clean natural gas–fired plants, coal-fire plants, biofuel-fired plants, and geothermal plants.

Alternatively, photothermal turboelectric or thermoelectric plants may be affixed to the stalks and bridges to provide some or all the necessary power to operate the system(s).

Such photothermal generators can include traditional steam effluent–driven turboelectric systems, steam-driven Pelton wheel motivation, Sterling generators, and other mechanisms. Moreover, multicycle-steam systems and regenerative heating may be used for the efficient conversion of solar heat to electrical power.

Nuclear-powered mechanical electric power stations may also be distributed along the stalks and bridges.

What is possible using Earth's orbit for beanstalks and bridges may also be duplicated for stalks on Mars.

Conceivably, a vast system of travel tubes could be assembled as more or less concentric ringlike shells that are centered on the sun.

For example, a system of radial, oblique to radial, and orthogonal accelerator tubes may be constructed at a radial coordinate between Earth's and Mars's orbits.

Another system might be constructed between the asteroid belt and Jupiter.

Yet other systems might be constructed between Jupiter and Saturn, Saturn and Uranus, Uranus and Neptune, and Neptune and Pluto.

The interplanetary sun-centered shells would rely on attainment of orbital velocity in order to reduce tidal stresses and to largely cancel centripetal-acceleration-based stresses.

For interplanetary subways, it is conceivable that relativistic velocities could be achieved for personal cars that would be accelerated by mass driver mechanisms. Linear induction braking could be used to slow the cars down. Such induction braking can be used to produce electrical energy, which would then be stored in batteries, flywheels, or capacitors for reuse. To the extent that the kinetic energy of a car could be completely reconverted to electrical energy, the energy could be repeatedly cycled. Thus travel about a given ring shell might effectively entail a virtual free-energy lunch.

Now, for cases where a much less massive body orbits a more massive body, the orbital velocity is defined as follows:

$$V_0 \approx [GM/r]^{1/2}$$

Where the orbit is circular.

Taking the mass of both orbiting bodies into account for a two-body system, the formula for orbiting velocity for the smaller body is as follows:

$$V_0 \approx [G(M_1 + M_2)/r]^{1/2}$$

Now, the orbital velocities of the planets around the sun are as follows:

47.4 km/s for Mercury

35.0 km/s for Venus

29.8 km/s for Earth

24.1 km/s for Mars

13.1 km/s for Jupiter

9.7 km/s Saturn

6.8 km/s for Uranus

5.4 km/s for Neptune

4.7 km/s for Pluto

So ring-worlds radially located between the planetary orbits would require revolutional velocities that can be easily interpolated as a first-order approximation upon mere inspection of the orbital velocities of the planets.

Rings with greater distances from the sun will require lower revolutional velocities.

However, tidal stresses exist for rings having a substantial radial thickness.

Now, tidal acceleration may be expressed as follows:

$$a_g = -\mathbf{r}\, G\, M/[(R \pm \Delta r)^2]$$

Extracting R^2, we obtain:

$$a_g = -r\, [GM/(R^2)]\{1/[1 \pm (\Delta r/R)]^2\}$$

Now, consider the Maclaurin series:

$$1/[(1 + X)^2] = 1 - 2X + 3X^2 - 4X^3 + 5X^4 - \ldots$$

So the series expansion for tidal acceleration is as follows:

$$a_g = -r\,[GM/(R^2)] \pm \{r\,[G(2M)/(R^2)](\Delta r/R)\} - r\,[G(3M)/(R^2)][(\Delta r/R)^2] \pm r\,[G(4M)/(R^2)][(\Delta r/R)^3] - r\,[G(5M)/(R^2)][(\Delta r/R)^4] \pm \ldots$$

Here, *R* is the distance between the center of the two bodies, *Δr* is the width of the body undergoing tidal stress from the larger mass (M), *G* is the universal gravitational constant, *-r* is the unit radial or tidal acceleration vector, which is parallel (and collinear) with *R*.

Now consider Newton's law: $\mathbf{F} = m\mathbf{a}$.

Thus, the tidal force pulling on the body will be approximately:

$$\mathbf{F}_{\mathbf{tidal}} = m\{-\mathbf{r}\, G\, M/[(R \pm \Delta r)^2]\} = m\{-r\,[GM/(R^2)]\{1/[1 \pm (\Delta r/R)]^2\}\} = m\{-r\,[GM/(R^2)] \pm \{r\,[G(2M)/(R^2)](\Delta r/R)\} - r\,[G(3M)/(R^2)][(\Delta r/R)^2] \pm r\,[G(4M)/(R^2)][(\Delta r/R)^3]$$

$$-r\,[G(5M)/(R^2)][(\Delta r/R)^4] \pm \ldots\}$$

We can surmise that ringlike interplanetary subways can have a radial width at least equal to that of the largest planets.

Since we are considering supermaterials as options for the construction of the subway systems, the radial extensity of the ring-words may be much greater than the width of Jupiter.

Let us consider a carbon nanotube truss member located 1 billion kilometers from the sun and in a stable orbit. Let us further consider that the member is radially oriented with respect to the center of the sun and which has a tensile strength of 100 billion newtons per square meter. We now consider the mass of the member commensurate with $\mathbf{F_{tidal}}$ being equal to 10^{11} N. We will assume $\Delta r = 10^{10}$ meters. We further consider the density of the carbon nanotube material is one metric ton per cubic meter.

So:

$$m = (10^{11}N) /\{- \mathbf{r}\ [6.67\times10^{-11}\ N\cdot(meter/kg)^2\ ][[1.989\times10^{30}]kg]/[[(10^{12}\ meters) \pm (10^{10}\ meters)]^2]\} = [7.6892\ x\ (10^{14})]\ kg = .\ [7.6892\ x\ (10^{11})]\ metric\ tons$$

Assuming a truss member with constant length specific mass, the material cross-sectional area of the member will be 76.892 m^2. The design factor is 76.892.

Let us consider a carbon nanotube truss member located 1 billion kilometers from the sun and in a stable orbit. Let us further consider that the member is radially oriented with respect to the center of the sun, and which has a tensile strength of 100 billion newtons per square meter. We now consider the mass of the member commensurate with $\mathbf{F_{tidal}}$ being equal to 10^{12} N. We will assume $\Delta r = 10^{11}$ meters. We further consider the density of the carbon nanotube material is one metric ton per cubic meter.

So:

$$m = (10^{12}N) /\{- \mathbf{r}\ [6.67\times10^{-11}\ N\cdot(meter/kg)^2\ ][[1.989\times10^{30}]kg]/[[(10^{12}\ meters) \pm (10^{11}\ meters)]^2]\} = [9.120628\ x\ (10^{15})]\ kg = .\ [9.120628\ x\ (10^{12})]\ metric\ tons$$

Assuming a truss member with constant length specific mass, the material cross-sectional area of the member will be 91.20628 m^2. The design factor is 91.20628.

Let us consider a carbon nanotube truss member located 10 billion kilometers from the sun and in a stable orbit. Let us further consider that the member is radially oriented with respect to the center of the sun and which has a tensile strength of

100 billion newtons per square meter. We now consider the mass of the member commensurate with $\mathbf{F_{tidal}}$ being equal to 10^{13} N. We will assume $\Delta r = 10^{11}$ meters. We further consider the density of the carbon nanotube material is 1 metric ton per cubic meter.

So:

m = (10^{13}N) /{- **r** [6.67×10^{-11} N·(meter/kg)^2][[1.989×10^{30}]kg]/[[(10^{13} meters) ± (10^{11} meters)]^2]} = [7.6892 x (10^{18})] kg = . [7.6892 x (10^{15})] metric tons

Assuming a truss member with constant length-specific mass, the material cross-sectional area of the member will be 768.92 m^2. The design factor is 768.92. Assuming a space-age alloy, or even ordinary steel with a density of 10 metric tons per cubic meter and which has 0.02 times the tensile strength of carbon nanotubes, we would have a design factor of 768.92/500 ≈ 1.5.

Let us consider a carbon nanotube truss member located 10 billion kilometers from the sun and in a stable orbit. Let us further consider that the member is radially oriented with respect to the center of the sun and which has a tensile strength of 100 billion newtons per square meter. We now consider the mass of the member commensurate with $\mathbf{F_{tidal}}$ being equal to 10^{13} N. We will assume $\Delta r = 10^{12}$ meters. We further consider the density of the carbon nanotube material is 1 metric ton per cubic meter.

So:

m = (10^{13}N) /{- **r** [6.67×10^{-11} N·(meter/kg)^2][[1.989×10^{30}]kg]/[[(10^{13} meters) ± (10^{12} meters)]^2]} = [9.120628 x (10^{18})] kg = . [9.120628 x (10^{15})] metric tons

Assuming a truss member with constant length-specific mass, the material cross-sectional area of the member will be 912.0628 m^2. The design factor is 912.0628. Assuming a space-age alloy, or even ordinary steel with a density of 10 metric tons per cubic meter and which has 0.02 times the tensile strength of carbon nanotubes, we would have a design factor of 912.0628/500 ≈ 1.8.

Let us consider a carbon nanotube truss member located 100 billion kilometers from the sun and in a stable orbit. Let us further consider that the member is radially oriented with respect to the center of the sun and which has a tensile strength of 100 billion newtons per square meter. We now consider the mass of the member commensurate with $\mathbf{F_{tidal}}$ being equal to 10^{13} N. We will assume $\Delta r = 10^{13}$

meters. We further consider the density of the carbon nanotube material is one metric ton per cubic meter.

So:

> m = (10^{13}N) /{- **r** [6.67×10^{-11} N·(meter/kg)^2][[1.989×10^{30}]kg]/[[(10^{14} meters) ± (10^{13} meters)]^2]} = [3.01508 x (10^{19})] kg = . [7.6892 x (10^{16})] metric tons

Assuming a truss member with constant length-specific mass, the material cross-sectional area of the member will be 7,689.2 m^2. The design factor is 7,689.2. Assuming a space-age alloy or even ordinary steel with a density of 10 metric tons per cubic meter and which has 0.02 times the tensile strength of carbon nanotubes, we would have a design factor of 7,689.2/500 ≈ 15.

We could build according to this latest scenario with sturdy concrete, and perhaps even compressed wood chemically treated to withstand the cryogenic temperatures of deep space without becoming too brittle.

Using a star having a mass five times the mass of the sun will result in *m* being five times greater, and all else remaining the same. In fact, since in the scenarios provided above where carbon nanotubes are used to construct the ring-world and we obtained an exemplar design factor of a minimum of 76.892, we may use any star having a mass in the range of 0.1 solar masses to 76.892 solar masses, all else being the same, and still not exceed the tensile strength of the ring-worlds.

Since graphene in its pure form is stronger than carbon nanotubes, we may be able to construct any of the above ring-worlds around any star including the heaviest and not exceed design limitations.

Taking Another Mathematical Approach

Now Earth travels once around the sun every Earth year. The distance of travel is about 2πR, where *R* is roughly 150,000,000,000 meters. Therefore, the orbital velocity of Earth is approximately (150,000,000,000 meters)(2π)/(31,000,000 seconds) = 30,400 meters per second = 30.4 km second.

The velocity vectors of Earth amid successive half-year intervals are antiparallel in orientation. Thus, the effective acceleration of Earth is approximately (4)(30,400 m/s)/(31,000,000 s) = 0.0039229 m/s^2, or 0.0003999 Earth Gs.

Thus, assuming such an acceleration, a cable made out of carbon nanotubes, graphene, carbon nitride fiber, diamond fiber, boron nitride nanotubes, carbon atom chains, and the like having a volumetric density of 1,000 kilograms per cubic meter which is 1 square centimeter in cross-section and 100 million kilometers long will have a tidal stress of (0.1 kg)(10^{11})(0.0039229 m/s^2) = 39,229,000 newtons, or 3,998,000 kilograms force.

In fact, the cable may be as long as roughly [(10,000,000)/(39,229,000)](100,000,000 km), or 25.49 million kilometers while undergoing an acceleration of 0.0039229 m/s^2 or 0.0003999 Earth Gs with the associated induced tidal stress without fatigue.

Thus, carbonaceous supermaterials-based cables can be used for transportation throughout the planetary solar system from orbital radii ranging from that of the planet Venus to locations well beyond that of Pluto because the centripetal accelerations associated with the orbital motions of the gas giants are radially progressively and strongly reduced with respect to that of Earth.

A large radially anchored ring-world infrastructure with a radial center of mass located about 1 billion kilometers from the sun and balanced by a radially external counterweight mechanism would have a circumference of about 6.28 billion kilometer. Assuming that each kilometer of the ring-world structure had a mass of 10 million metric tons, the entire ring would have a mass of 6.28 x 10^{16} metric tons.

Further assume that each kilometer arcuate element of the ring could house and support 100,000 persons simultaneously, the ring world could support 628 trillion persons.

More than one ring world may in theory concentrically disposed. Such ring-worlds may be radially connected by travel tubes in the form of electromagnetic mass drivers. Travel tubes may span the entire circumference of the ring-worlds as well. The mass drivers may optionally be configured similarly to modern-day Maglev trains. In one scenario, the tubes would pull the train or cars forward by electromagnets that would serially pull on ferromagnetic elements affixed to the cars. In another scenario, the cars would include permanent magnets to assist in the propulsion. In still another example, the cars would include electromagnetics for propulsion. For cases where the cars contain electromagnetics, the tubes may contain permanent magnets and/or iron or steel elements for which the cars would react against.

Assuming a comfortable 1-G of acceleration, the cars may obtain a velocity of [9.81 m/s^2](100,000 s) or 981,000 m/s in 100,000 seconds, which is about one day.

The distance traveled will be 49.05 billion meters or 49,050,000 kilometers in about one day. Increasing the acceleration to 10 Gs and the terminal velocity will be 9,810,000 m/s. The distance traveled in one day will be 490,500,000 kilometers. Increase the acceleration to 100 Gs, and the terminal velocity will be 98,100,000 m/s, or 98,100 kilometers per second in a Newtonian approximation. This is one-third of the velocity of light! The relativistic kinetic energy of the spacecraft will be about 0.0606 times that of its total invariant energy. Such a terminal velocity can be achieved for roughly one revolution around a ring-world having a radius of 1 billion kilometers, assuming roughly 100 Gs of tangential acceleration.

Assume ten complete revolutions around a solar system ring-world of the above radius where the spacecraft gains about 0.0606 times its total invariant energy in relativistic kinetic energy for each revolution. The spacecraft Lorentz factor will be about 1.606 with a velocity of 0.7827 C.

Assume 100 complete revolutions around the above solar system ring world where the spacecraft gains about 0.0606 times its total invariant energy in relativistic kinetic energy for each revolution. The spacecraft Lorentz factor will be about 1 + 6.06 or 7.06, with a velocity of 0.98994 C.

Assume 1,000 complete revolutions around the solar system ring world where the spacecraft gains about 0.0606 times its total invariant energy in relativistic kinetic energy for each revolution. The spacecraft Lorentz factor will be about 1 + 60.6 or 61.6, with a velocity of 0.999864 C.

Assume 10,000 complete revolutions around the solar system ring world where the spacecraft gains about 0.0606 times its total invariant energy in relativistic kinetic energy for each revolution. The spacecraft Lorentz factor will be about 1 + 606 or 607 with a velocity of 0.99999864 C.

Assume 100,000 complete revolutions around the solar system ring-world where the spacecraft gains about 0.0606 times its total invariant energy in relativistic kinetic energy for each revolution. The spacecraft Lorentz factor will be about 1 + 6,060 or 6,061 with a velocity of 0.99999864 C.

At a Lorentz factor of 100,000, the car occupants would experience a forward time travel of 100,000 years with respect to the stationary ring world frame. In 10 years, the occupants would experience a forward time travel of 1,000,000 years. In 100 years, the occupants would experience a forward time travel of 10 million years. Assuming human life expectancy can be augmented to over 1,000 years, human persons could experience alive and awake forward time travel of 100 million years.

Cryogenic sleep preservation and/or near freezing hibernation states can enable much further forward time travel for the car occupants.

For reduced radial acceleration and centripetal G-forces, ring-worlds having a radius of 10 billion kilometers, or perhaps even 100 billion kilometers, can be used. Accordingly, the energy gain per revolution could be much higher than that for the 1-billion-kilometer-radius ring-world.

Alternatively, once the cars reach the desired terminal Lorentz factors, the cars could be released into interstellar space. At a Lorentz factor of 100,000, the car occupants would experience a forward time travel of 100,000 years with respect to the stationary ring-world frame and travel 100,000 light-years through intergalactic space. In 10 years, the occupants would experience a forward time travel of 1,000,000 years and travel 1,000,000 light-years through intergalactic space. In 100 years, the occupants would experience a forward time travel of 10 million years and travel 10,000,000 light-years through intergalactic space. Assuming human life expectancy can be augmented to over 1,000 years, persons could experience alive and awake forward time travel of 100 million years and travel 100,000,000 light-years through intergalactic space. Cryogenic sleep preservation and/or near freezing hibernation states can enable much further forward time travel for the car occupants.

The ratio of spacecraft kinetic energy gain and its total invariant energy can vary from one ring-world to another, as well as within a given ring-world. Larger ring-worlds are more facilitative of high ratios as such than ring-worlds of smaller radii. This is true for three main reasons. First, the acceleration circumferential path per revolution is greater for larger rings, thus enabling longer acceleration paths. Secondly, the radial acceleration for a given Lorentz factor is less for cars traversing larger ring-worlds. Thirdly, the tangential acceleration for a very large ring-world can be significantly less than that for a much smaller ring world but yet enable the car traversing one revolution of the larger ring world to obtain a significantly higher Lorentz factor than that which is practical for one loop around the smaller ring-world.

We now consider a locally interstellar subway system.

An interstellar subway system as such may reasonably extend out to a distance of about 20 light-years from ol' Sol. The numbers of stars and brown dwarfs within 20 light-years of Earth is estimated to be about 135. Most of this mass within the local interstellar neighborhood exist as hydrogen and helium, and perhaps as much as 0.0001 of the mass exists as heavy metals. In astronomic literature, heavy metals are elements and isotopes heavier than helium atoms or nuclei.

Assume that the heavy metal composition is conducive for the formation of carbonaceous super-materials, super-alloys, rock, concretes and the like in the amount of 0.00001 of the total baryonic mass content within the local interstellar neighborhood. Since the free interstellar gas is generally somewhere around two to ten times more plentiful than the stellar matter in a given local region of interstellar space, we can expect that perhaps the useful heavy metal composition of the local interstellar neighborhood would be roughly equal to 0.00002 to 0.0001 times the combined mass embodied in the interstellar objects within the local interstellar neighborhood. Since the average star is about 1/5 of the mass of the sun, the total available construction material ranges from (0.00002)(0.2)(135) to (0.0001)(0.2)(135) solar masses, or 0.00054 to 0.0027 solar masses. This would be about [2 x (10^{27})](0.00054) to [2 x (10^{27})](0.0027) metric tons or [1.08 x (10^{24})] to [5.4 x (10^{24})] metric tons.

We now assume that each person needs 1,000 metric tons for life support. The total number of simulateously supportable persons will be [1.08 x (10^{21})] to [5.4 x (10^{21})]. Since human metabolism radiates about 80 watts, the above human population will produce a mere (80)[1.08 x (10^{21})] to (80)[5.4 x (10^{21})] watts of radiant power, or about [8.64 x (10^{22})] to [4.32 x (10^{23})] watts of radiative power. This is equivalent to that of a red dwarf star or 0.0002 to 0.001 times the solar output.

Since the total stellar output within the interstellar neighborhood or 20 light-years radial extension from the sun is several orders of magnitude greater than the human metabolic power for the above conjectured population, stellar power alone can plausibly provide life support for such a huge human population.

Even the CMBR light that passes through the local interstellar neighborhood can provide enough power for such life-support activities. In fact, since the effective temperature of starlight alone with respect to the surface of Earth is of the same order of magnitude as that of the incident CMBR, capturing either source of radiant energy would enable easy support of the above conjectured populations.

It is possible that a Dewar type of multihull system can capture human metabolic waste heat and other industrial thermal and other electromagnetic emission. The entire hab and tube structure of the local interstellar system may accordingly be enclosed in conductive multihull shells to capture almost all waste heat and magnetic emmisions from the subway tubes.

The travel tubes can include linear and curvilinear forms. For extreme spacecraft Lorentz factor capable cars, mass driver tubes can take any of the former topologies.

Now, for linear or curvilinear mass drivers, the accrued Lorentz factor for the cars is equal to:

$\gamma = \cosh(aT/c) = [1 + [(at/C)^2]]^{1/2} = [ad/(C^2)] + 1$

Thus, assuming a tangential acceleration of 1-G in the car frame and a curvilinear travel path of 100 light-years, or about 10^{18} meters, the accrued Lorentz factor will be $\gamma = \{[9.81\ m/s^2][10^{18}\ m]/\{[9 \times (10^{16})]\ m^2\ s^{2}\}\} + 1 = 110$.

Assuming a tangential of 1-G in the car frame, and a curvilinear travel path of 10,000 light-years, or about 10^{20} meters, the accrued Lorentz factor will be $\gamma = \{[9.81\ m/s^2][10^{20}\ m]/\{[9 \times (10^{16})]\ m^2\ s^{2}\}\} + 1 = 10{,}900$.

Assuming a tangential of 1-G in the car frame, and a curvilinear travel path of 1,000,000 light-years, or about 10^{22} meters, the accrued Lorentz factor will be $\gamma = \{[9.81\ m/s^2][10^{22}\ m]/\{[9 \times (10^{16})]\ m^2\ s^{2}\}\} + 1 = 1{,}090{,}000$.

Assuming a tangential of 1-G in the car frame, and a curvilinear travel path of 100,000,000 light-years or about 10^{24} meters, the accrued Lorentz factor will be $\gamma = \{[9.81\ m/s^2][10^{24}\ m]/\{[9 \times (10^{16})]\ m^2\ s^{2}\}\} + 1 = 109{,}000{,}000$.

Assuming a tangential acceleration of 10-Gs in the car frame and a curvilinear travel path of 100 light-years, or about 10^{18} meters, the accrued Lorentz factor will be $\gamma = \{[98.1\ m/s^2][10^{18}\ m]/\{[9 \times (10^{16})]\ m^2\ s^{2}\}\} + 1 = 1{,}091$.

Assuming a tangential of 10-Gs in the car frame, and a curvilinear travel path of 10,000 light-years, or about 10^{20} meters, the accrued Lorentz factor will be $\gamma = \{[98.1\ m/s^2][10^{20}\ m]/\{[9 \times (10^{16})]\ m^2\ s^{2}\}\} + 1 = 109{,}000$.

Assuming a tangential of 10-Gs in the car frame, and a curvilinear travel path of 1,000,000 light-years, or about 10^{22} meters, the accrued Lorentz factor will be $\gamma = \{[98.1\ m/s^2][10^{22}\ m]/\{[9 \times (10^{16})]\ m^2\ s^{2}\}\} + 1 = 10{,}900{,}000$.

Assuming a tangential of 10-Gs in the car frame, and a curvilinear travel path of 100,000,000 light-years, or about 10^{24} meters, the accrued Lorentz factor will be $\gamma = \{[98.1\ m/s^2][10^{24}\ m]/\{[9 \times (10^{16})]\ m^2\ s^{2}\}\} + 1 = 1{,}090{,}000{,}000$.

Assuming a tangential acceleration of 10-Gs in the car frame, and a curvilinear travel path of 100 light-years, or about 10^{18} meters, the accrued Lorentz factor will be $\gamma = \{[98.1\ m/s^2][10^{18}\ m]/\{[9 \times (10^{16})]\ m^2\ s^{2}\}\} + 1 = 1{,}091$.

Assuming a tangential of 100-Gs in the car frame, and a curvilinear travel path of 10,000 light-years, or about 10^{20} meters, the accrued Lorentz factor will be $\gamma = \{[981\ m/s^2][10^{20}\ m]/\{[9 \times (10^{16})]\ m^2\ s^{2}\}\} + 1 = 1{,}090{,}000$.

Assuming a tangential acceleration of 100-Gs in the car frame and a curvilinear travel path of 1,000,000 light-years, or about 10^{22} meters, the accrued Lorentz factor will be $\gamma = \{[981\ m/s^2][10^{22}\ m]/\{[9 \times (10^{16})]\ m^2\ s^{2}\}\} + 1 = 109{,}000{,}000$.

Assuming a tangential acceleration of 100-Gs in the car frame, and a curvilinear travel path of 100,000,000 light-years, or about 10^{24} meters, the accrued Lorentz factor will be $\gamma = \{[981\ m/s^2][10^{24}\ m]/\{[9 \times (10^{16})]\ m^2\ s^{2}\}\} + 1 = 10{,}900{,}000{,}000$.

Assume that an acceleration profile could be maintained such that the craft could tangentially accelerate at a constant 1 G, ship's reference frame over a path length equal to that of the circumference of the currently observable universe.

Substituting 10 m/(s^2) for g to simplify calculations, 3×10^8 m/s for C, and $[76 \times (10^9)]\ [3.1 \times (10^7)]$ seconds for t, we obtain:

$$T_o = \{[3 \times (10^8)]/(10)\}\ ln\ \{[3 \times (10^8)]\ \{[[[3 \times (10^8)]^2] + [[(10)[2.356 \times 10^{18}]]^2]^{1/2}] + [(10)[2.356 \times 10^{18}]]\}\ /[[3 \times (10^8)]^2]\}\ \text{seconds}$$

$$= [3 \times (10^7)]\ ln\ \{[3 \times (10^8)]\{\{\{[9 \times (10^{16})] + [[2.356 \times (10^{19})]^2]\}^{1/2}\} + [2.356 \times (10^{19})]\}/[9 \times (10^{16})]\}\ \text{seconds}$$

$$= \{[3 \times (10^7)]\ ln\ [1.57 \times (10^{11})]\}\ \text{seconds} = 773\ \text{million seconds} = 24.9\ \text{years}$$

The ship would have aged only 24.9 years' ship time for an overall averaged Lorentz factor or time dilation of 3.05 billion in 76×10^9 years background reference frame time. For slowing down within the tube, the spacecraft could use electrodynamic braking to cover 152 billion light-years' travel path- length in 49.8 years' ship time.

Now assume that a tangential acceleration profile could be maintained at a constant 0.1 G, ship's reference frame.

Substituting 1.0 m/(s^2) for g, 3×10^8 m/s for C, and $(76 \times 10^9)\ (3.1 \times 10^7$ seconds) for t, we obtain:

$$To = \{[3 \times (10^8)]/(1.0)\}\ ln\ \{[3 \times (10^8)]\ \{[[[3 \times (10^8)]^2] + [[(1.0)[2.356 \times 10^{18}]]^2]^{1/2}] + [(1.0)[2.356 \times 10^{18}]]\}\ /[[3 \times (10^8)]^2]\ \}\ \text{seconds}$$

$$= [3 \times (10^8)]\ ln\ \{[3 \times (10^8)]\{\{\{[9 \times (10^{16})] + [[2.356 \times (10^{18})]^2]\}^{1/2}\} + [2.356 \times (10^{18})]\}/[9 \times (10^{16})]\}\ \text{seconds} = [3 \times (10^8)]\ ln\ [1.57 \times (10^{10})]\ \text{seconds}$$

$$= 7.043\ \text{billion seconds} = 227.2\ \text{years}$$

The ship would have aged only 227.2 years' ship time for an overall averaged Lorentz factor or time dilation of 334.5 million in 76 x 10^9 years background reference frame time. For slowing down, the spacecraft could use electrodynamic braking to cover a 152 billion light-years' travel path-length in 454.4 years' ship time.

Assume that a tangential acceleration profile could be maintained at a constant 10 G, car's reference frame.

We substitute 100 m/(second^2) for *g*, 3 x 10^8 m/s for C, and [76 x (10^9)] [3.1 x (10^7)] seconds for *t*, and obtain:

$$To = \{[3 \times (10^8)]/(100)\}\ ln\ \{[3 \times (10^8)]\ \{[[[3 \times (10^8)]^2] + [[(100)[2.356 \times 10^{18}]]^2]^{1/2}] + [(100)[2.356 \times 10^{18}]]\}\ /[[3 \times (10^8)]^2]\}\ \text{seconds}$$

$$= [3 \times (10^6)]\ ln\ \{[3 \times (10^8)]\{\{\{[9 \times (10^{16})] + [[2.356 \times (10^{20})]^2]\}^{1/2}\} + [2.356 \times (10^{20})]\}/[9 \times (10^{16})]\}\text{seconds} = \{[3 \times (10^6)]\ ln\ [1.57 \times (10^{12})]\}\ \text{seconds}$$

$$= 84.25 \text{ million seconds} = 2.7176 \text{ years}$$

In 76 x 10^9 years' background reference frame time, the car would have aged only 2.7176 years car time for an overall averaged Lorentz factor or time dilation of 27.97 billion. For deceleration, the car could use electrodynamic braking to cover a 152 billion light-years' travel path-length in 5.435 years' ship time.

The travel velocity of the cars at high mass-driver Lorentz factors will induce very large centripetal forces within the contents of the cars. Thus, a means to reduce the effective G forces for the centripetal motions in circular mass-drivers is required.

Among these is the enclosure of crew members' bodies in hydrostatically sealed breathable oxygenated liquid-containing vessels. Alternatively, perhaps nanotechnology types of pressure suits could completely encase the crew members' bodies. The pressure suits might optionally pump high pressure air into the lungs of the crew members wearing them and gradually relax the lung pressure as the rate of acceleration was reduced to more manageable levels. The crew members may be enclosed in smart fabric types of whole body pressure suits such as might optionally be constructed from rheo-elastic or other electro-elastic materials.

Another possibility includes placing the crew and any passengers in either a cryogenic form of sleep. The bulk modulus of the sleeping human bodies can approach that of hard steel for freezing temperatures near absolute zero.

Perhaps some sort of G-force cancellation techniques can be used such as an instilled electrical charge within the car and where a reactive electrical field emitted within the car or externally would cancel out the forces due to centripetal acceleration. The charge created within human bodies can be deposited by nanotechnology mechanisms for precise control.

Alternatively, a magnetic-field-generation mechanism within the cars or external to the cars can induce a dipole moment within the passengers' bodies thus magnetizing the bodies. The magnetic field can be adjusted in order to effect nearly complete G-force cancellation. Alternatively, a gravity field generator or antigravity field generator can be used to cancel out the centripetal acceleration experienced by car passengers. Furthermore, the cars may optionally utilize two or more of the above centripetal force cancellation mechanisms simultaneously.

Now, how on Earth is a 1-metric-ton car traveling a 20 light-year-radius loop going to be contained within a circular accelerator without rupturing the accelerator via centripetally induced tensile stress! The simple answer may involve none other than 10-kilometer-thick wall tubes made of solid carbonaceous supermaterials. Such tubes when square in cross-section would have a tensile strength of $(10^7)(10^{12})(4)$ newtons = $[4 \times 10^{19}]$ newtons. A 1-kilometer long section would have a mass of 400 billion metric tons. A 120 light-year long circular track would have a mass of $(120)(10^{13})[4 \times 10^{11}]$ metric tons = $[4.8 \times 10^{26}]$ metric tons.

Now consider the formula:

$$\gamma = \cosh (aT/c) = [1 + [(at/C)^2]]^{1/2} = [ad/(C^2)] + 1 \approx 30$$

Where we replace d with 3×10^{17} meters, or 30 light-years. We assume an acceleration of $1\ G = 9.81\ m/s^2$.

Thus,

$$[ad/(C^2)] + 1 = \{[9.81\ m/s^2]\ [3 \times 10^{17}\ m] / \{[9 \times 10^{16}]\ m^2/s^2\}\} + 1 = 33.7 \approx 30$$

Since one complete circuit involves the car going from a Lorentz factor of 33.7 in one direction to a Lorentz factor of 33.7 in an antiparallel direction and then back to a Lorentz factor of 33.7 in the former direction, we assume that the centripetal acceleration of the spacecraft would likewise impose a G force of 1 G for a spacecraft revolving in the tube at a constant Lorentz factor of 33.7. However, a spacecraft orbiting at a Lorentz factor of 33.7 will have a total relativistic mass with respect to the tubular stationary frame of 33.7 M_{rest}. Therefore, the tube will need to resist a centripetal force of 33.7 metric tons where the car has an invariant mass of one metric ton for which the revolutional Lorentz factor of the car will be 33.7.

Now consider the formula:

$\gamma = \cosh (aT/c) = [1 + [(at/C)^2]]^{1/2} = [ad/(C^2)] + 1$

Where we replace *d* with 3 x 10^{17} meters, or 30 light-years, and assume an acceleration of 10,900,000 Gs = [(10,900,000) (9.81)] m/s^2 = 1.06929 x 10^8 m/s^2.

Therefore:

$$\gamma = \cosh (aT/c) = [1 + [(at/C)^2]]^{1/2} = [ad/(C^2)] + 1 = \{\{[1.06929 \times 10^8] m/s^2\} [3 \times 10^{17} m] / \{[9 \times 10^{16}] m^2/s^2\}\} + 1$$

$$= 3.5643 \times 10^8$$

Consider one complete circuit involves the car going from a Lorentz factor of 3.5643 x 10^8 in one direction to a Lorentz factor of 3.5643 x 10^8 in an antiparallel direction, and then back to a Lorentz factor of 3.5643 x 10^8 in the former direction. We assume that the centripetal acceleration of the spacecraft would likewise impose an acceleration force of 10,900,000 Gs on the tube for a car revolving in the tube at a constant Lorentz factor of 3.5643 x 10^8 under the condition where there would be no relativistic mass increase. However, a spacecraft orbiting at a Lorentz factor of 3.5643 x 10^8 will have a total relativistic mass with respect to the tubular stationary frame of 3.5643 x 10^8 M_{rest}. Therefore, the tube will need to resist a centripetal force of [3.5643 x 10^8] [10,900,000] metric tons or [3.885 x 10^{15}] metric tons where the car has an invariant mass of one metric ton for which the revolutional Lorentz factor of the car will be 3.5643 x 10^8. This is equal to 3.811 x 10^{19} newtons. Therefore, the loop should handle the load with a small safety margin.

Now consider again the formula:

$\gamma = \cosh (aT/c) = [1 + [(at/C)^2]]^{1/2} = [ad/(C^2)] + 1$

Where we replace *d* with 3 x 10^{17} meters, or 30 light-years and assume an acceleration of 10,900,000,000 Gs = [(10,900,000,000) (9.81)] m/s^2 = 1.06929 x 10^{11}.

Therefore:

$$\gamma = \cosh (aT/c) = [1 + [(at/C)^2]]^{1/2} = [ad/(C^2)] + 1 = \{\{[1.06929 \times 10^{11}] m/s^2\} [3 \times 10^{17} m] / \{[9 \times 10^{16}] m^2/s^2\}\} + 1 = 3.5643 \times 10^{11}$$

Since one complete circuit involves the car going from a Lorentz factor of 3.5643×10^{11} in one direction to a Lorentz factor of 3.5643×10^{11} in an antiparallel direction and then back to a Lorentz factor of 3.5643×10^{11} in the former direction, we assume that the centripetal acceleration of the spacecraft would likewise impose an acceleration force of 10,900,000,000 Gs on the tube for a car revolving in the tube at a constant Lorentz factor of 3.5643×10^{11} under the condition where there would be no relativistic mass increase. However, a spacecraft orbiting at a Lorentz factor of 3.5643×10^{11} will have a total relativistic mass with respect to the tubular stationary frame of 3.5643×10^{11} M_{rest}. Therefore, the tube will need to resist a centripetal force of [3.5643×10^{11}] [10,900,000,000] metric tons or [3.885×10^{21}] metric tons where the car has an invariant mass of one metric ton for which the revolutional Lorentz factor of the car will be 33.7. This is not a trivial force at 3.8112×10^{25} newtons. So we need something stronger than carbonaceous super materials.

Neutronium to the rescue!

We now assume tube construction out of extreme materials such as neutroniums, quarkoniums, higgsiniums, and the like super-nuclear density materials.

Now the force of attraction between nucleons is about 10,000 newtons. Therefore, the tensile strength of neutronium would be approximately {(10,000 newtons)[$(10^{15})^2$] = 10^{34} newtons per square meter. Since carbonaceous super materials have a tensile strength that is roughly 10^{11} newtons per square meter, but which have a density that is only 10^{-15} times that of neutronium, neutronium tubes have a strength to invariant mass ratio that is 100 million times that of carbonaceous super-materials. Consequently, neutronium subway tubes may have a mass that is 100 million times less than carbonaceous supermaterials having the same tensile strength.

Nature already has precedence for neutronium in the context of neutron stars. However, the neutronium in such stars is continuously being regenerated under the enormous gravitationally self-induced pressures within neutron stars.

Large assemblages of neutrons and protons have been proven to exist in the form of the atomic nuclei of heavy periodic table elements.

The yield strength of protons and neutrons is about one hundred times greater than the force of attraction among these nucleons within the composition of the atomic nucleus. The latter force of attraction is about 10,000 newtons. Therefore, the yield strength of protons and neutrons is about 1,000,000 newtons. A differential cross-sectional area of a column of up-down quarkonium where the column is $1{,}000^{1/2}$

neutrons by $1,000^{1/2}$ neutrons wide is about $(3)[1,000^{1/2}]$ up and/or down quarks wide by $(3)[1,000^{1/2}]$ up and/or down quarks thick. The tensile strength of typical up-down quarkonium will thereby be about $\{\{[(3)^{1/2}]\}^2\}(100)$ times greater than that of neutronium or on the order of 1,000 times greater than that of neutronium. There are likely a large number of possible up-down quarkonium types, considering the range of plausible crystalline patterns, quarkonium excited or isomer states and the like. Thus, we may have some freedom in the designed strength of specific quarkoniums by up to 3, or perhaps even 4, orders of magnitude. Note the analogous cases of comparison for the element carbon, which includes soft pencil graphite or activated charcoal or the much stronger forms of carbonaceous supermaterials. Here, the yield strength of the materials spans a range of several orders of magnitude.

Non-hadronically differentiated up-down quarkonium having the same density as neutronium may in theory have a strength to weight ratio that is 10 billion times that of carbonaceous supermaterials.

Quarkoniums in the form of quark nuggets left over as relics from the early stages of the Big Bang are plausible and have been the subject of serious theoretical studies. The quark nuggets would theoretically be on the order of one to two meters in size and have a mass of roughly that of Earth's moon.

Some theories of stellar evolution involving intermediate stages between neutron stars and black holes have posited the existence of quark stars. Either way, many theoretical considerations regarding the interior of neutron stars suggest the existence of a quarkonium core within at least some neutron stars.

The ratio $[3.8112 \times 10^{25}]/ [3.8112 \times 10^{19}]$ is 1,000,000. Therefore, about $\{(10^{-8})(10^{6}) [4.8 \times 10^{26}]/[2 \times 10^{27}]\}$ solar masses, or 0.0024 solar masses of neutronium, should suffice.

For non-hadronically differentiated up-downium having a mass-specific yield of that of the proton, only 0.000024 solar masses of neutronium would suffice.

Using 0.24 solar masses of neutronium to construct the ring should enable 1 metric ton car Lorentz factors of $(10)[3.5643 \times 10^{11}] = 3.5643 \times 10^{12}$. Using 0.24 solar masses of non-hadronically differentiated up-downium should enable Lorentz factors of 3.5643×10^{13}.

Now, here comes the zinger.

Consider communications lines set up between the circular tube and thus the extreme Lorentz factor cars and a repeater located near or at the very axial center of the tube ring, in other words, no further than 10 light-years from the car at any time. Further consider that an identical system is eventually assembled at a distance of 1 billion light-years from the first habitat, and indeed, yet further on, where quintillions of identical systems are assembled in a spherical patterns having a radius of 1 billion light-years but where some such systems shadow in part systems located closely behind the former systems.

Further consider scenarios where the communications from each car of a first constructed ring system centered near the Sun can be uploaded in micro-second long pulses or shorter pulses, which may be digitally modulated, amplitude modulated, frequency modulated, polarity modulated, and/or super-chirality modulated and then which are broadcast to each of the spherically disposed sub-way worlds located a distance of 1 billion light-years from the sun.

Assume each car of the sun-centered ring-world is time dilated by a factor of 3.5643×10^{11} with respect to the ring and consequently with respect to the ring's central repeater. Further consider that each of the 1 billion light-year distant spherically disposed identical systems have revolving cars that are also traveling at a Lorentz factor of 3.5643×10^{11} with respect to its own host ring. The two-way communications time experienced by such car passengers sending out messages to the sun-centered system would be a mere 0.005611 years, or about 2.04949 days in the car's frame.

In fact, the longest complete signal time between locations along the 1-billion-light-year-radius sphere that are diametrically opposed to each other would be only 4.09898 days in the car's frame.

For a similar sphere with a central ring hab-centered around the Sun, but which has a radius of 10 billion light-years, and which contains on the order of 10^{21} sub-way hab-tube systems identical to the ones proposed previously, the two way radial communication time will be a mere 20.4949 days car time. For diametrically opposed ring-worlds, the two-way communication will be a mere 40.9898 days or 1.36 months car time.

Now, assume each car of the Sun-centered ring world is time dilated by a factor of 3.5643×10^{12} with respect to the ring and consequently with respect to the ring's central repeater. Further consider that each of the 1 billion light-year distant spherically disposed identical systems have revolving cars that are also traveling at a Lorentz factor of 3.5643×10^{12} with respect to its own host ring. The two-way communications time experienced by such car passengers sending out messages to the sun-centered system would be a mere 0.0005611 years, or about 0.204949 days in the car's frame car time.

In fact, the longest complete signal time between locations along the 1-billion-light-year-radius sphere that are diametrically opposed to each other would be only 0.409898 days car time in the car's frame.

For a similar sphere with a central ring hab-centered around the Sun, but which has a radius of 10 billion light-years, and which contains on the order of 10^{21} sub-way hab-tube systems identical to the ones proposed previously, the two way radial communication time will be a mere 2.04949 days car time. For diametrically opposed ring-worlds, the two-way communication will be a mere 4.09898 days car time.

Assume 1-billion-light-year-radius rings so-constructed that support one metric ton cars revolving at a Lorentz factor of 3.5643 x 10^{13} with respect to the rings and its central transceiving devices. The two-way communications time experienced by such car passengers sending out messages to the Sun-centered system would be a mere 0.00005611 years or about 0.49187 hours in the car's frame.

In fact, the longest complete signal time between locations along the 1 billion light year radius sphere which are diametrically opposed to each other would be only 0.983755 hours' car time in the car's frame.

For a similar sphere with a central ring hab-centered around the sun but which has a radius of 10 billion light-years, and which contains on the order of 10^{21} subway hab-tube systems identical to the ones proposed previously, the two-way radial communication time will be a mere 0.204949 days' car time. For diametrically opposed ring-worlds, the two-way communication will be a mere 0.409898 hours' car time.

My friend Adam Crowl has verified the correctness of the above reasoning in principle but kindly noted that the carbonaceous super-material rings capable of supporting cars having a Lorentz factor of 3.5643 x 10^{8} would have a mass beyond the Jeans limit and thereby collapse under its own weight into a disposition of sausage-like links, which would then collapse into a collection of spheres, which would in turn collapse into a sphere.

What if we could set the ring rotating at a sufficient Keplerian velocity so that it does not self-collapse? This rotational motion would not be enough to significantly alter the Lorentz factor capacities of the cars nor the communication up-load and down-load time but it could be employed to increase the effective Jeans limit of the ring.

Alternatively, repulsive laser like magnetic fields or electric field flux patterns concentrated radially toward paired portions of the ring-worlds may provide the necessary force to overcome tendencies for ring collapse. The ring may be suitably magnetically and/or electrically charged for the associated self-repulsion.

The neutronium rings would likely be self-supporting even while non-rotating, and the neutronium density, non-hadronically differentiated up-downium, would certainly be stable against collapse even for non-rotating systems.

Assuming a neutronium ring having a mass of 2,400 solar masses and the ring could support a one metric ton car revolving at a Lorentz factor of 3.5643×10^{14}. For non-hadronically differentiated nuclear density up-downium, the ring could support a car revolving at gamma = 3.5643×10^{15}!

Considering once again the above 1-billion-light-year-radius format but this time for the neutronium 2,400 solar mass ring case except where the number of ring-word tube systems is reduced by a factor of 10,000 so as to avoid gravitational collapse, the two-way communication between the cars and the hab-ring-world centered near the sun would be a mere $[2 \times 10^9]/ [3.5643 \times 10^{14}]$ years or 2.9512 minutes car time. The communication time between diametrically opposed locations on the sphere would be a mere 5.9025 minutes car time. For the 10-billion-light-year-radius ring, the two respective communications times would be a mere 29.512 minutes and 59.025 minutes, car time, respectively.

Considering once again the above 1-billion-light-year-radius format but this time for the up-downium 2,400 solar mass ring case, the two-way communication between the cars and the hab-ring-world centered near the sun would be a mere $[2 \times 10^9]/ [3.5643 \times 10^{15}]$ years or 17.394 seconds. The communication time between diametrically opposed locations on the sphere would be a mere 34.788 seconds car time. For the 10-billion-light-year-radius ring, the two respective communications times would be a mere 173.94 seconds' and 347.88 seconds' car time, respectively.

Should the big bang's expansion rate slow to a relative distance specific crawl over the next trillion years of so, within say about 2 trillion years from now, perhaps a little longer, we should be able to construct 1-trillion-light-year-radius sphere analogues consisting of $[10^{21}][10^{-4}][10^3]$ 2,400 solar mass rings. This is a whopping 10^{20} ring worlds. For neutronium ring-world tubular systems, the two-way communication time between a spherically disposed hab and the sun-centered hab will still be very short at a mere 4.8316 hours' car time. Two-way communications entirely across the sphere will take place in 9.6633 days car time. For the up-downium cases, the two-way communication time between a spherically disposed hab and the Sun-centered hab will still be very short at a mere

0.48316 hours car time. Two-way communications entirely across the sphere will take place in only 0.96633 days' car time.

We cannot permit too great of mass concentration in the construction of the sphere assemblage; otherwise, we will reach the Jeans limit analogue to gravitating masses disposed over the associated cosmic distance scales. Black-hole radius grows in proportion to black-hole mass, and so we would be wise to limit the number of such 2,400 solar mass ring-worlds in constructed spheres to a number that scales with the diameter of the sphere.

Consider the prospects of extending a subway system to the scale of the Milky Way galaxy. The subway tubes may be distributed to coincide with the rotation of the spiral arms of the Milky Way. Additionally, one or more rotating rings of roughly the same radius of the Milky Way may be concentrically disposed. Other rings may be located at fractional values of the radius of the galaxy.

Since the average star is about 1/5 of the mass of the sun, the total available construction material ranges from (0.00002)(0.2)[4 x 10^{11}] to (0.0001)(0.2)[4 x 10^{11}] solar masses or 1,600,000 to 8,000,000 solar masses. This would be about [2 x (10^{27})](1,600,000) to [2 x (10^{27})](8,000,000) metric tons, or [3.2 x (10^{33})] to [1.6 x (10^{34})] metric tons. Assuming that a single human person only requires 1,000 metric tons of mass for life support, [3.2 x (10^{30})] to [1.6 x (10^{31})] human persons may be simultaneously supported by the Milky Way. Assuming that most stars are red dwarfs with a lifetime on the order of $10^{12.5}$ years, in cases where the human life expectancy can be increased to 1,000 years, the Milky Way alone could nurture 1.0119 x 10^{43} to 5.0596 x 10^{43} persons. This can be considered an ensemble of souls. This huge number is roughly equal to the number of atoms in 50 cubic kilometers of liquid water at STP conditions.

Assume circulinear travel tube circumference equal to the 76,000 light-years or about the 0.000001 times the circumference of the observable universe. Assume that an acceleration profile could be maintained such that the craft could accelerate at a constant 1 G, ship's reference frame so that the craft can make 1 million revolutions around the ring.

Substituting 10 m/(s^2) for g for computational simplicity, 3 x 10^8 m/s for C, and [76 x (10^9)] [3.1 x (10^7)] seconds for t, we obtain:

$$T_o = \{[3 \text{ x } (10^8)]/(10)\}\ ln\ \{[3 \text{ x } (10^8)]\ \{[[[3 \text{ x } (10^8)]^2] + [[(10)[2.356 \text{ x } 10^{18}]]^2]^{1/2}] + [(10)[2.356 \text{ x } 10^{18}]]\}\ /[[3 \text{ x } (10^8)]^2]\} \text{ seconds}$$

$= [3 \times (10^7)]\ ln\ \{[3 \times (10^8)]\{\{\{[9 \times (10^{16})] + [[2.356 \times (10^{19})]^2]\}^{1/2}\} + [2.356 \times (10^{19})]\}/[9 \times (10^{16})]\}$ seconds

$= \{[3 \times (10^7)]\ ln\ [1.57 \times (10^{11})]\}$ seconds = 773 million seconds = 24.9 years

The car would have aged only 24.9 years' ship time for an overall averaged Lorentz factor or time dilation of 3.05 billion in only 76 x 10^9 years' background reference frame time. For slowing down, the car could use electrodynamic braking such as reverse mass-driver modes to travel a total path length of 152 billion light-years' travel distance in 49.8 years' ship time. Now assume that an acceleration profile could be maintained at a constant 0.1 G, ship's reference frame.

Substituting 1.0 m/(s^2) for *g* for computational simplicity, 3 x 10^8 m/s for C, and (76 x 10^9) (3.1 x 10^7 seconds) for *t*, we obtain:

$To = \{[3 \times (10^8)]/(1.0)\}\ ln\ \{[3 \times (10^8)]\ \{[[[3 \times (10^8)]^2] + [[(1.0)[2.356 \times 10^{18}]]^2]^{1/2}] + [(1.0)[2.356 \times 10^{18}]]\}\ /[[3 \times (10^8)]^2]\ \}$ seconds

$= [3 \times (10^8)]\ ln\ \{[3 \times (10^8)]\{\{\{[9 \times (10^{16})] + [[2.356 \times (10^{18})]^2]\}^{1/2}\} + [2.356 \times (10^{18})]\}/[9 \times (10^{16})]\}$ seconds $= [3 \times (10^8)]\ ln\ [1.57 \times (10^{10})]$ seconds

= 7.043 billion seconds = 227.2 years

The car would have aged only 227.2 years' car time for an overall averaged Lorentz factor or time dilation of 334.5 million in 76 x 10^9 years background reference frame time. For slowing down, the car could use electrodynamic braking such as reverse mass-driver modes to cover a total path length of 152 billion light-years travel distance in 454.4 years' ship time. Assume that an acceleration profile could be maintained at a constant 10 G, ship's reference frame.

We substitute 100 m/($second^2$) for *g* for computational simplicity, 3 x 10^8 m/s for C, and [76 x (10^9)] [3.1 x (10^7)] seconds for *t*, and obtain:

$To = \{[3 \times (10^8)]/(100)\}\ ln\ \{[3 \times (10^8)]\ \{[[[3 \times (10^8)]^2] + [[(100)[2.356 \times 10^{18}]]^2]^{1/2}] + [(100)[2.356 \times 10^{18}]]\}\ /[[3 \times (10^8)]^2]\}$ seconds

$= [3 \times (10^6)]\ ln\ \{[3 \times (10^8)]\{\{\{[9 \times (10^{16})] + [[2.356 \times (10^{20})]^2]\}^{1/2}\} + [2.356 \times (10^{20})]\}/[9 \times (10^{16})]\}$seconds $= \{[3 \times (10^6)]\ ln\ [1.57 \times (10^{12})]\}$ seconds

= 84.25 million seconds = 2.7176 years

In 76×10^9 years' background reference frame time, the car would have aged only 2.7176 years' ship time for an overall averaged Lorentz factor or time dilation of 27.97 billion. For deceleration, the spacecraft could use electrodynamic braking such as reverse mass-driver modes, to cover a total of 152 billion light-years' travel distance 5.435 years' ship time. Were the car traveling as a free spacecraft in a straight path, the actual distance of travel from the Milky Way galaxy would be considerably greater yet by roughly 2 orders of magnitude. The recessional velocity of the spacecraft would be roughly 2 orders of magnitude greater than C due to the expansion of space-time, assuming that the expansion of the universe continues, on average, at its current rate over the next 152 billion years.

Now yet again consider the formula:

$\gamma = \cosh(aT/c) = [1 + [(at/C)^2]]^{1/2} = [ad/(C^2)] + 1$

Where we replace d with 1.9×10^{26} meters, or 19,000,000,000 light-years, and assume an tangential acceleration of $1G = 9.81\ m/s^2$.

Therefore:

$$\gamma = \cosh(aT/c) = [1 + [(at/C)^2]]^{1/2} = [ad/(C^2)] + 1 = \{[9.81\ m/s^2][1.9 \times 10^{26}\ m] / \{[9 \times 10^{16}]\ m^2/s^2\}\} + 1$$

$$= 2.071 \times 10^{10}$$

The actual radial centripetal force exerted by the 1 metric ton car were it not for special relativistic mass increase would be (10^6) metric tons force. However, the car is traveling at a Lorentz factor of with respect to the travel tube reference frame, and so the centripetal force acting on the ring due to the car will be $(10^6)[2.071 \times 10^{10}](9{,}810)$ newtons, or 2.03165×10^{20} newtons.

A track made of 45.0738-kilometer-thick wall tubes constructed of solid carbonaceous super-materials should provide bursting resistance. Such tubes when square in cross-section would have a tensile strength of $(10^7)(10^{10})(45.0738^2)$ newtons = $[2.03165 \times 10^{20}]$ newtons. A 1-kilometer-long section would have a mass of 2.03165×10^{12} metric tons. A 76,000 light-year long circular track would have a mass of $(76{,}000)(10^{13})[2.03165 \times 10^{12}]$ metric tons = $[1.54405 \times 10^{30}]$ metric tons, or about 772 solar masses.

However, we would prefer that perhaps a population of 100 million cars could revolve around the tube simultaneously so as to transport humans from our cosmic era well into the future. Such humans may remain alive and awake during the transport process. Therefore, travel tubes made of neutronium but having the same mass would suffice. In order to support 10 billion cars, non-hadronically differentiated up-downium may suffice.

Alternatively, since one complete circuit involves the car going from a Lorentz factor of 2.071×10^{10} in one direction to a Lorentz factor of 2.071×10^{10} in an antiparallel direction and then back to a Lorentz factor of 2.071×10^{10} in the former direction, we assume that the centripetal acceleration of the spacecraft would likewise impose an acceleration force of 1,000,000 Gs on the tube for a car revolving in the tube at a constant Lorentz factor of 2.071×10^{10} under the condition where there would be no relativistic mass increase. However, a spacecraft orbiting at a Lorentz factor of 2.071×10^{10} will have a total relativistic mass with respect to the tubular stationary frame of 2.071×10^{10} M_{rest}. Therefore, the tube will need to resist a centripetal force of $[2.071 \times 10^{10}]$ [1,000,000] metric tons or $[2.071 \times 10^{16}]$ metric tons where the car has an invariant mass of one metric ton for which the revolutional Lorentz factor of the car will be 2.071×10^{10}. This is not a trivial force at 2.03165×10^{20} newtons. So we need something stronger than carbonaceous super materials for the case where the car's Lorentz factor is 10,000 times greater.

Once again consider the formula:

$$\gamma = \cosh(aT/c) = [1 + [(at/C)^{2}]]^{1/2} = [ad/(C^{2})] + 1$$

Where we replace *d* with 1.9×10^{26} meters, or 19,000,000,000 light-years, and assume an tangential acceleration of 10,000 Gs = 98,100 m/s^2.

Therefore:

$$\gamma = \cosh(aT/c) = [1 + [(at/C)^{2}]]^{1/2} = [ad/(C^{2})] + 1 = \{[98{,}100\ m/s^{2}][1.9 \times 10^{26}\ m] / \{[9 \times 10^{16}]\ m^{2}/s^{2}\}\} + 1$$

$$= 2.071 \times 10^{14}$$

Thus, the radial centripetal force exerted by the 1-metric-ton car on the 76,000 light-year-circumference loop were it not for special relativistic mass increase is 10^{10} metric tons force. However, the car is traveling at a Lorentz factor of 2.071×10^{14} with respect to the travel tube reference frame, and so the centripetal force acting on the ring due to the car will be $(10^{10})[2.071 \times 10^{14}](9{,}810)$ newtons, or 2.03165×10^{28} newtons. This is 100 million times the centripetal force exerted on the previous carbonaceous supermaterial system, which is logical, given the 100-millionfold mass-specific strength of neutronium over that of carbonaceous supermaterials. For the case of the up-downium, the maximum supportable Lorentz factor is 2.071×10^{15}, or ten times greater than in the latter neutronium case.

Now consider neutronium tubes having a mass of 772,000,000 solar masses and a circumference of 76,000 light-years. These tubes would support a 1-metric-ton car revolving at a Lorentz factor of 2.071×10^{17}. Tubes constructed of the up-downium having the same mass and radius would support a Lorentz factor of 2.071×10^{18}. Thus, the radial centripetal force exerted by the 1 metric ton car on the 76,000 light-year circumference loop were it not for special relativistic mass increase is 10^{13} metric tons force.

One-billion-light-year-radius spherical dispositions of 21.4 billion such neutronium galactic ring-worlds centered around an identical Milky Way ring-world, could send and receive two-way communications to the Milky Way ring-world in only $\{[2 \times 10^9]/[2.071 \times 10^{17}]\}[31{,}000{,}000]$ seconds, or 0.2993 seconds. Two-way communications diametrically spanning the sphere could be completed in only 0.5987 seconds. For an otherwise-identical up-downium sphere, the respective communication times are 0.02993 seconds and 0.05987 seconds.

One-trillion-light-year-radius spherical dispositions of 21.4 trillion such neutronium galactic ring-worlds centered around an identical Milky Way ring-world, could send and receive two-way communications to the Milky Way ring-world in only $\{[2 \times 10^{12}]/[2.071 \times 10^{17}]\}[31{,}000{,}000]$ seconds, or 299.3 seconds. Two-way communications diametrically spanning the sphere could be completed in only 598.7 seconds. For an otherwise-identical up-downium sphere, the respective communication times are 29.93 seconds and 59.87 seconds.

One-quadrillion-light-year-radius spherical dispositions of 21.4 quadrillion such neutronium galactic ring-worlds centered around an identical Milky Way ring-world, could send and receive two-way communications to the Milky Way ring-world in only $\{[2 \times 10^{15}]/[2.071 \times 10^{17}]\}[31{,}000{,}000]$ seconds, or 3.464 days. Two-way communications diametrically spanning the sphere could be completed in only 6.929 days. For an otherwise-identical up-downium sphere, the respective communication times are 0.3464 days and 0.6929 days.

Such 76,000-light-year-radius ring-worlds could each support rotating habitats traveling at a Lorentz factor on the order of millions to billions while still housing many, many trillions of forward-time-traveling citizens as ambassadors to our cosmic future descendants.

Still once again, consider the formula:

$$\gamma = \cosh(aT/c) = [1 + [(at/C)^2]]^{1/2} = [ad/(C^2)] + 1$$

Except this time, we replace *d* with 1.9×10^{24} meters, or 190,000,000 light-years, and assume a Lorentz factor of 2.071×10^{17}.

Therefore:

$\gamma = \cosh (aT/c) = [1 + [(at/C)^{2}]]^{1/2} = [ad/(C^2)] + 1 = \{(a)[1.9 \times 10^{24} m] / \{[9 \times 10^{16}] m^2/s^2\}\} + 1$

$= 2.071 \times 10^{17}$

Therefore, by solving for *a*, we obtain an acceleration equal to 10^9 Gs.

The radial centripetal force exerted by the 1-metric- ton car were it not for special relativistic mass increase is 10^9 metric tons. However, the car is traveling at a Lorentz factor of 2.071×10^{17} with respect to the travel tube reference frame and so the centripetal force acting on the ring due to the car will be $(10^9)[2.071 \times 10^{17}](9,810)$ newtons, or 2.03165×10^{30} newtons. Now consider neutronium tubes having a mass of 772,000,000 solar masses and a circumference of 760,000,000 light-years. These tubes would support a one metric ton car revolving at a Lorentz factor of 2.071×10^{17}. Tubes constructed of the up-downium having the same mass and radius would support a Lorentz factor of 2.071×10^{18}.

Now consider neutronium tubes having a mass of 772,000,000,000,000 solar masses and a circumference of 760,000,000 light-years. These tubes would support a 1-metric-ton car revolving at a Lorentz factor of 2.071×10^{20}. The radial centripetal force exerted by the 1-metric-ton car were it not for special relativistic mass increase is 10^{12} metric tons. Tubes constructed of the up-downium having the same mass and radius would support a Lorentz factor of 2.071×10^{21}. The radial centripetal force exerted by the 1-metric-ton car were it not for special relativistic mass increase is 10^{13} metric tons.

One-billion-light-year-radius spherical dispositions of 214 such neutronium galactic ring-worlds centered around an identical Milky Way ring-world could send and receive two-way communications to the Milky Way ring-world in only $\{[2 \times 10^9]/[2.071 \times 10^{20}]\}[31,000,000]$ seconds, or 0.0002993 seconds. Two-way communications diametrically spanning the sphere could be completed in only 0.0005987 seconds. For an otherwise-identical up-downium sphere, the respective communication times are 0.00002993 seconds, and 0.00005987 seconds.

One-trillion-light-year-radius spherical dispositions of 21,400 such neutronium galactic ring-worlds centered around an identical Milky Way ring-world, could send and receive two-way communications to the Milky Way ring-world in only $\{[2 \times 10^{12}]/[2.071 \times 10^{20}]\}[31,000,000]$ seconds or 0.2993 seconds. Two-way communications diametrically spanning the sphere could be completed in only 0.5987 seconds. For an otherwise-identical up-downium sphere, the respective communication times are 0.02993 seconds and 0.05987 seconds.

One-quadrillion-light-year-radius spherical dispositions of 21,400,000 such neutronium galactic ring-worlds centered around an identical Milky Way ring-world could send and receive two-way communications to the Milky Way ring-world in only $\{[2 \times 10^{15}]/[2.071 \times 10^{20}]\}[31{,}000{,}000]$ seconds, or 4.9883 minutes. Two-way communications diametrically spanning the sphere could be completed in only 9.9766 minutes. For an otherwise-identical up-downium sphere, the respective communication times are 0.49883 minutes and 0.99766 hours.

One-quintillion-light-year-radius spherical dispositions of 21,400,000,000 such neutronium galactic ring-worlds centered around an identical Milky Way ring-world, could send and receive two-way communications to the Milky Way ring-world in only $\{[2 \times 10^{18}]/[2.071 \times 10^{20}]\}[31{,}000{,}000]$ seconds, or 3.4641 days. Two-way communications diametrically spanning the sphere could be completed in only 6.92819 days. For an otherwise-identical up-downium sphere, the respective communication times are 0.34641 weeks and 0.692819 weeks.

Still yet again, consider the formula:

$$\gamma = \cosh(aT/c) = [1 + [(at/C)^{2}]]^{1/2} = [ad/(C^{2})] + 1$$

Except this time, we replace *d* with 1.9×10^{30} meters or 190,000,000,000,000 light-years and assume a Lorentz factor of 2.071×10^{23}.

Therefore:

$$\gamma = \cosh(aT/c) = [1 + [(at/C)^{2}]]^{1/2} = [ad/(C^{2})] + 1 = \{(a)[1.9 \times 10^{30}\ m] / \{[9 \times 10^{16}]\ m^{2}/s^{2}\}\} + 1$$

$$= 2.071 \times 10^{23}$$

Therefore, by solving for *a*, we obtain an acceleration equal to 10^{9} Gs.

The radial centripetal force exerted by the 1 metric ton car were it not for special relativistic mass increase is 10^{9} metric tons. However, the car is traveling at a Lorentz factor of 2.071×10^{23} with respect to the travel tube reference frame and so the centripetal force acting on the ring due to the car will be $(10^{9})[2.071 \times 10^{23}](9{,}810)$ newtons, or 2.03165×10^{36} newtons. Now consider neutronium tubes having a mass of 772,000,000,000,000,000,000 solar masses or 7.72×10^{20} solar masses and a circumference of 760,000,000,000,000 light-years, or 7.6×10^{14} light-years. These tubes would indeed support a 1-metric-ton car revolving at a Lorentz factor of 2.071 x 10^{23}. Tubes constructed of the up-downium having the same mass and radius would support a Lorentz factor of 2.071 x 10^{24}.

Now consider neutronium tubes having a mass of 7 x 10^{26} solar masses and a circumference of 7.6 x 10^{14} light-years. These tubes would support a 1-metric-ton car revolving at a Lorentz factor of 2.071 x 10^{26}. Tubes constructed of the up-downium having the same mass and radius would support a Lorentz factor of 2.071 x 10^{27}.

So 10^{16} light-year-radius spherical dispositions of 21,472 such neutronium galactic ring-worlds centered around an identical Milky Way ring-world could send and receive two-way communications to the Milky Way ring-world in only {[2 x 10^{16}]/[2.071 x 10^{23}]}[31,000,000] seconds or 2.993 seconds. Two-way communications diametrically spanning the sphere could be completed in only 5.987 seconds. For an otherwise-identical up-downium sphere, the respective communication times are 0.2993 seconds and 0.5987 seconds.

So 10^{19} light-year-radius spherical dispositions of 21,472,000 such neutronium galactic ring-worlds centered around an identical Milky Way ring-world could send and receive two-way communications to the Milky Way ring-world in only {[2 x 10^{19}]/[2.071 x 10^{23}]}[31,000,000] seconds, or 49.895 minutes. Two-way communications diametrically spanning the sphere could be completed in only 99.79 minutes. For an otherwise-identical up-downium sphere, the respective communication times are 4.9895 minutes and 9.979 minutes.

So, 10^{22} light-year-radius spherical dispositions of 21,472,000,000 such neutronium galactic ring-worlds centered around an identical Milky Way ring-world could send and receive two-way communications to the Milky Way ring-world in only {[2 x 10^{22}]/[2.071 x 10^{23}]}[31,000,000] seconds, or 34.649 days. Two-way communications diametrically spanning the sphere could be completed in only 69.299 days. For an otherwise-identical up-downium sphere, the respective communication times are 3.4649 days and 6.9299 days.

So 10^{25} light-year-radius spherical dispositions of 21,472,000,000,000 such neutronium galactic ring-worlds centered around an identical Milky Way ring-world could send and receive two-way communications to the Milky Way ring-world in only {[2 x 10^{25}]/[2.071 x 10^{23}]}[31,000,000] seconds, or 94.863 years. Two-way communications diametrically spanning the sphere could be completed in only 189.727 years. For an otherwise-identical up-downium sphere, the respective communication times are 9.4863 years and 18.9727 years.

We can consider still more extreme cases in such detail; however, by now the reader likely has an intuitive understanding of the mathematical functions describing systems covered herein.

Thus I have provided a thought problem solution with first order mathematical analysis indicating in simple terms the plausibility of effectively extraordinarily superluminal communication over cosmic distance scales. Special Relativity ironically makes this possible.

One major caveat for such extreme Lorentz factors includes virtually complete evacuation of thermal emissions from within the travel tubes. CMBR photons would be blueshifted to extremely hard gamma radiation and may annihilate a travel car body made of neutronium and quarkonium panels. A second caveat is actually being able to fabricate such extreme materials on such massive scales. A third caveat is the assembly such huge quantities of nuclear density materials into the order and controlled extremely extensive dispositions. A fourth caveat is the need for a suitable power source.

Now, the radius of a black hole is

$R_{bh} = 2GM/c^2 \approx 2.95\ (M/M_{Sun})$ km

A black-hole with the radius of the observable universe would have a mass of

> [13.75 ± 0.11 billion light-years][10^{13} km/1 light-year](1 solar mass)/(2.95 km) = 4.66 x 10^{22} solar masses. The mass of a critically open universe is about [10^{53} kg]/[2 x 10^{30} kg] solar mass = 5 x 10^{22} solar masses.

A marginally open universe will have a mass of about 10^{53} kilograms. Therefore, any black-hole travel tubes having a radius equal to that of the observable universe will be limited to a mass of about 5 x 10^{22} solar masses.

In fact, any circulinear travel tube car will be limited in radius $R > 2GM_{tt}/c^2 = 2.95\ (M_{tt}/M_{Sun})$ km.

Recall the following simple points.

The radial centripetal force exerted by the 1-metric-ton car were it not for special relativistic mass increase is 10^9 metric tons. However, the car is traveling at a Lorentz factor of 2.071 x 10^{23} with respect to the travel tube reference frame, and so the centripetal force acting on the ring due to the car will be (10^9)[2.071 x 10^{23}](9,810) newtons, or 2.03165 x 10^{36} newtons. Now consider neutronium tubes having a mass of 7,720,000,000,000,000,000,000 solar masses, or 7.72 x 10^{21} solar masses and a circumference of 760,000,000,000,000 light-years or 7.6 x 10^{14} light-years. These tubes would support a 1-metric-ton car revolving at a Lorentz factor of 6.549 x 10^{23}. Tubes constructed of the up-downium having the same mass and radius would support a Lorentz factor of 2.071 x 10^{24}.

Consider neutronium tubes having a mass of 7.72 x 10^{27} solar masses solar masses and a circumference of 7.6 x 10^{14} light-years. These tubes would support a 1-metric-ton car revolving at a Lorentz factor of 6.549 x 10^{26}. Tubes constructed of the up-downium having the same mass and radius would support a Lorentz factor of 2.071 x 10^{27}.

Now the radius of the observable universe is 1.375 x 10^{10} light-years and a black hole having the same radius has a mass of 4.66 x 10^{22} solar masses.

Thus, we would expect the following boundary condition to hold.

> [1.375 x 10^{10} light-year]/[4.66 x 10^{22} solar mass] = [2.419 x 10^{14} light-year]/M_{tt}
>
> → M_{tt} = [2.419 x 10^{14} light-year] [4.66 x 10^{22} solar mass]/[1.375 x 10^{10} light-year] = 8.1982 x 10^{26} solar masses

Thus, M_{tt} is a little less than the Jean's limit mass for the previously computed exemplar ring, and so it is plausible that the ring can be upheld against collapse into a useless black hole.

We can provide a means by which the ring's tensile strength can remain constant for progressively larger rings, thereby permitting constant minor cross-section of the ring. Thus, for every factoral increase, *n*, in ring growth beyond a circumference of 7.6 x 10^{14} light-years, we can support a factoral increase in car Lorentz factor by $n^{1/2}$. Thus, for an up-downium linear mass density–maximized ring having a circumference of 7.6 x 10^{18} light-years, we could support a Lorentz factor of [2.071 x 10^{29}]. For an up-downium linear mass density–maximized ring having a circumference of 7.6 x 10^{22} light-years, we could support a Lorentz factor of [2.071 x 10^{31}]. For an up-downium linear mass density–maximized ring having a circumference of 7.6 x 10^{26} light-years, we could support a Lorentz factor of [2.071 x 10^{33}]. For an up-downium linear mass density maximized ring having a circumference of 7.6 x 10^{30} light-years, we could support a Lorentz factor of [2.071 x 10^{35}], and so on.

Now the length specific mass of an 8.1982 x 10^{26} solar mass ring having a circumference of7.6 x 10^{14} light-years is {[8.1982 x 10^{26}][2 x 10^{27}]}/{[7.6 x 10^{14}][10^{16}]} kg/m = 2.1574 x 10^{23} kg/m. Assuming an up-downium density equal to neutronium at 10^{18} kg/m^3, the travel tube material cross-sectional area will be

215,740 square meters. This is the maximum possible linear mass density for the up-downium considered herein. The tensile strength of the tube as specified will be 2.1574 x 10^{39} newtons.

Now what can be done with mass-driver travel tubes can in principle be duplicated with relativistic rockets and the relativistic Lorentz turning force. The environments around large stars are an ideal starting point for such considerations. The next levels of such a rocket mechanism might include globular clusters, galaxies, galaxy clusters, followed by galaxy superclusters. Eventually, nucleations of galactic matter larger than superclusters may occur, which have more or less radially and circumferentially symmetrical and uniform polodial magnetic fields.

In our last exemplar Lorentz factor of 2.071 x 10^{35}, we are only at a relativistic mass of about 15 orders of magnitude less than that of the currently observable universe. In theory, for 1-metric-ton cars, a Lorentz factor of about 10^{50} should be possible for revolutional systems. For revolutional systems having a circumference of 7.6 x 10^{30} light-years, a Lorentz factor as high as about 10^{70} should be possible. From here on, the maximum possible Lorentz factor for revolutional systems should scale with the radius of revolution.

Now for closed universes, it is interesting to note that a spacecraft traveling in one direction long enough may eventually come back to its rough starting position much as an aircraft flying around Earth along a meridian will eventually fly back to or over the location whence it began its flight. The relativistic mass limit of the universally circumnavigating craft may well be limited to that of the maximum non-black-hole mass concentration for a maximally expanded closed universe. In such a case, substantially all of the real mass content within the universal closed space-time would have been converted into spacecraft kinetic energy.

Given the considerations in the previous paragraph, it may be possible to increase the rate of universal expansionary slowdown and collapse simply by powering the exemplar spacecraft up to super-general-relativistic masses via extraction of propulsion energy from the zero-point energy fields. Such a mechanism would result in the immediate formation of a black hole where the enclosing universe was not already one.

On a smaller scale, such a mechanism may enable the formation of black holes of cosmic radius scales within either a marginally open or open universe. For open universes, it would be of great help if the rate of universal expansion would non-trivially decrease over cosmic time scales. Such a slowing may be required so that the space-time revolution process of the spacecraft would not be diluted in its associated relativistic mass energy content as a result of space-time expansion.

We have an interesting result and point to ponder here. Assuming that black holes can be formed by such Lorentz turning force effects from spacecraft accelerated to super-general-relativistic kinetic energies, perhaps the spacecraft can be interpreted as forming its own black hole. Should wormhole travel be possible in the external vicinity of the extremely stressed space-time near-black-hole-event horizons, or if not, within the black-hole event horizon itself, perhaps such spacecraft accelerated to super-relativistic kinetic energies may be interpreted as entering its own black hole. Since most of the energy of the spacecraft and the space-time enclosing the spacecraft revolutional path would be contained within the spacecraft, perhaps due to special relativistic mass increase of the spacecraft, the spacecraft would enter a black hole formed immediately by its relativistic mass thereby causing the spacecraft to disconnect from the space-time or external universe whence it accelerated up to speed. Such a self-imposed black-hole state may be interpretable as inter-universal travel, extra-dimensional travel, hyperspatial travel, wormhole travel, and the like.

Another caveat is the need to actually send communications over cosmic distances for which the signal strength would not be diluted by relativistic Doppler effects. This will not be a problem in cases where the phase of rotational motion is in tandem and the plane of rotation is substantially precisely parallel even where the spacecraft Lorentz factors are limited but also extreme.

References and Suggested Readings

Ad Astra Rocket Company. ASPL Director Franklin Chang Díaz, 2005. Retrieved August 27, 2011, from http://www.adastrarocket.com/HiResImagesForPublicRelease/Franklin-ASPLHiRes.jpg

Ad Astra Rocket Company. VASIMR® Operating Principles. Retrieved March 28, 2011, http://www.adastrarocket.com/HiResImagesForPublicRelease/VASIMR_operating_principles.jpg

Ad Astra Rocket Company. Retrieved August 27, 2011, from http://www.adastrarocket.com/HiResImagesForPublicRelease/VX-200-FullPowerBothStagesHiRes.jpg

Ad Astra Rocket Company. (2009). Technology. Retrieved March 28, 2011, from http://www.adastrarocket.com/aarc/Technology

Ad Astra Rocket Company. (2009). VX-200. Retrieved March 28, 2011, from http://www.adastrarocket.com/aarc/VX200

Bekuo. Human Mission to Mars with 10 MW Nuclear Powered VASIMR. Retrieved August 27, 2011, from http://www.adastrarocket.com/HiResImagesForPublicRelease/BekuoHiRes.jpg

Bekuo. Human Mission to Mars with 200 MW Nuclear Powered VASIMR. Retrieved August 27, 2011, from http://www.adastrarocket.com/HiResImagesForPublicRelease/Bekuo-200MW-1-MidRes.jpg

Boeing Company. (1995). Defense, space & security Xenon ion propulsion center. Retrieved March 28, 2011 from http://www.boeing.com/defense-space/space/bss/factsheets/xips/xips.html (Boeing Company 1995)

Braeunig, R. A. (2011). Rocket & Space Technology, Rocket Propellants: Retrieved July 28, 2011, from http://www.braeunig.us/space/propel.htm

Centauri Dreams. http://www.centauri-dreams.org.

Close, F. E. (2009). *Antimatter*. Oxford: Oxford University Press.

Crawford, I. (2010). Targets for Icarus: Planets within 15 light-years of the Sun. Retrieved July 1, 2011, from the Project Icarus website: http://www.icarusinterstellar.org/blog/category/astronomical-target/

Crowl, A. (2009). Deuterium fusion Starships II. Retrieved December 5, 2011, from http://crowlspace.com/?p=589

Crowl, A. (2011). Starflight on the Cheap II: Beating the Fuel Costs. Retrieved July 1, 2011, from the Project Icarus website: http://www.icarusinterstellar.org/blog/category/fuel-2/

Crowl, A. Starflight on the Cheap II: Beating the Fuel Costs. Retrieved September 30, 2011, from the Project Icarus website: http://www.icarusinterstellar.org/blog/starflight-cheap-ii-beating-fuel-costs/#more-429

Enzmann Starship. Retrieved April 18, 2011, from the Enzmann Starship website: http://enzmannstarship.com/

Erich von Däniken. (2005). Erich von Däniken. Retrieved March 30, 2011, from the Erich von Däniken website: http://www.evdaniken.com/

Feder, T. (2009). Need for clean energy, waste transmutation revives interest in hybrid fusion–fission reactor. Sweet solution or pie in the sky? Hybrids get new attention. Physics Today, 62, 24-27.

Franzen, C. (2011). TPM, How The World's Lightest Material Was Made: Retrieved December 1, 2011, from http://idealab.talkingpointsmemo.com/2011/11/making-the-worlds-lightest-material.php?ref=fpnewsfeed_beta

Gilster, P. (2004). An antimatter-driven sail to the Kuiper Belt. Retrieved March 28, 2011, from the Centauri Dreams website: http://www.centauri-dreams.org/?p=28

Gilster, P. (2007). A Note on the Enzmann Starship. Retrieved December 5, 2011, from the Centauri Dreams website: http://www.centauri-dreams.org/?p=1142

Gilster, P. (2009). LightSail: A Near-Term Space Sail. Retrieved July 28, 2011, from the Centauri Dreams website: http://www.centauri-dreams.org/?p=10190

Gilster P. (2011). Progress Toward the Dream of Space Drives and Stargates. Retrieved November 12, 2011, from the Centauri Dreams website: http://www.centauri-dreams.org/?p=18076

Gilster, P. (2010). The Enigma of Contact. Retrieved September 15, 2011, from the Centauri Dreams website: http://www.centauri-dreams.org/?p=12229

Gilster, P. (2009) The Problem with Warp Drive. Retrieved October 17, 2011, from the Centauri Dreams website: http://www.centauri-dreams.org/?p=10826

Jamieson, V. (2010). Starship pilots: speed kills, especially warp speed. New Scientist. Retrieved March 30, 2011, from http://www.newscientist.com/article/dn18532-starship-pilots-speed-kills-especially-warp-speed.html

Lezec, H. (2009) Left-handed metamaterials operating in the visible: negative refraction and negative radiation pressure. 2009 APS March Meeting Volume 54, Number 1

Long, K. Slowing Down The Icarus Probe & Induced Deceleration. Retrieved July 1, 2011, from the Project Icarus website: http://www.icarusinterstellar.org/blog/slowing-icarus-probe-induced-deceleration/

M. Alan Kazlev, Richard Baker, David Dye, Mauk Mcamuk and Chris Shaeffer. Encyclopedia Galactica, Chemical Rocket. Retrieved July 28, 2011 from the Orion's Arm Website: http://www.orionsarm.com/eg-article/493687ff373fd

Misner, C.W., Thorne, K. S., & Wheeler, J. A. (1973). *Gravitation*. San Francisco: W. H. Freeman

NASA. (2010). Reaching for the stars. Retrieved March 30, 2011, from http://science.nasa.gov/science-news/science-at-nasa/1999/prop12apr99_1/

Obousy, R. *The Daedalus Propulsion System*. Retrieved July 1 2011, from the Project Icarus website: http://www.icarusinterstellar.org/blog/daedalus-propulsion-system/#more-449

Padilla, W. J., Dimitri, N. B., Smith, D. R. (2006). Negative Refractive Index Metamaterials. Materials Today 9, 28-35 .

Planetary Society. (2011). What we do Projects lightsail - solar sailing. Retrieved March 28, 2011, from http://www.planetary.org/programs/projects/solar_sailing/

Powell, D. (2011). Science News, Diamond cousin proposed: Retrieved November 25, 2011, from http://www.sciencenews.org/view/generic/id/70420/title/Diamond_cousin_proposed

Project Icarus. (2011). Current Research: Retrieved July 1, 2011, from http://icarusinterstellar.org/currentresearch.php

Project Icarus. (2011). Project Icarus: Retrieved July 1, 2011, from http://icarusinterstellar.org/icarus_project.php

Project Icarus.(2011). Team Designers: Retrieved July 1, 2011, from http://icarusinterstellar.org/team_members.php

Sagan, C. (1980). *Cosmos*. New York: The Random House Publishing Group.

The Nuclear Weapon Archive. (2001). The B61 (Mk-61) Bomb,Intermediate yield strategic and tactical thermonuclear bomb: Retrieved July 28, 2011, from http://nuclearweaponarchive.org/Usa/Weapons/B61.html

The Nuclear Weapon Archive. (2001). The W87 Warhead, Intermediate yield strategic ICBM MIRV warhead: Retrieved July 28, 2011, from http://nuclearweaponarchive.org/Usa/Weapons/W87.html

The Nuclear Weapon Archive. (1997). The W88 Warhead, Intermediate yield strategic SLBM MIRV warhead: Retrieved July 28, 2011, from http://nuclearweaponarchive.org/Usa/Weapons/W88.html

Thorne, K. S. (1994). *Black Holes and Time Warps: Einstein's Outrageous Legacy*. New York: W. W. & Norton Company, Inc.

University Of Alaska-Fairbanks, Physics Department. "Antimatter fission/fusion drive." Retrieved March 30, 2011, from http://ffden-2.phys.uaf.edu/213.web.stuff/Scott%20Kircher/fissionfusion.html

Wikipedia. (2011). Beam-powered propulsion. Retrieved September 23, 2011, from http://en.wikipedia.org/wiki/Beam-powered_propulsion

Wikipedia. (2011). Carbon nanotubes. Retrieved March 28, 2011, from http://en.wikipedia.org/wiki/Carbon_nanotube

Wikipedia. (2011) Fusion rocket. Retrieved November 22, 2011, from http://en.wikipedia.org/wiki/Fusion_rocket.

Wikipedia. (2011). Nuclear pulse propulsion. Retrieved March 30, 2011, from http://en.wikipedia.org/wiki/Nuclear_pulse_propulsion

Wikipedia. (2012). Solar sail. Retrieved October 8, 2011, from http://en.wikipedia.org/wiki/Solar_sail

Wikipedia (2011). Solar thermal rocket. Retrieved July 28, 2011, from http://en.wikipedia.org/wiki/Solar_thermal_rocket

Printed in the United States
by Baker & Taylor Publisher Services